The FOS and JUN Families
of Transcription Factors

Edited by
Peter E. Angel, Ph.D
Peter A. Herrlich, M.D., Ph.D.
Institute for Genetics and Toxicology
Karlsruhe, Germany

CRC Press
Taylor & Francis Group
Boca Raton London New York

CRC Press is an imprint of the
Taylor & Francis Group, an **informa** business

First published 1994 by CRC Press
Taylor & Francis Group
6000 Broken Sound Parkway NW, Suite 300
Boca Raton, FL 33487-2742

Reissued 2018 by CRC Press

A Library of Congress record exists under LC control number: 94013477

Publisher's Note
The publisher has gone to great lengths to ensure the quality of this reprint but points out that some imperfections in the original copies may be apparent.

Disclaimer
The publisher has made every effort to trace copyright holders and welcomes correspondence from those they have been unable to contact.

ISBN 13: 978-1-138-50650-3 (hbk)
ISBN 13: 978-1-138-56216-5 (pbk)
ISBN 13: 978-0-203-71003-6 (ebk)

Visit the Taylor & Francis Web site at http://www.taylorandfrancis.com and the
CRC Press Web site at http://www.crcpress.com

PREFACE

This book attempts to summarize our current understanding of a particularly important transcription factor family: AP-1. Is this the right time to give an interim review? In a way, yes, since AP-1 has become one of the prototype factors regulated by physiological and pathological signals from the microenvironment of cells. Much of what we currently explore with other transcription factors has been primed by work on AP-1.

Nevertheless, the AP-1 area still seems to expand exponentially; for example, the leucine zipper dimerization principle and the expanding number of putatively interacting subunits may "generate" the existence of some 100 to 200 different AP-1 factors. It is, therefore, a difficult area to review with any confidence of completeness. We have asked several experts in this field to contribute chapters. In order to keep chapters in an independently readable form, we have not rigorously eliminated repetitions. Rather, we have compromised by inserting connecting paragraphs and allowing for some redundancies, particularly if topics were approached from different angles. Furthermore, we have allowed a variety of interpretations.

We wish to thank all contributors and Ingrid Kammerer for help in preparing this book.

THE EDITORS

Peter E. Angel, Ph.D., is a junior staff member of the Institute for Genetics, Kernforschungszentrum Karlsruhe, Germany. He studied biology at the University of Karlsruhe, obtaining an M.S. degree in 1983. The diploma thesis was completed at the Institute for Genetics in 1983, the Ph.D. thesis (summa cum laude) in 1987. Dr. Angel then worked as a research associate at the University of California, San Diego, for almost three years. In 1990 he obtained a cancer fellowship at the Deutsche Forschungsgemeinschaft (Heisenberg Stipendium). He chose the Institute for Genetics in Karlsruhe as his home institution in 1990 and assumed his present position as a principal investigator in 1993. His research is supported, in addition to Kernforschungszentrum Karlsruhe, by Deutsche Forschungsgemeinschaft and the European Community. Dr. Angel is the author of 50 original papers, some of which became citation classics. His research interests relate to transcription factors, protein-protein interactions, and the function and regulation of matrix metalloproteinases.

Peter A. Herrlich, M.D., Ph.D., is chairman of the Institute for Genetics, Kernforschungszentrum Karlsruhe, and professor of genetics at the University of Karlsruhe. Dr. Herrlich obtained his M.D. degree at the University of Munich in 1964, and, after an internship at Cook County Hospital in Chicago, a medical license. After his postdoctoral years at the Max Planck Institute for Biochemistry, Munich, and at Rockefeller University, New York, he obtained a "Habilitation" (Ph.D.) degree at the Free University of Berlin in 1972. Dr. Herrlich was a senior staff member (associate professor) at the Max Planck Institute for Molecular Genetics, Berlin (1973-1977) and assumed his present position in 1977. In 1986-1987 he was a visiting professor at the University of California, San Diego. Dr. Herrlich is a member of the European Molecular Biology Organization, the European Science Foundation, and several other scientific associations. He is a co-editor of *Carcinogenesis* and a member of the editorial boards of *New Biologist, Molecular Carcinogenesis*, and the *International Journal of Cancer*. He has received the Meyenburg and Acker Prizes for Cancer Research. Dr. Herrlich serves on various agencies and study sections, e.g., as a member of the Kuratorium of Boehringer Ingelheim Fonds. His research has been supported, in addition to Kernforschungszentrum Karlsruhe, by Deutsche Forschungsgemeinschaft, the European Community, the Fonds der Chemischen Industrie, and Boehringer Ingelheim. Dr. Herrlich is an author on more than 300 papers and two books, and has co-edited one book. Examples of his current research interests are: genes involved in metastatic properties of cancer, DNA repair, signal transduction, and transcription factors.

INTRODUCTION

The vast array of behaviors and functions exhibited by a living organism are the result of induced changes of gene expression, especially at the level of transcription. Genes from the large complement of inherited genetic information are selected and their transcription turned on or off according to the needs that a given new condition imposes. Changes in transcription occur in response to a great variety of microenvironmental cues, including the supply of nutrients, stress agents, hormones, growth factors, position of cells (or nuclei) within a morphogen gradient, extracellular matrix, and components on neighboring cells. Indeed, understanding the control of gene expression is equivalent to understanding many of the fundamental properties that define life. If we set our goals modestly, we can say that there has been enormous progress in the last 10 years with respect to understanding the molecules participating in gene regulation. We can, for instance, answer questions such as: What are the biological structures through which a cell measures the conditions outside? How is this "knowledge" transformed into short- and long-term changes in transcription? And, finally, how are such responses integrated into crucial internal cellular programs, such as the cell cycle?

SOME BASICS OF THE TRANSCRIPTIONAL PROCESS IN EUCARYOTES

Transcription is the function of a multisubunit enzyme, RNA polymerase, which finds, with the help of auxiliary factors, the beginning of the gene and the noncoding strand. All of the proteins involved at the stage of transcription form the initiation complex at the promoter. In most of the genes transcribed by RNA polymerase II, a sequence conserved among most genes, the TATA box, determines where this complex assembles. In TATA-less promoters, an initiator sequence substitutes. Figure I-1 shows the multiprotein complex in position. Evidence for the constitution of the basic transcriptional initiation complex stems from *in vitro* transcription studies and purification from nuclear extracts of the minimal set of components required for the start reaction. Work on polymerase II was guided by previous studies on bacterial transcription and by sequence comparisons of promoter regions revealing the conserved features of the TATA-like elements.

When such defined start sequences in conjunction with a measurable reporter gene are incorporated into cells, only extremely low rates of expression will occur, regardless of how the cells are treated. None of the environmental cues listed above can turn on transcription. Promoter activity is only established by linking additional sequences to the TATA–reporter construct. In general terms, such sequences could be named enhancers since they "enhance" the activity of the basic initiation complex. The TATA–reporter construct would serve as an "enhancer trap," since it indicates the presence of sequences from anywhere in the genome that are able to enhance transcription. In a given gene such sequences are most often found 5' of the TATA box (with respect to the direction of transcription; the 5' is determined by the first nucleotide of the transcript). In fact, one or the other sequence needs to be in close proximity (e.g., within 100 bp) to make a promoter functional. However, many enhancing sequences have been found at a considerable distance (e.g., 3.6 kb) and also within the gene (in introns and exons) and 3' of the gene. To assemble all of the sequence elements necessary for completely "normal" behavior of a gene, rather large segments of the genome (possibly 300 kb or more) are needed. Evidence of this latter conviction comes from transgene technology. It appears that only the large yeast artificial chromosome (YAC) clones ensure that a gene carries all the intrinsic information needed for transcriptional regulation through ontogeny and adult life.

Enhancer sequences (and their opposite: silencers) are recognized by specialized proteins: transcription factors. One common property of these factors is their ability to bind DNA in a highly specific manner. Each factor binds with preference to the specific

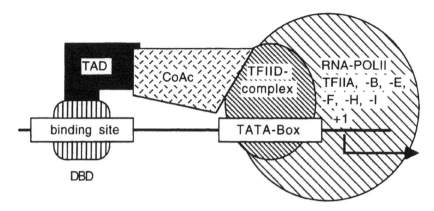

Figure I-1 The RNA polymerase II preinitiation complex. The transcription factor, binding to specific upstream enhancer elements, is composed of a DNA binding domain (DBD) and a transactivation domain (TAD). The TAD interacts with the components of the basal transcription machinery (RNA polymerase II plus basal factors such as TFIIA, -B, -D, etc). TFIID stands for a protein complex that consists of the TATA-binding protein (TBP) binding to the TATA sequence and associated factors (TAFs) that do not bind to DNA but interact directly or indirectly with TBP. The individual subunits of TFIID as well as their putative spatial arrangement are described in: Drapkin et al., *Curr. Opinion Cell Biol.*, 5, 469, 1993; and Weinzierl et al., *Nature*, 362, 511, 1993. The TAD of the transcription factor can interact either directly with TBP or TAFs or through the action of a bridging protein (co-activator) that mediates the link between the TAD of the transcription factor and TBP or TAFs.

enhancer sequence it is "built" for. The affinity calculated is in the range of $10^{-10}\ M$. The transcription factors this book is devoted to are among the best characterized and have contributed considerably to the discovery of transcription factor properties.

Transcription factors carry several functional domains in addition to DNA binding. One of these, transactivation domain (TAD), is responsible for making contact with the basic transcriptional machinery. For an enhancer factor, this contact results in elevated transcriptional initiation. Transcription factors seem to be able to interact directly with the TATA-binding protein (TBP), at least *in vitro*, or indirectly with the TBP via interaction with TBP-associated factors (TAFs). Thus, the "active" initiation complex possesses at least two anchors on the DNA: one at the enhancer binding site, the other through TBP binding to the TATA box. To permit formation of such contacts over considerable distances, enhancer sequences may be located away from the initiation complex. It is thought that the DNA stretch between the initiation complex and the enhancer loops to bring the enhancer sequence and its transcription factor in close proximity to the preinitiation complex (Figure I-1).

The most interesting problem of transcription factor function concerns the modulation of the transactivating function. Transactivation is the interaction between the components of the basic preinitiation complex, and the interacting portion of the transcription factor is referred to as the transactivation domain (Figure I-1). The activity of many, if not all, transcription factors is influenced in response to the conditions of the microenvironment. They are responsible for changing transcription when appropriate stimuli act on cells. These activations are rather specific for given agents, ensuring that changes in the genetic program occur only in response to the specific stimulus. The molecular mechanism of such activation has been identified in only a few examples. Steroid hormone receptors, for example, are converted to active transcription factors by the binding of the specific hormone ligand. Stimulation at the outer surface of the plasma membrane often results in a process of signal transduction via protein kinase–phosphatase cascades. Transcription factors, as the ultimate targets of these cascades, are posttranslationally modified (e.g., directly by phosphorylation at serines/threonines). Probably only very few of the transactivation-modulating mechanisms have yet been identified. It is conceivable, furthermore, that transcription factors are responsive to more than one modifying reaction and, in particular, to other proteins by protein–protein interaction.

THE AP-1 FAMILY

Much of our present knowledge about transcription factors comes from the discovery and study of the AP-1 factor family. One of its members, the heterodimer Fos–Jun, was found as early as 1982, as a protein complex containing the viral oncogene product Fos, without a clue of its function. The term AP-1 (activating protein-1) was coined for an activity that supports basal level transcription *in vitro* at the metallothionein gene promoter (1987). AP-1 was also immediately recognized as the decisive control element of the collagenase promoter *in vivo*, and it was demonstrated that it could be activated by external stimulation with the tumor promoter TPA.

The classical control site where AP-1 binds in the collagenase promoter reads 5'–TGAGTCA–3'. Even in the first purification, AP-1 looked like a family of related proteins. This interpretation was later proven to be true by the identification of multiple cJun- and cFos-related proteins encoded by distinct genes. This family has relative affinity to the classical AP-1 binding sequence. It appears now, however, that each member of the family has a strong preference for one very specific DNA sequence that is only related to the classical one. For instance, one portion of the family prefers the consensus 5'–TGANNTCA–3', with an additional nucleotide in the center of the recognition sequence. This sequence is preferentially bound by members of the cAMP responsive element binding protein (CREB/ATF) family. In fact, distinct members of the CREB/ATF family dimerize with the various Jun proteins, resulting in a change of sequence specificity. Moreover, in addition to the core element 5'–TGA(N)$_{1 or 2}$TCA–3', adjacent nucleotides can influence specificity.

Each AP-1 factor is composed of two subunits. It is not known, however, how many AP-1 factors exist. If each subunit could combine with each other subunit (which is not strictly possible), we could predict some 100 members of the family. Obviously, an understanding of the functional significance of such a large family lags behind our ability to characterize each member.

Most data on promoters cannot distinguish which AP-1 factor is active under a given *in vivo* condition. Many assignments are, therefore, arbitrary unless the need for a particular factor has been proven by antisense or knockout techniques. An added level of complexity is rendered to this system by the observation that different AP-1 dimers seem to affect DNA structure differently (DNA bending). DNA may also affect transcription factor conformation, as has been shown for steroid receptors. Furthermore, individual dimeric complexes can function in opposite directions. Fos, for example, is active in proliferation, differentiation, and apoptosis. Clearly, the context of activation as well as the variety of dimerizing partners present determine the ultimate response of the cell.

Why a book on the AP-1 family and not on other families, such as SP-1 or the octamer binding factors? We believe that a unique treatment is justified by the central role AP-1 plays both in mammalian organisms and in the investigation of transcription factors in general. AP-1 has served to detect one of the decisive DNA binding motifs and the bZip interaction. It is the AP-1 family (and NFκB) for which inducible activation was first detected and carried into molecular exploration. Its members are engaged in the control of cell proliferation as well as various types of differentiation, and, further, in neural function and stress responses. AP-1 is one of the key factors that translates external stimuli into both short- and long-term changes of gene expression. Several subunits of AP-1 are transforming proteins and required for transformation by other oncogenes. Work on AP-1 has also led to the discovery of cross-talk and cross-coupling between transcription factors.

This monograph attempts, therefore, to describe the "state of the art" with respect to Fos and Jun in detail. Topics will range from the structure of AP-1 and the principles of its DNA binding and function, to the exploration of its functions in physiology and pathology. The reader will realize that the story of AP-1 focuses on most of the current fronts of transcription research, including coordination with the cell cycle and the role of modifying enzymes in transcriptional regulation.

CONTENTS

I. Characteristics of AP-1 Subunits

II. Regulation of AP-1 Activity and Synthesis

A. Protein–Protein Interactions

B. Posttranslational Modification

C. Nuclear Uptake

D. Promoter Regulation

III. Functions in Physiological Processes

A. AP-1-Regulated Target Genes

B. Neural Cell Function

C. Differentiation

D. Stress Response

IV. Functions in Pathology

A. Fos and Jun Oncogenes

B. Oncogenes Acting Through Fos and Jun

C.   Adenovirus E1A–AP-1 Connection

D.   Tumor Promotion and Carcinogenesis

E.   Tumor Progression

Section I

CHARACTERISTICS
OF AP-1 SUBUNITS

Chapter 1

General Structure of AP-1 Subunits and Characteristics of the Jun Proteins

Peter Angel and Peter Herrlich

CONTENTS

GENERAL STRUCTURE OF THE AP-1 SUBUNITS

According to its function in controlling gene expression, the prototype of a transcription factor has to comprise two, or possibly three, properties: a region of the protein responsible for binding to a specific DNA recognition sequence (DNA binding domain); a region required for transcriptional activation (transactivation domain) once the protein is bound to DNA; and, possibly, a modulation domain. These functions are encoded by separate regions (modules) of the protein that generally function independently; however, domain swapping experiments have shown that the transactivation domain can be fused to a heterologous DNA binding domain to form a potent transcription factor of new promoter specificity.

In Jun proteins, the transactivation domain is located within the *N*-terminal half of the protein while the DNA binding domains of Jun and CREB/ATF are located at the *C*-terminus (Figure 1-1A). The transactivation domains of Fos and CREB/ATF have not been determined precisely, but, in the case of Fos, transactivation seems to be influenced by amino acids at both the *N*-terminus and *C*-terminus of the protein (see Chapters 2 and 8). The DNA binding domain of Fos is located near the center of the protein.

In vivo mutation analysis of the Jun proteins has identified three subdomains that together form the transactivation domain (Angel et al., 1989; Hirai et al., 1990). The subdomains are characterized by an abundance of acidic amino acids that are essential for activity (Angel et al., 1989). In addition, *in vitro*, a fourth region near the DNA binding domain has been identified (Bohmann and Tjian, 1989). These regions are thought to be responsible for the link between the transcription factor bound to DNA and the RNA polymerase II preinitiation complex. The link is established by either direct or indirect interaction with components of the basal transcription machinery. Based on these amino acid sequences of the transactivation domain, the Jun proteins belong to the family of "acidic-blob"-type transcription factors (Angel et al., 1989; Oehler and Angel, 1992). However, in contrast to other members of the acidic-blob group, for example, the yeast transcription factor GAL4 (Ma and Ptashne, 1987) or the VP16 protein of herpes simplex virus (Trietzenberg et al., 1988), the transactivation domains of the Jun proteins do not seem to form an extensive α-helical structure, since these regions also contain single or multiple proline residues known to disrupt α-helices.

4

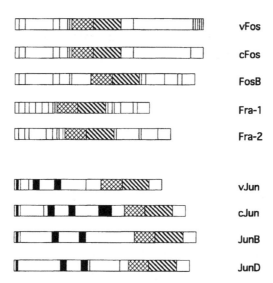

Figure 1-1A Structural organization of the Jun and Fos proteins. The hatched boxes (▨) represent the "leucine zipper" region. The cross-hatched boxes (▩) indicate the "basic domain." The box marked by stripes (▥) indicates the part of vFos generated by a frame shift mutation in cFos. In the Fos proteins the shaded boxes (☐) denote highly conserved regions of unknown functions; vertical lines represent deletions (comparing members of either the Fos or Jun families). In the Jun proteins the solid boxes (■) represent the transactivation regions; the shaded boxes (▦) in cJun show the glutamine- and proline-rich region of unknown function (present in human but not in mouse or avian cJun).

In contrast to the transactivation domains, whose structural properties are poorly understood, a large body of information on the DNA binding domains of Jun, Fos, and other AP-1 factors has been collected (Vogt and Bos, 1990; Busch and Sassone-Corsi, 1990; Ransone and Verma, 1990). Mutation analysis has revealed characteristic properties that are evolutionarily conserved between the Fos, Jun, and CREB proteins, thus defining the protein family called "bZip proteins" (Landschulz et al., 1988). bZip (see Figure 1-1B) stands for the amino acid sequences of the two independently acting subregions of the DNA binding domain: the "basic domain," rich in basic amino acids responsible for contacting the DNA, and the "leucine zipper" region, characterized by heptad repeats of leucine, which is responsible for dimerization that in turn is a prerequisite for DNA binding (Kouzarides and Ziff, 1988; Sassone-Corsi et al., 1988; Gentz et al., 1989; Turner and Tjian, 1989; Neuberg et al., 1989b). While amino acid substitutions or deletions in the leucine zipper abolish dimerization, mutations in the basic domain abolish DNA binding without affecting dimerization. Domain swapping experiments have shown that both domains are interchangeable among the different bZip proteins without loss of their physical properties (Neuberg et al., 1989a; Sellers and Struhl, 1989; Kouzarides and Ziff, 1989; Cohen and Curran, 1990).

Based on the potential formation of an α-helical structure of the leucine zipper region and the need for dimerization, dimer formation was proposed to be mediated by interdigitation (zipper formation) of the leucines, which are located in a linear fashion at the inner phase of the helix (Landschulz et al., 1988). Since proteins exist that contain a

├─ "basic region" ─┤ ├──────── "leucine zipper" ────────┤

```
EEKRRIRRERNKMAAAKCRNRRRELTDT  LQAETDQLEDEKSALQTEIANLLKEKEKLEFILAAH   cFos  (aa 137-200)
EEKRRVRRERNKLAAAKCRNRRRELTDR  LQAETDQLEEEKAELESEIAELQKEKERLEFVLVAH   FosB  (aa 155-218)
EERRRVRRERNKLAAAKCRNRRKELTDF  LQAETDKLEDEKSGLQREIEELQKQKERLEFMLVAH   Fra-1 (aa 105-168)
EEKRRIRRERNKLAAAKCRNRRRELTEK  LQAETEELEEEKSGLQKEIAELQKEKEKLEFMLVAH   Fra-2 (aa 124-187)
RIKAERKRMRNRIAASKCRKRKLERIAR  LEEKVKTLKAQNSELASTANMLREQVAQLKQKVMNH   cJun  (aa 252-315)
RIKVERKRLRNRLAATKCRKRKLERIAR  LEDKVKTLKAENAGLSSAAGLLREQVAQLKQKVMTH   JunB  (aa 265-328)
RIKAERKRLRNRIAASKCRKRKLERISR  LEEKVKTLKSQNTELASTASLLREQVAQLKQKVLSH   JunD  (aa 262-325)
DEKRRKFLERNRAAASRCRQKRKVWVQS  LEKKAEDLSSLNGQLSEVTLLRNEVAQLKQ        ATF-2 (aa 283-341)
ARKREVRLMKNREAARECRRKKKEYVKC  LENRAVLENQNKTLIEELKALKDLYCHKSD         CREB  (aa 283-341)
```

Figure 1-1B Amino acid sequences of the "bZip" region of the Fos, Jun, and CREB/ATF proteins. The positions of the leucines in the "leucine zipper" are designated by shaded boxes; basic amino acids in the "basic region" are indicated by bold letters.

heptad repeat of leucine residues but do not dimerize with Fos, CREB/ATF, or Jun (e.g., the Myc protein), and because other amino acids within the leucine zipper region were also found to be critical for dimer formation, the "zipper model" was revised (O'Shea et al., 1989a, 1989b), proposing that the leucine zipper region of both Fos and Jun forms an α-helical structure of 4-3 repeats (see Figure 5 in Chapter 2). The location of a hydrophobic amino acid at position a and the leucines at position d favor the formation of dimers in parallel orientation by a so-called "coiled coil" interaction, with the leucines and hydrophobic amino acids positioned toward the contact interface (see Figure 2-5). It has been proposed that the stability of dimers is determined by charged amino acids at positions e and g of the heptad repeat, forcing either electrostatic repulsion or stabilization of the dimer. Amino acid residues at positions b, c, and f are not involved in dimer formation (Schuermann et al., 1991; O'Shea et al., 1992). On alignment of two Fos monomers in a coiled coil structure, the e and g positions are both covered by acidic amino acids that are likely to cause electrostatic repulsion, which may explain why Fos is not able to form stable homodimers. In contrast, efficient dimerization between the Jun and Fos proteins is likely to be promoted by ionic interactions between the positively charged residues at the e and g positions of Jun with the negatively charged residues in Fos (see also Chapter 2). The number of such putative interactions is lower in Jun/Jun homodimers, resulting in reduced stability (O'Shea et al., 1989). The same type of interaction rules are probably responsible for complex formation of cJun with specific members of the CREB/ATF protein family (ATF-2), whereas Fos does not efficiently form Fos/CREB heterodimers (Benbrook and Jones, 1990; McGregor et al., 1990; Hai and Curran, 1991; Müller et al., unpublished results).

We have tested the validity of these rules by individually modifying charged amino acids at positions e and g of the leucine zippers of cJun and ATF-2, and have obtained contradictory results: cJun homodimer formation in solution was not significantly affected after replacing two lysines with two glutamines, although this change was expected to result in strong repulsion. In contrast, however, the replacement of two glutamines in the ATF-2 leucine zipper with two lysines did have a negative effect on dimer formation (van Dam et al., unpublished results). Thus, although electrostatic interactions between amino acids at positions e and g may contribute to dimer stability, additional interactions, including association with DNA, appear to be important as well. Another exception to the e–g rule has been found for homodimerization of JunB. Two amino acids at positions b and c strongly affect homodimer formation by an unknown mechanism (Deng and Karin, 1993), most likely a general destabilization of the α-helical structure of the leucine zipper of JunB (Deng and Karin, 1993).

Most studies have been performed in solution *in vitro* in the absence of DNA. It is conceivable that DNA exerts influence on dimer formation by preference for specific dimers or by participating in the complex formation. In turn, as will be described below, Jun/Jun and Jun/Fos each cause different bending of DNA (opposite bending; see Kerppola and Curran, 1991a, 1991b).

Coiled coil formation as a major mechanism of protein–protein interaction is not without precedence; this type of interaction was first described for the structure of the α-class of fibrous proteins, such as keratin, myosin, and fibrinogen (Cohen and Parry, 1986). A special feature of transcription factors, the ability to form dimers, is combined with the ability to bind to DNA in a sequence-specific manner. Sequence specificity in the basic domain is encoded by the amino acid sequence that interacts with the major groove of the DNA (Vinson et al., 1989; O'Neil et al., 1990). Based on studies on C/EBP, sequence specificity is established in such a way that both α-helices of the dimerized proteins bifurcate beyond the leucine zipper as a consequence of the positive charges of the juxtaposed basic regions, forcing the two basic regions to track along the major groove of the DNA (scissor grip model; Vinson et al., 1989). Because the Fos, Jun, and CREB/

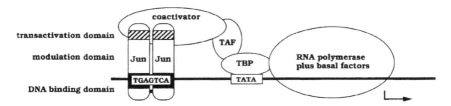

Figure 1-2 Transactivation of AP-1-dependent genes. The hatched (▨) and solid boxes (■) signify the transactivation and DNA binding domains, respectively, of Jun. The existence of putative modulation domain(s) of Jun interacting with other cellular proteins still must be confirmed. The TATA binding protein (TBP), TBP-associated factors (TAFs), RNA polymerase II, and basal transcription factors (TFIIA, -B, -E, -F, -H, and -I) make up the RNA polymerase II preinitiation complex (see also Figure 1-3 and Figure I-1). In Jun homodimers, the transactivation domain of Jun interacts with a specific coactivator (p52/54) to make contact with the preinitiation complex. The receiving target of coactivator interaction (which type of TAF?) remains to be determined. The mechanism by which the members of the Fos and CREB/ATF protein families interact with the components of the basal transcription machinery is still largely unknown.

ATF proteins exhibit some differences in sequence specificity and structure of the DNA–protein complexes (Hai and Curran, 1991; Ryseck and Bravo, 1991), the combined ability of dimerization and DNA binding allows for multiple combinations of dimers that differ in their biological properties (see also Chapter 2).

In addition to the DNA binding and transactivation domains, transcription factors may also contain specific regions responsible for: (1) interaction with other cellular proteins distinct from components of the RNA polymerase initiation complex (modulation domain, Figure 1-2), (2) nuclear translocation, and (3) regulation of protein stability. In both the Fos and Jun proteins, regions have been identified that interact with other cellular proteins; for example, steroid hormone receptors (see Chapters 3 and 4). While the nuclear localization signal in both Fos and Jun has been identified within the basic domain (Tratner and Verma, 1991; Chida and Vogt, 1992), the specific regions involved in the rapid degradation of the Fos and Jun proteins, possibly regulated by phosphorylation and ubiquitination (Papavasiliou et al., 1992; D. Bohmann, personal communication), have not yet been found.

CHARACTERISTICS OF THE JUN PROTEINS

Originally thought to be encoded by a single type of protein, AP-1 is generated by a series of dimers of products of the Fos, Jun, and CREB/ATF protein families (Bohmann et al., 1987; Angel et al., 1988a; Distel et al., 1987; Rauscher et al., 1988a, 1988b; Hai et al., 1988), as well as other bZip proteins (e.g., LRF-1), which have not yet been characterized (Hsu et al., 1991). In addition, associations have been observed between Fos or Jun and the p65 subunit of NFκB (Stein et al., 1993), and ATF-2 and p50-NFκB (Du et al., 1993). Combinatorial association can draw on three Jun genes (c-jun, junB, junD), four Fos genes (c-fos, fosB, fra-1, fra-2) and several CREB/ATF genes (Benbrook and Jones, 1990; Hai and Curran, 1991; Ryseck and Bravo, 1991; Vogt and Bos, 1990). Despite the high degree of homology in the overall structural features described in the previous section, the different members of the Fos, Jun, and CREB families exhibit significant differences, which lead to subtle differences in DNA binding and transcriptional activation (Ryseck and Bravo, 1991; Hai and Curran, 1991; Chiu et al., 1989; Hirai et al., 1989;

Deng and Karin, 1993) and which suggest specific functions in gene regulation for individual dimers. This chapter focuses primarily on the Jun proteins. Characteristics of the Fos proteins will be described in Chapter 2.

THE DNA BINDING DOMAIN

Considerable differences in selectivity and stability among the bZip proteins have been found with regard to dimerization. As described above, the efficiency of stable dimer formation depends primarily on the charged amino acid residues at positions e and g of the α-helical structure of the leucine zipper region (Schuermann et al., 1991; O'Shea et al., 1992). With respect to these sites, the amino acid sequences between individual members of each family (e.g., cJun, JunB, and JunD) are fairly conserved. They differ greatly, however, when comparing the subfamilies Jun, Fos, and CREB/ATF (Figure 1B). These differences in amino acid composition of the contact interface of the dimer probably determine the specificity rules of dimerization. Such specificity of coiled coil interaction has been proved by domain-swapping experiments (Sellers and Struhl, 1989; Neuberg et al., 1989a) and has served to screen cDNA libraries for interacting subunits (Benbrook and Jones, 1990; McGregor et al., 1990; Ivashkiv et al., 1990). Most importantly, Fos cannot bind to DNA and regulate gene expression on its own, but depends on the presence of Jun or other proteins to form heterodimers. In contrast, both CREB/ATF and Jun proteins can bind to DNA on their own by forming homodimers (Bohmann and Montminy, 1987; Angel et al., 1988a; Hoeffler et al., 1988; Maekawa et al., 1989; Gonzalez and Montminy, 1989), although with lower stability (Jun/Jun as compared to Jun/Fos heterodimers).

The existence of a large number of possible combinations of partners would only by relevant physiologically if the individual dimers differed in DNA affinity or sequence specificity of the combined DNA binding domains of the subunits. Putative core sequence elements recognized by AP-1 factors are quite variable and fall into two classes, 7 bp (TPA responsive element, TRE) and 8 bp (cAMP responsive element, CRE) long. Site-directed mutagenesis experiments have shown that the amino acids in the basic regions of Fos and Jun required for interaction with the DNA are highly conserved between the two proteins (see also Chapter 2). *In vitro* binding studies of chimeric bZip proteins (e.g., of Jun background in which the basic domain of Jun was replaced by the equivalent sequence of Fos) revealed differences in sequence specificity (Nakabeppu and Nathans, 1989): dimers with two basic regions of Jun (or Fos) bind to the CRE with somewhat greater affinity than to the "classical" AP-1 site (TRE). In contrast, heterodimers of Fos and Jun bind preferentially to the TRE (Nakabeppu and Nathans, 1989). CREB homodimers prefer the CRE but also exhibit significant affinity for the TRE (Masquilier and Sassone-Corsi, 1992). There is little difference in the binding site specificity between the Fos and Jun basic regions since some base pair substitutions of the binding site resulted in a marked decrease in binding whereas others had a lesser effect or resulted in enhanced binding (Risse et al., 1989). Nevertheless, UV cross-linking experiments have suggested that both Jun and Fos make direct contact with DNA in a favored orientation of the complex on the TRE (Nakabeppu and Nathans, 1989). This interpretation is in line with the finding that Jun/Fos heterodimers prefer binding to asymmetrical, 7-mer TREs while Jun homodimers prefer binding to symmetrical, 8-mer CRE sequences (Nakabeppu and Nathans, 1989; Ryseck and Bravo, 1991). Exchanging Fos in a Fos/Jun heterodimer against ATF-2 abolishes binding to the TRE but allows efficient interaction with the CRE (Benbrook and Jones, 1990; McGregor et al., 1990; Ivashkiv et al., 1990; Hai and Curran, 1991; van Dam et al., 1993). In addition to the core nucleotide sequence of the TRE or CRE, sequence specificity is also affected by the flanking nucleotides (Deutsch et al., 1988; Ryseck and Bravo, 1991).

Models of bZip protein/DNA interaction propose that contacts between the straight α-helix of the basic domain and the major groove of straight B-DNA are limited to a maximum of 12 contiguous amino acids, which can contact a maximum of 5 bp of DNA (Vinson et al., 1989; O'Neil et al., 1990). In contrast, the basic region extends over 20 amino acid residues (Nakabeppu and Nathans, 1989), and the DNA binding site consists of at least 7 nucleotides (Angel et al., 1987; Risse et al., 1989). Thus, the basic region of the protein and/or the DNA must be bent or distorted to allow for the identified regions of contact between the molecules. Circular dichroism spectroscopy has demonstrated that the basic domain undergoes a conformational transition to a structure of high α-helix content in the presence of the cognate DNA binding site (O'Neil et al., 1990; Patel et al., 1990). This conformational change of the protein induced by DNA binding enforces DNA bending, as measured by the anomalous electrophoretic mobility of bent DNA fragments (Kerppola and Curran, 1991a, 1991b). Based on circular permutation and phasing analysis, Jun/Jun homodimers and Jun/Fos heterodimers induce bends in opposite directions: Fos bends DNA away from the dimer interface, causing the major groove to extend over the recognition site, and Jun/Jun homodimers bend DNA toward the dimer interface (Kerppola and Curran, 1991a, 1991b). While these data have been obtained by analyzing Jun/Jun homodimers and Jun/Fos heterodimers, it is not known whether this is true for other dimeric complexes, such as Fra-1/Jun or Jun/ATF-2. In addition, posttranslational modifications (e.g., phosphorylation) within the DNA binding domain of Jun (see Chapter 5) may affect the ability of Jun to bend DNA. Regardless of dimer specificity, differences in DNA bending might have important consequences on the transactivation function of Jun/Jun homodimers or Jun/Fos heterodimers: protein-induced bending has been proposed to be a mechanism that allows the interaction of factors bound to separate sites on the DNA and selection among different potential protein–protein interactions in a region containing multiple factors bound to DNA (Moitoso de Vargas et al., 1989).

THE TRANSACTIVATION DOMAIN

In vivo competition (squelching) experiments (Figure 1-3) have suggested the necessary participation of an intermediary protein (coactivator) interacting with the transactivation domain of cJun, in order to link Jun to TBP or TAFs (Angel et al., 1989; Oehler and Angel, 1992). In this type of experiment, transcriptional activation of an AP-1-dependent promoter by overexpression of cJun is strongly reduced by coexpression of a Jun mutant that contains the transactivation domain but lacks the DNA binding domain (schematically illustrated in Figure 1-3). Overexpression of the Jun mutant represses the Jun-induced transcription but does not affect basal activity of the reporter plasmid or the activity of other promoters whose expression is regulated by DNA binding proteins distinct from AP-1. These results suggest that the Jun mutant competes for binding of an intermediary factor to the transactivation domain of Jun required for transcriptional activation. *In vivo* competition correlates with the binding of a cluster of proteins with a molecular weight of 52, 53, and 54 kDa (p52/54; Oehler and Angel, 1992) in that transactivation domain-negative Jun mutants neither squelch nor bind the 52- to 54-kDa proteins. Transcriptional interference of cJun-specific transactivation as well as reduced physical interaction between Jun and p52/54 is also observed in the presence of other "acidic-blob"-type transactivators, including GAL4 and VP16, suggesting that the intermediary factor(s), including p52/54, are shared by transcription factors whose transactivation domains are characterized by an abundance of acidic amino acids. In contrast, transcription factors that are characterized by a different type of transactivation domain (e.g., Sp1, CTF, GHF1/Pit-1, or the estrogen receptor) do not interfere with Jun-specific transactivation, showing that these proteins either use a different type of bridging protein or interact directly with TBP or TAFs (Oehler and Angel, 1992).

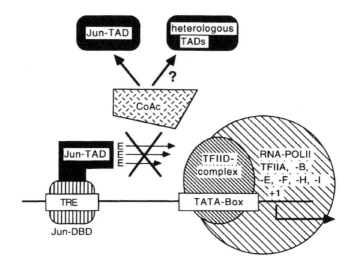

Figure 1-3 *In vivo* competition of Jun-dependent transactivation. To demonstrate the requirement and specificity of a bridging protein (coactivator, CoAc), transactivation of a TRE-dependent promoter is measured in the presence of an excess of truncated forms of Jun (Jun-TAD) or other transcription factors (heterologous TADs) that contain the transactivation domain (TAD) but lack the DNA binding domain (DBD). Diminution of promoter activity is explained by titrating out the coactivator through competitive binding of the mutants. "E" stands for the presence of glutamate residues required for TAD activity (and, most likely, interaction with the coactivator protein).

Diminution of cJun-specific transactivation is also observed by coexpression of the transactivation domain of JunB, suggesting that cJun and JunB make use of the same bridging protein (Oehler and Angel, 1992). In fact, the amino acid sequences of the three subdomains of the cJun and JunB transactivation domains share high degrees of homology (Vogt and Bos, 1990), and both proteins are efficient activators of promoters containing multiple binding sites (Chiu et al., 1989). Nevertheless, analysis of transactivation of AP-1-dependent promoters containing a single binding site has clearly shown that nonconserved amino acids outside these three subdomains have an important function in establishing a functional transactivation domain: while cJun homodimers strongly activate such promoters, neither JunB nor JunD homodimers efficiently stimulate transcription but, rather, suppress stimulation of these promoters by cJun, possibly by forming inactive cJun/JunB heterodimers (Deng and Karin, 1993). These distinct behaviors of cJun and JunB with respect to their transactivation potential are probably due to differences in their transactivation domains. Studies with chimeric proteins containing the DNA binding domain of GAL4 and a minimal transactivation domain, cJun or JunB (amino acids 5–89), revealed that GAL4–cJun is a potent transactivator while GAL4–JunB fails to activate transcription (Franklin et al., 1992). In the case of JunB, a switch from the inactive to the active state of the transactivation domain might occur because of an unknown mechanism of cooperativity established by binding of JunB to multiple binding sites in a given promoter. This switch might also occur by heterodimerization of JunB or JunD with Fos, resulting in potent transactivators of promoters containing either single or multiple AP-1 binding sites. Interestingly, two of the three regions that constitute the transactivation domain of the Jun proteins (domains II and III) are also found in Fos and are required for transactivation by Fos (see also Chapter 2). These motifs, homology boxes 1 and 2 (HOB1, HOB2), can be exchanged; e.g., HOB1 of Fos can be replaced by

HOB1 of Jun (Sutherland et al., 1992). Since interaction of Jun and p52/54 depends on the presence of Jun's transactivation domain, it is possible that p52/54, or a closely related protein, also interacts with Fos. So far, however, we have not detected stable physical interactions between Fos and p52/54. As a tentative interpretation, Fos does not carry domain I (amino acids 6–12 of Jun), which was found to be essential for the interaction between Jun and p52/54 (Oehler and Angel, 1992).

In summary, the formation of specific homo- and heterodimers in a given cell depends on the relative abundance of individual members of the bZip protein family. The mix of dimers will change with any change of a given subunit's abundance. Due to the unique properties of the specific domains of individual bZip proteins (for the Fos proteins, see Chapter 2) genes will be activated according to the subset of single or multiple TRE or CRE sequences they carry in their promoters, and according to the prevalent AP-1 dimers present.

ACKNOWLEDGMENTS

We would like to thank Thomas Oehler for preparing some of the figures and Dr. Larry Sherman for critical reading of the manuscript.

REFERENCES

Angel, P., Imagawa, M., Chiu, R., Stein, B., Imbra, R. J., Rahmsdorf, H. J., Jonat, C., Herrlich, P., and Karin, M., Phorbol ester-inducible genes contain a common cis element recognized by a TPA-modulated trans-acting factor, *Cell*, 49, 729, 1987.

Angel, P., Allegretto, E. A., Okino, S. T., Hattori, K., Boyle, W. J., Hunter, T., and Karin, M., Oncogene jun encodes a sequence-specific trans-activator similar to AP-1, *Nature*, 332, 166, 1988a.

Angel, P., Hattori, K., Smeal, T., and Karin, M., The jun proto-oncogene is positively autoregulated by its product, Jun/AP-1, *Cell*, 55, 875, 1988b.

Angel, P., Smeal, T., Meek, J., and Karin, M., Jun and v-Jun contain multiple regions that participate in transcriptional activation in an interdependent manner, *New Biol.*, 1, 35, 1989.

Benbrook, D. M. and Jones, N. C., Heterodimer formation between CREB and JUN proteins, *Oncogene*, 5, 295, 1990.

Bohmann, D. and Tjian, R., Biochemical analysis of transcriptional activation by Jun: Differential activity of c- and v-Jun, *Cell*, 59, 709, 1989.

Bohmann, D., Bos, T. J., Admon, A., Nishimura, T., Vogt, P. K., and Tjian, R., Human proto-oncogene c-jun encodes a DNA binding protein with structural and functional properties of transcription factor AP-1, *Science*, 238, 1386, 1987.

Busch, S. J. and Sassone-Corsi, P., Dimers, leucine-zippers and DNA binding domains, *Trends Genet.*, 6, 36, 1990.

Chida, K. and Vogt, P. K., Nuclear translocation of viral Jun but not of cellular Jun is cell cycle dependent, *Proc. Natl. Acad. Sci. U.S.A.*, 89, 4290, 1992.

Chiu, R., Angel, P., and Karin, M., Jun-B differs in its biological properties from, and is a negative regulator of, cJun, *Cell*, 59, 979, 1989.

Cohen, C. and Parry, D. A. D., α-Helical coiled coils — a widespread motif in proteins, *Trends Biochem. Sci.*, 11, 245, 1986.

Cohen, D. R. and Curran, T., Analysis of dimerization and DNA binding functions in Fos and Jun by domain-swapping: involvement of residues outside the leucine zipper/basic region, *Oncogene*, 5, 929, 1990.

Deng, T. and Karin, M., JunB differs from cJun in its DNA-binding and dimerization domains, and represses c-Jun by formation of inactive heterodimers, *Genes Dev.*, 7, 479, 1993.

12

Deutsch, P. J., Hoeffler, J. P., Jameson, J. L., and Habener, J. F., Cyclic AMP and phorbol ester-stimulated transcription mediated by similar DNA elements that bind distinct proteins, *Proc. Natl. Acad. Sci. U.S.A.*, 85, 7922, 1988.

Distel, R. J., Ro. H.-S., Rosen, B. S., Groves, D. L., and Spiegelman, B. M., Nucleoprotein complexes that regulate gene expression in adipocyte differentiation: Direct participation of c-fos, *Cell*, 49, 835, 1987.

Du, W., Thanos, D., and Maniatis, T., Mechanisms of transcriptional synergism between distinct virus-inducible enhancer elements, *Cell*, 74, 887, 1993.

Franklin, C. C., Sanchez, V., Wagner, F., Woodgett, J. R., and Kraft, A. S., Phorbol ester-induced amino-terminal phosphorylation of human JUN but not JUNB regulates transcriptional activation, *Proc. Natl. Acad. Sci. U.S.A.*, 89, 7247, 1992.

Gentz, R., Rauscher, F. J., III, Abate, C., and Curran, T., Parallel association of Fos and Jun leucine zippers juxtaposes DNA binding domains, *Science*, 243, 1695, 1989.

Gonzalez, G. A. and Montminy, M. R., A cluster of phosphorylation sites on the cyclic AMP-regulated nuclear factor CREB predicted by its sequence., *Nature*, 337, 749, 1989.

Hai, T. and Curran, T., Cross-family dimerization of transcription factors Fos/Jun and ATF/CREB alters DNA binding specificity, *Proc. Natl. Acad. Sci. U.S.A.*, 88, 3720, 1991.

Hai, T., Liu, F., Allegretto, E. A., Karin, M., and Green, M. R., A family of immunologically related transcription factors that includes multiple forms of ATF and AP-1, *Genes Dev.*, 2, 1216, 1988.

Hirai, S.-I., Ryseck, R.-P., Mechta, F., Bravo, R., and Yaniv, M., Characterization of junD: A new member of the jun proto-onocogene family, *EMBO J.*, 8, 1433, 1989.

Hirai, S.-I., Bourachot, B., and Yaniv, M., Both Jun and Fos contribute to transcription activation by the heterodimer, *Oncogene*, 5, 39, 1990.

Hoeffler, J. P., Meyer, T. E., Yun, Y., Jameson, J. L., and Habener, J. F., Cyclic AMP-responsive DNA-binding protein: Structure based on a cloned placental cDNA, *Science*, 242, 1430, 1988.

Hsu, J.-C., Laz, T., Mohn, K. L., and Taub, R., Identification of LRF-1, a leucine-zipper protein that is rapidly and highly induced in regenerating liver, *Proc. Natl. Acad. Sci. U.S.A.*, 88, 3511, 1991.

Ivashkiv, L. B., Liou, H.-C., Kara, C. J., Lamph, W. W., and Verma, I. M., mXBP/CRE-BP2 and c-Jun form a complex which binds to the cyclic AMP, but not to the 12-O-tetradecanoylphorbol-13-acetate, response element, *Mol. Cell. Biol.*, 10, 1609, 1990.

Kerppola, T. K. and Curran, T., Fos-Jun heterodimers and Jun homodimers bend DNA in opposite orientations: Implications for transcription factor cooperativity, *Cell*, 66, 317, 1991a.

Kerppola, T. K. and Curran, T., DNA bending by Fos and Jun: The flexible hinge model, *Science*, 254, 1210, 1991b.

Kouzarides, T. and Ziff, E., The role of the leucine zipper in the fos–jun interaction, *Nature*, 336, 646, 1988.

Kouzarides, T. and Ziff, E., Leucine zippers of fos, jun and GCN4 dictate dimerization specificity and thereby control DNA binding, *Nature*, 340, 568, 1989.

Landschulz, W. H., Johnson, P. F., and McKnight, S. L., The leucine zipper: A hypothetical structure common to a new class of DNA binding proteins, *Science*, 240, 1759, 1988.

Ma, J. and Ptashne, M., Deletion analysis of GAL4 defines two transcriptional activating segments, *Cell*, 48, 847, 1987.

McGregor, P. F., Abate, C., and Curran, T., Direct cloning of leucine zipper proteins: Jun binds cooperatively to the CRE with CRE-BP1, *Oncogene*, 5, 451, 1990.

Maekawa, T., Sakura, H., Kanei-Ishii, C., Sudo, T., Yoshimura, T., Fujisawa, J.-I., Yoshida, M., and Ishii, S., Leucine zipper structure of the protein CRE-BP1 binding to the cyclic AMP response element in brain, *EMBO J.*, 8, 2020, 1989.

Masquilier, D. and Sassone-Corsi, P., Transcriptional cross-talk: Nuclear factors CREM and CREB bind to AP-1 sites and inhibit activation by Jun, *J. Biol. Chem.* 267, 22460, 1992.

Moitoso de Vargas, L., Kim, S., and Landy, A., DNA looping generated by DNA bending protein IHF and the two domains of lambda integrase, *Science*, 244, 1457, 1989.

Nakabeppu, Y. and Nathans, D., The basic region of Fos mediates specific DNA binding, *EMBO J.*, 8, 3833, 1989.

Neuberg, M., Adamkiewicz, J., Hunter, J. B., and Müller, R., A Fos protein containing the Jun leucine zipper forms a homodimer which binds to the AP1 binding site, *Nature*, 341, 243, 1989a.

Neuberg, M., Schuermann, M., Hunter, J. B., and Müller, R., Two functionally different regions in Fos are required for the sequence-specific DNA interaction of the Fos/Jun protein complex, *Nature*, 338, 589, 1989b.

Oehler, T. and Angel, P., A common intermediary factor (p52/54) recognizing "acidic-blob"-type domains is required for transcriptional activation by the Jun proteins, *Mol. Cell. Biol.*, 12, 5508, 1992.

O'Neil, K. T., Hoess, R. H., and DeGrado, W. F., Design of DNA-binding peptides based on the leucine zipper motif, *Science*, 249, 774, 1990.

O'Shea, E. K., Rutkowski, R., and Kim, P. S., Evidence that the leucine zipper is a coiled coil, *Science*, 243, 538, 1989a.

O'Shea, E. K., Rutkowski, R., Stafford, W. F., III, and Kim, P. S., Preferential heterodimer formation by isolated leucine zippers from Fos and Jun, *Science*, 245, 646, 1989b.

O'Shea, E., Rutkowski, R., and Kim, P. S., Mechanism of specificity in the Fos-Jun oncoprotein heterodimer. *Cell*, 68, 699, 1992.

Papavasiliou, A. G., Treier, M., Chavrier, C., and Bohmann, D., Targeted degradation of c-Fos, but not v-Fos, by a phosphorylation-dependent signal on c-Jun, *Science*, 258, 1941, 1992.

Patel, L., Abate, C., and Curran, T., DNA binding by Fos Jun results in altered protein conformation. *Nature*, 347, 572, 1990.

Ransone, L. J. and Verma, I. M., Nuclear proto-oncogenes Fos and Jun, *Annu. Rev. Cell. Biol.*, 6, 539, 1990.

Rauscher, F. J., III, Cohen, D. R., Curran, T., Bos, T. J., Vogt, P. K., Bohmann, D., Tjian, R., and Franza, B. R., Jr., Fos-associated protein p39 is the product of the jun proto-oncogene, *Science*, 240, 1010, 1988a.

Rauscher, F. J., III, Sambucetti, L. C., Curran, T., Distel, R. J., and Spiegelman, B. M., Common DNA binding site for Fos protein complexes and transcription factor AP-1, *Cell*, 52, 471, 1988b.

Risse, G., Jooss, K., Neuberg, M., Brüller, H.-J., and Müller, R., Asymmetrical recognition of the palindromic AP-1 binding site (TRE) by Fos protein complexes, *EMBO J.*, 8, 3825, 1989.

Ryseck, R.-P. and Bravo, R., c-Jun, JunB and JunD differ in their binding affinities to AP-1 and CRE consensus sequences: Effect of Fos proteins, *Oncogene*, 6, 533, 1991.

Sassone-Corsi, P., Ransone, L. J., Lamph, W. W., and Verma, I. M., Direct interaction between fos and jun nuclear oncoproteins: Role of the "leucine zipper" domain. *Nature*, 336, 692, 1988.

Schuermann, M., Hunter, J. B., Hennig, G., and Müller, R., Non-leucine residues in the leucine repeats of Fos and Jun contribute to the stability and determine the specificity of dimerization, *Nucleic Acids Res.*, 19, 739, 1991.

Sellers, J. W. and Struhl, K., Changing Fos oncoprotein to a Jun-independent DNA-binding protein with GCN4 dimerization specificity by swapping 'leucine zipper', *Nature*, 341, 74, 1989.

Stein, B., Baldwin, A., Ballard, D., Greene, W., Angel, P., and Herrlich, P., Cross-coupling of the NFκBp65 and Fos/Jun transcription factors produces potentiated biological functions, *EMBO J.*, 12, 3879, 1993.

Sutherland, J. A., Cook, A., Bannister, A. J., and Kouzarides, T., Conserved motifs in Fos and Jun define a new class of activation domain, *Genes Dev.*, 6, 1810, 1992.

Tratner, I. and Verma, I. M., Identification of a nuclear targeting sequence in the Fos protein, *Oncogene*, 6, 2049, 1991.

Trietzenberg, S. J., Kingsbury, R. C., and McKnight, S. L., Functional dissection of VP16, the transactivator of herpes simplex virus immediate early gene expression, *Genes Dev.*, 2, 718, 1988.

Turner, R. and Tjian, R., Leucine repeats and an adjacent DNA binding domain mediate the formation of functional cFos-cJun heterodimers, *Science*, 243, 1689, 1989.

van Dam, H., Duyndam, M., Rottier, R., Bosch, A., de Vries-Smits, L., Herrlich, P., Zantema, A., Angel, P., and van der Eb, A., Heterodimer formation of cJun and ATF-2 is responsible for induction of c-jun by the 243 amino acid adenovirus E1A protein, *EMBO J.*, 12, 479, 1993.

Vinson, C. R., Sigler, P. B. and McKnight, S. L. Scissor-grip model for DNA recognition by a family of leucine zipper proteins, *Science*, 246, 911, 1989.

Vogt, P. K. and Bos, T. J., Jun: Oncogene and transcription factor, *Adv. Cancer Res.*, 55, 1, 1990.

Chapter 2

The Fos Family: Gene and Protein Structure, Homologies, and Differences

Marcus Schuermann

CONTENTS

INTRODUCTION

The c-*fos* proto-oncogene was first isolated as the cellular homologue of two viral *fos* oncogenes encoded by the Finkel-Biskis-Jenkins (FBJ) and Finkel-Biskis-Reilly (FBR) murine sarcoma viruses (MuSV) both of which induce osteosarcomas in rats and mice (Van Beveren et al., 1983; Finkel et al., 1966; Finkel et al., 1975). It has since been learned that expression of c-*fos* is rapidly and transiently inducible by the addition of growth factors, such as PDGF and EGF (see Chapter 8, this volume; Greenberg and Ziff, 1984; Kruijer et al., 1984; Müller et al., 1984). This stimulation involves direct transcriptional activation at the promoter level and places the c-*fos* gene among the first cellular "immediate early genes" described. Subsequent work has shown that the encoded product, the c-Fos protein, is a nuclear phosphoprotein, associated with chromatin (Renz et al., 1987; Sambucetti and Curran, 1986) and complexed to a second protein in the range of 39 kDa (hence, termed p39) (Curran and Teich, 1982a). Several important discoveries followed, which have laid the foundation for our current view of Fos as a transcription factor subunit (Müller, 1986; Curran and Franza, 1988; Herrlich and Ponta, 1989; Morgan and Curran, 1989; Lucibello and Müller, 1991; Angel and Karin, 1991). Among these discoveries were the following: (1) p39 is identical to the product of the proto-oncogene *jun* (Rauscher et al., 1988a; Angel et al., 1988; Sassone-Corsi et al., 1988a), (2) both proteins are major components of the transcription factor AP-1 (Bohmann et al., 1987; Lee et al., 1987; Chiu et al., 1988), and (3) a specific palindromic recognition element, $TGA^G/_CTCA$, described as AP-1 binding site or TRE (derived from *T*PA-*r*esponsive *e*lement), is the major DNA target for both proteins (Angel et al., 1987; Distel et al., 1987).

0-8493-4573-1/94/$0.00+$.50
© 1994 by CRC Press, Inc.

Similarly, three additional *fos*-related genes were identified, either by cross-hybridization of conserved DNA fragments (leading to isolation of the *fos*B gene) (Zerial et al., 1989) or through immunological cross-reactions of their encoded gene products, so-called Fos related antigens (Franza et al., 1987), giving rise to the isolation of the *fra*-1 (Cohen and Curran, 1988) and *fra*-2 genes (Matsui et al., 1990; Nishina et al., 1990).

All *fos* members exhibit a number of similarities, with respect to the regulation of their expression and the level of protein function. In most differentiated tissues, *fos* genes are expressed at low levels but are inducible by a variety of extracellular stimuli working through common signal transduction pathways (Franza et al., 1987; Cohen and Curran, 1988; Matsui et al., 1990; Nishina et al., 1990, see also Chapter 8, this volume). This rapid increase in the rate of transcription is counteracted by a number of mechanisms acting at the different transcriptional and posttranscriptional levels that help in the negative regulation of gene expression, which also implies cross-regulation among the Fos members (Lucibello and Müller, 1991).

At the protein level, all products encoded by *fos* genes show a number of similar properties, such as sequence-specific binding to the TRE, complex formation with Jun proteins, and transactivation from TRE-containing promoter sequences, which can be explained by the considerable degree of structural conservation within this protein family (Franza et al., 1987; Cohen and Curran, 1988; Matsui et al., 1990; Nishina et al., 1990; Schuermann et al., 1991b). These common aspects make it difficult to assess a particular role for the Fos protein in the process of cell proliferation or differentiation. Only recently has there been evidence that shows that individual members may also diverge in their presumptive biological functions. This evidence is based, on the one hand, on observed differences in protein function, such as the transregulatory influence on distinct promoter segments (Gizang-Ginsberg and Ziff, 1990; Suzuki et al., 1991; Wisdom et al., 1992), the transrepression of transcription from promoters containing serum-responsive elements (SREs) (Nakabeppu and Nathans, 1991; Mumberg et al., 1991; Yen et al., 1991), or the interference with members of the steroid receptor family (Lucibello et al., 1990). On the other hand, this assumption is also supported by the observation of differential expression of the individual *fos* genes in certain tissues (Morgan and Curran, 1989) and during certain stages of development (Dony and Gruss, 1987; Redemann-Fibi et al., 1991). Since many of these features are discussed elsewhere in this volume, it is the purpose of this chapter to summarize primarily the basic structural features of the *fos* genes and proteins as well as the influence of structure on protein function.

ORGANIZATION OF THE *fos* GENES AND GENE EXPRESSION

While the cDNA sequence of all *fos* genes has been determined, less is known about the genomic structure of the respective genes. To date, aside from the organization of the viral genes, only the structure of the c-*fos* and *fos*B genes has been sufficiently documented (Van Beveren et al., 1983; Van Straaten et al., 1983), and the structure of the *fra*-2 gene only in part (Nishina et al., 1990).

C-*fos*

The c-*fos* gene is 4 kbp long, including the proximal promoter sequences, and is interrupted by three introns (see Figure 2-1). From this gene, a single 2.2-kbp mRNA is transcribed, which is in agreement with the position mapped for the transcriptional start and the polyadenylation site. The mouse c-*fos* gene has been assigned to the [E-D] region of chromosome 12 (D'Eustachio, 1984), and the human gene assigned to chromosome 14q21–q31 (Barker et al., 1984). Sequence analysis of c-*fos* genes from mouse, man, and chicken reveals that the gene is highly conserved among vertebrate species; cDNAs

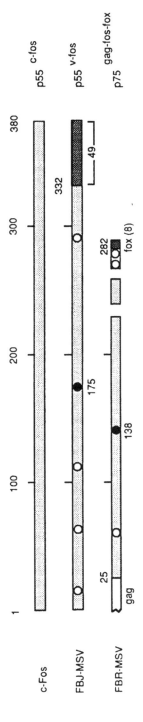

Figure 2-1 Structure of the viral and cellular c-Fos proteins (Curran and Teich, 1982b; Van Beveren et al., 1983, 1984). Hatched bars show areas of almost conserved structural identity, stippled and open bars in FBJ–Fos and FBR–Fos denote sequences not homologous to c-Fos. Amino acid positions are indicated relative to the c-Fos protein. Circles indicate point mutations leading to the substitution of individual amino acids in both proteins (positions 15, 67 110, 175, and 291 in FBJ–Fos and positions 64, 138, 268, 279, and 280 in FBR–Fos, relative to the c-Fos protein). The mutations affecting functional properties are indicated by a filled circle.

derived from mouse and chicken show 79% identity at the protein level (Mölders et al., 1987; Fujiwara et al., 1987). Underlining the high degree of evolutionary conservation, a gene related to c-*fos* (d*FRA*) exists in *Drosophila*. The encoded protein, harbors in its central region a domain, the "bZip region", that is conserved in all Fos proteins and is needed to bind to a palindromic TRE sequence and allow formation of dimeric complexes with a Jun-like protein (dJRA). Nevertheless, with respect to the rest of the protein, there are considerable deviations indicating that, in *Drosophila*, dFRA might also serve other functions (Perkins et al., 1990).

The c-*fos* promoter has been analyzed extensively, and a number of regulatory elements mediating the basal activity, response to growth factors, oncogenes, cAMP pathways and G_1-specific transcription factors have been identified (see Chapter 8, this volume). There are also indications that some of these relevant promoter elements as well as the exon–intron boundaries may be conserved in evolution, since these features are completely conserved between mouse and man (Van Straaten et al., 1983).

In most tissues, c-*fos* expression is tightly controlled. mRNA is expressed at only relatively low levels, but can be rapidly and transiently induced as early as 10 to 15 min after the addition of growth factors, phorbol esters, cytokines, or a number of compounds activating different intracellular pathways. Following a brief peak of expression, mRNA levels are reduced efficiently and kept low in the absence of external stimuli. At least three mechanisms have been shown to cooperate in this downregulation of c-*fos* expression: transcriptional shutoff through negative autoregulation of the c-*fos* promoter (König et al., 1989; Lucibello et al., 1989; Sassone-Corsi et al., 1988b; see also Chapter 8), premature termination of nascent RNA transcripts (Lamb et al., 1990), and, posttranscriptionally, rapid turnover of RNA because of destabilizing sequences located in both the 3′-untranslated and protein-coding regions (Meijlink et al., 1985; Shyu et al., 1989).

V-*fos*

It is generally believed that many proto-oncogenes were acquired by retroviruses during evolution and, hence, have been structurally modified to facilitate viral survival (Varmus, 1982). Underscoring this concept, two viral *fos* genes have been determined from the genome of two viral strains, both of which induce osteosarcomas in mice: the FBJ-MuSV and the FBR-MuSV (Curran and Teich, 1982b; Van Beveren et al., 1984; see also Chapter 14). The discovery of a virus-borne *fos* oncogene led to the identification of the corresponding cellular homologue, the c-*fos* proto-oncogene (Van Beveren et al., 1983).

Both viral *fos* genes show distinct structural deviations from the c-*fos* gene. FBJ-*fos* diverges at the 3′ end of the coding region resulting from a frame-shift mutation caused by a deletion of 104 nucleotides. This affects the translation of the last 48 amino acids following Pro332, which are replaced by 49 residues provided by the different reading frame (see Figure 2-2). Thus, the FBJ-MuSV derived *fos* gene codes for a protein of 381 amino acids, approximately 55 kDa in size, which is comparable to the molecular weight of the c-Fos protein. Because of the C-terminal alterations, however, important residues including regulatory phosphorylation sites are missing in FBJ-Fos, leading to increased protein stability, impaired negative autoregulation potential and, probably, release from nuclear export control as described by Roux et al. (1990; see also Chapter 7). Apart from the C-terminal alterations, the rest of FBJ-Fos is nearly identical to c-Fos, with the exception of five amino acid substitutions that are of no or only marginal relevance for protein function. To date, only the Glu–Lys substitution at position 175 has been shown to have a detectable influence on the interaction with Jun proteins (Schuermann et al., 1991a).

Contrary to FBJ, the FBR-MuSV derived *fos* gene diverges considerably from c-*fos*, owing to large structural alterations in the FBR-MuSV genome, including truncation at both termini: 25 amino acids of c-Fos at the amino-terminus and 98 amino acids at the

c-fos (chr. 12)

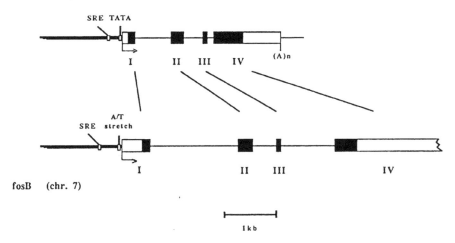

fosB (chr. 7)

1 k b

Figure 2-2 Genomic organization of the mouse c-*fos* and *fos*B genes (Lazo et al., 1992; Mumberg and Schuermann, unpublished results). Shown are the conserved elements in the promoter of both genes and the common exon–intron structure. SRE, serum response element.

C-terminus have been truncated. The *C*-terminus has been replaced by eight residues derived from cellular sequences (termed *fox)* (Müller, 1986). Moreover, due to the lack of premature splice and termination signals, the FBR-*fos* gene is expressed only as a Gag–Fos–Fox fusion protein with a molecular weight of 75 kDa. In addition to these gross alterations, two small internal deletions, spanning 13 and 10 amino acids, respectively, and five point mutations leading to amino acid substitutions have been found (see Figure 2-2). At present, it is not clear to what extent these individual deletions affect the functional properties of the FBR-Fos protein. Most noticeable, however, is the increased potential of FBR-Fos to morphologically transform (see Chapter 14) and to promote the establishment of mouse connective tissue cells. The latter was shown to be due to a single point mutation in the FBR-*fos* gene, leading to a change of Glu138 to Val (Jenuwein and Müller, 1987).

*fos*B

The *fos*B gene is similar to c-*fos* in both its genomic structure and with respect to the coding part (Zerial et al., 1989; Lazo et al., 1992; Mumberg and Schuermann, unpublished results). The *fos*B gene covers a region of approximately 8 kbp on mouse chromosome 7 (A1–B1 region) (Lazo et al., 1992). As can be deduced from Figure 2, the mouse *fos*B gene contains three introns and has identical exon–intron boundaries to c-*fos*. In addition, a serum response element (SRE) and an adjacent AP-1-like sequence (FAP) found in the c-*fos* promoter are also present in the *fos*B promoter. The promoter also contains several overlapping "TATA"-like elements. The start site of transcription has been located 40 nucleotides downstream (Lazo et al., 1992). A polyadenylation signal has not been mapped so far. The high degree of structural conservation between c-*fos* and *fos*B suggests that these genes have originated from a common ancestor form. Like c-*fos*, expression of *fos*B is also transiently inducible in most adult tissue by a number of growth factors and shows a comparable pattern with respect to its subsequent down regulation.

Contrary to c-*fos*, however, the primary transcript derived from the *fos*B gene is subjected to alternative splicing. The alternative splice eliminates 140 nucleotides within exon IV due to the presence of appropriate splice signals therein. Thus, two mature mRNA forms of approximately 3.9 and 4 kbp in length occur in the cytoplasm (Nakabeppu and Nathans, 1991; Mumberg et al., 1991; Yen et al., 1991; Dobrzanski et al., 1991). The mRNAs code for two versions of the same protein, both of which differ in their molecular properties, particularly in their transforming and transrepressing abilities (see below and Chapter 14, this volume). Furthermore, the two forms of *fos*B mRNA are also expressed at different times following serum stimulation, with the long form preceding the short form by approximately 30 min, thus pointing to a novel and potentially interesting way to regulate *fos*-dependent gene expression (Mumberg et al., 1991).

fra-1 AND *fra*-2

While the genomic structure of *fra*-1 has not been reported, two cDNA sequences derived from either rat embryo fibroblasts or human U937 monocytic cells have been isolated (Cohen and Curran, 1988; Matsui et al., 1990). Like c-*fos*, both *fra*-1 genes are highly conserved and cross-hybridize with avian genomic DNA. The rat *fra*-1 clone is a nearly full-length cDNA version and corresponds in size with the major 1.6-kbp transcript observed in most cells on serum or TPA stimulation. In addition, a minor 3.3-kbp transcript also has been observed in TPA-stimulated U937 cells (Matsui et al., 1990).

With respect to genomic organization, more is known about the *fra*-2 gene. As deduced from a partial sequence of chicken *fra*-2, this gene contains four exons with exon–intron boundaries identical to those of c-*fos* and *fos*B genes (Nishina et al., 1990). A *fra*-2 cDNA clone also has been obtained from human U937 cells. Both genes show a considerable degree of conservation (Matsui et al., 1990). Three mRNA forms can be distinguished following TPA treatment of U937 cells: two minor forms of 1.7 and 6 kbp and a major transcript of 2.3 kbp. In chicken embryo fibroblasts stimulated by serum, the 6 kbp form seems to be the most prominent. Whether these different mRNA transcripts result from the use of alternative start, splice, or polyadenylation signals remains to be shown. Unlike the rapid induction of c-*fos* and *fos*B mRNA, accumulation of *fra*-1 and *fra*-2 transcripts is not seen before 30 to 60 min following serum stimulation and remains elevated for approximately 1 to 3 hours thereafter. This initial lag phase in the kinetics of induction may reflect the additional requirement of factors that are newly synthesized and/or activated in response to mitogenic signals (e.g., c-Fos-regulated gene products; see Chapter 10, this volume), which may help to augment and maintain the level of transcription.

FOS PROTEIN STRUCTURE: A LESSON IN
MODULAR ARCHITECTURE

In support of the concept that all *fos* genes may have been derived from a common ancestor, the coding sequences of all *fos* genes show remarkable areas of almost complete conservation on the nucleic acid and protein levels. If one omits the two viral Fos proteins, FBJ-Fos and FBR-Fos, which are c-Fos homologues that have undergone different point mutations and structural alterations, the products of the *fos*-related genes show five major stretches of significant homology: two short areas comprising about 10 and 20 amino acids in the *N*-terminal part, a central domain of about 85 amino acids, and two stretches of approximately 25 amino acids each at the *C*-terminal end (see Figure 2-3). These areas of relatively conserved sequences are interrupted by stretches of amino acids showing little or no homology at all. The nonconserved segments are of variable length and are responsible for the different protein sizes, ranging from 380 amino acids for c-Fos (mouse/human) to 275 amino acids for Fra-2 (human; see Figure 2-3). The clustering of conserved and nonconserved areas suggests that all Fos proteins are structured in a modular way. These modules either serve common functions exerted by all proteins or are specific for the function of individual members. Concluding from the conserved exon–intron boundaries of the c-*fos, fos*B, and *fra*-2 genes, the divergency may have evolved gradually, rather than through mechanisms involving exon shuffling.

Several domains have been localized within the Fos protein that serve both common as well as separate molecular functions. The most important of these domains is a region in the central part of the protein that is indispensable for TRE-dependent transcriptional activation and morphological transformation (Jenuwein and Müller, 1987; Schuermann et al., 1989), two properties of Fos that seem to be closely linked. This region harbors the DNA binding domain (Halazonetis et al., 1988; Kouzarides and Ziff, 1988; Nakabeppu et al., 1988; Rauscher et al., 1988b; Gentz et al., 1989; Turner and Tjian, 1989; Neuberg et al., 1989a, 1989b), a region allowing complex formation with Jun proteins (Schuermann et al., 1989; Kouzarides and Ziff, 1988; Gentz et al., 1989; Turner and Tjian, 1989; Neuberg et al., 1989a; Landschulz et al., 1988b; Sassone-Corsi et al., 1988c; Ransone et al., 1989), and a nuclear translocation sequence (see below; Tratner and Verma, 1991; Jooss and Müller, unpublished observations). Common to nearly all cellular Fos proteins are sequences at the *C*-terminal end that, in combination with Jun proteins, are involved in the transrepression of SRE-dependent transcription (Neuberg et al., 1989b). Repression concerns the c-*fos* promoter (autorepression-type) (König et al., 1989; Lucibello et al., 1989; Sassone-Corsi et al., 1988b; Schönthal et al., 1989; see also Chapter 8) and other genes, e.g., the *egr*-1 and *egr*-2 genes (Gius et al., 1990). The exception to this rule is the FosB-S protein, which, as a result of alternative splicing (see above), lacks 101 amino acids of the FosB *C*-terminus and which, consequently, has lost its transrepressor function. A third molecular mechanism, which implies negative regulation of transcriptional control mediated by the glucocorticoid receptor, may involve sequences within the *N*-terminal part of c-Fos, located between amino acids 40 and 111. This domain is unique to c-Fos and may explain why this transrepressor function is not seen with other Fos proteins, such as FosB (Lucibello et al., 1990; see also Chapter 4). Finally, a number of potential phosphorylation sites are conserved within the Fos protein family that could serve as substrates for different protein kinases, including casein kinase II (CKII), cdc2, and protein kinase A (see Chapter 6, this volume). Because the c-Fos protein generally becomes extensively phosphorylated following treatment of cells with either serum or TPA (Barber and Verma, 1987; Müller et al., 1987), the analysis of this kind of protein modification is able to reveal important mechanisms that control the activity of the different Fos proteins.

Although, as indicated above, some regions involved in the different molecular functions of Fos have been characterized in more detail, there is still no clear concept concerning the relevance of other *N*- and *C*-terminal sequences in Fos, parts of which are conserved within this protein family and which, therefore, may serve important functions. Evidence demonstrating the existence of distinct transactivation or transrepressor domains is preliminary, and the mapping of posttranslational modification sites is still developmental. Our current understanding of the role of these additional protein domains, therefore, does not go far beyond comparing structural relationships among different members of the Fos family.

THE DNA CONTACT SITE (BASIC REGION)

As explained in Chapter 1, the complex formation between Fos, Jun, and the TRE is brought about by a conserved motif located in the central part of the protein. This region, which exists in an α-helical conformation, contains two clusters of basic amino acids

```
c-fos mouse   MMFSGFNADYEASSSRCSSASPAGDSLSYYHSPADSFSSMGSPVNTQDFCADLSVSSANF   60
fosB  mouse   M-FQAFPGDYDSGSRCSSSPSAESQ----YLSSVDSFGSPTAAASQE-CAGLGEMPGSF   54
fra-1 rat     M-YRDFGEPGPSSGAGSA----------YGRPAQPQQAQTQTVQQQKFHL--------   39
fra-2 human   M-YQDYPGNFDTSSRGSS----------GSPAHAESYSSGGGQQKFRVDMPGSGSAF   47
              *  + +++  +        +++ ++                         ++

c-fos mouse   IPTVTAISTSPDLQWLVQPTLVSSVA------------------PSQTRAPHY----GLPTQS   102
fosB  mouse   VPTVTAITTSQDLQWLVQPTLISSMAQSQGQPLASQPPAVDPYDMPGTSYST-PGLSAYS   113
fra-1 rat     VPSINAVSGSQELQWMVQPHFLGP----------------SGYPR-PLTYPQYS-----P   77
fra-2 human   IPTINAITTSQDLQWMVQPTVITSMS--------------NPYPR-SHPYSPLPGLASVP   92
              +++++*+++++++++++*  +++                    + +*

c-fos mouse   AG-AYARAGMVKT--------VSGGRAQSIGRGKVEQLSPEEEKRRIRRERNKMAAAKCR   155
fosB  mouse   TGGASGSGGPSTSTTTSGPVSARPARARPRRPREETLTPEEEKRRVRRERNKLAAAKCR   173
fra-1 rat     PQ---PRPGVIRA------LGPPPGV----RRRPCEQISPEEEERRVRRERNKLAAAKCR   125
fra-2 human   GHMALPRPGVIKT------IGTTVGR---RRRD-EQLSPEEEEKRRIRRERNKLAAAKCR   142
              +*         ++ +    **   * ++******+*+********+**+********

c-fos mouse   NRRRELTDTLQAETDQLEDEKSALQTEIANLLKEKEKLEFILAAHRPACKIPDDLGFPEE   215
fosB  mouse   NRRRELTDRLQAETDQLEEEKAELESEIAELQKEKERLEFVLVAHKPGCKIP----YEEG   229
fra-1 rat     NRRKELTDFLQAETDKLEDEKSGLQREIEELQKQKERLELVLEAHRPICKIP------EED   180
fra-2 human   NRRRELTEKLQAETEELEEEKSGLQKEIAELQKEKEKLEFMLVAHGPVCKIS------PEE   197
              ********+  *** ++++++++*  *+++  *********+*** *** **

c-fos mouse   MSVASLDLTGGLPEASTPESEEAFTLPLLNDPEPKPSLEPVKSISNVELKAEPFDDFLFP   275
fosB  mouse   PGPGPLAEVRDLPG-STSAKEDGFGWLL--PPPPPPL----------PFQ-------PPL   267
fra-1 rat     KKD-----TGGTSSTSGAGS--------------------------PPG-------P   199
fra-2 human   RRSPP---APGLQPMRSGGG-----------------------SVGAVVVKQEPLEE-DSP   231
              +  +   ++ +

c-fos mouse   ASSRPSGSETSRSVPD-VDLSGSFYAADWEPLHSNSLGMGPMVTELEPLCTPVVTCTPGC   334
fosB  mouse   ------------SSRDAPPNLTASLFTH---------SEVQVLGDPFPVVSPS-   299
fra-1 rat     CRPVPCISLSPGPVLEP---------------EALHTPTL----------MT-TPSL   230
fra-2 human   SSSAGLDKAQRSVIKPISIAGGFYGE--EPLHTPIV----------VTSTPAV   273
              +          + +*+
```

```
c-fos mouse   TTYTSSFVFTYP-----EADSFPS--CAAAHRKGSSSN-EPSSDSLSSPTLLAL   380
fosB  mouse   --YTSSFVLTCP-----EVSAFAG-----AQR---TSGSEQPSDPLNSPSLLAL   338
fra-1 rat     TPFTPSLVFTYPST-------PEPCSSAHRKSSSSSGDPSSDPLGSPTLLAL     275
fra-2 human   TPGTSNLVFTYPSVLEQESPASPSESCSKAHRRSSSS-GDQSSDSLNSPTLLAL   326

              ++++++++*            **  +*  ++++++++++*++*****
```

Figure 2-3 Sequence alignment of the c-Fos, FosB, Fra-1, and Fra-2 proteins. Amino acids identical in all four proteins are indicated beneath by an asterix; + denotes residues similar in character. Data are taken from Van Beveren et al., 1983; Cohen and Curran, 1988; Zerial et al., 1989; Matsui et al., 1990.

followed by a periodic array of five leucines and a histidine. The predominant basic character of the first motif and the original assumption on the function of the leucine repeat lead to generation of the term bZip (*basic* region + leucine *zip*per) (Kerppola and Curran, 1991). Apart from members of the Fos and Jun family, this motif was also found in several other nuclear factors, including members of the ATF/CREB family, C/EBP and GCN4, and hence denominates these proteins as a common subclass of transcription factors (bZip proteins; see Figure 2-4). Interaction with DNA first depends on the two basic regions and the alanine spacer residing therein (Gentz et al., 1989; Turner and Tjian, 1989; Neuberg et al., 1989b). When basic amino acid clusters are substituted with uncharged residues or the charge density is altered, DNA binding is almost completely abolished (Gentz et al., 1989; Turner and Tjian, 1989; Neuberg et al., 1989b; Ransone et al., 1989). Moreover, the recognition of a TRE requires the exact alignment of the two basic stretches since the introduction of two additional alanines in the spacer region disrupts binding (Neuberg et al., 1989b). This indicates that the alanine region may indeed help to position the flanking residues next to specific bases in the TRE. Apart from an intact basic region, DNA binding by Fos requires complex formation with Jun proteins (Neuberg et al., 1989a, 1989b), which is mediated by residues in the adjacent leucine zipper region (see Figure 2-4). Based on the presence of bZip motifs in many nuclear proteins, including Fos and Jun, the identical spacing of the leucine repeat, and the nearly symmetrical nature of the TRE, several models have been proposed to explain how the dimers are located with respect to the DNA recognition sequence. One idea (scissor grip model) has received support from a number of experiments (see also Chapter 1), including work on other bZip proteins (C/EBP; Vinson et al., 1989). The scissor grip model predicts that the helices of the two interacting proteins bifurcate beyond the leucine zipper as a consequence of the positive charges of the two juxtaposed basic regions. The two basic domains would then track along the major groove of the DNA in opposite directions (BR-II domain), bending in the vicinity of the conserved asparagine residue in order to "grip" around the DNA (with the help of the BR-I domain). Evidence supporting this model of symmetrical interaction resembling a "Y"-shaped structure has been provided by a number of experiments using circular dichroism spectroscopy and proteolytic cleavage on preformed protein–DNA complexes (O'Neil et al., 1990; Shuman et al., 1990; Talanian et al., 1990). Results obtained from substitution experiments in which grouped or individual amino acids within the basic region have been exchanged are also in line with this hypothesis (Neuberg et al., 1991).

Interestingly, the DNA binding potential of many transcription factors can be controlled through posttranslational modification, thus allowing its modulation by specific regulatory enzymes (see also subsequent chapters). In Jun, phosphorylation of three residues, Thr231, Ser243, and Ser249, was shown to negatively influence DNA binding (Boyle et al., 1991; Karin and Smeal, 1992). This phosphorylation is brought about by different protein kinases, either through CKII on Thr231 and Ser249 or, at least *in vitro*, through kinases identical or similar to glycogen synthase kinase-3 (GSK-3) and extracellular signal-regulated kinase (ERK-2). There is, however, no evidence that phosphorylation can modulate the DNA binding capacity of Fos in a similar way. Apart from protein phosphorylation, *in vitro* DNA binding studies have revealed that both proteins seem to be controlled by an unusual mechanism of protein modification that involves reduction and oxidation of a single cysteine residue located in the basic domain of both Fos and Jun proteins (Abate et al., 1990a, 1990b). *In vitro*, oxidation interferes with the binding activity, which can be restored by treatment with high concentrations of reducing agents or with a cellular factor identical to an endonuclease involved in DNA repair (Xanthoudakis and Curran, 1992). There is no clear evidence for the existence of such redox regulation *in vivo*. The Cys154 residue nevertheless seems to be important. When replaced by serine,

```
              BR-I      spacer  BR-II           L1             L2            L3            L4         L5

c-Fos     ekrirrerNkmAaakcBnrrreltdt l qaetdq l edeksa Lqteian l lkekek l efilaah
FosB      ekrivrrerNlAaakcBnrrreltdr l qaetdq l eeekae Leseiae l qkeker l efvlvah
Fra-1     errrvrrerNk l AaakcBnrrkeltdf l qaetdk l edeksg Lqreiee l qkqker l elvleah
Fra-2     ekrirrerNk l AaakcRnrrreltek l qaetee l eeeksg Lqkeiae l qkekek l efmlvah

c-Jun     ikaerkrmqNriAaskcRkrk l eriar l eekvkt l kaqnse Lastanm l reqvaq l kqkvmnh
JunB      ikverkrlrNrlAatkcRkrk l eriar l edkvkt l kaenag Lssaagl l reqvaq l kqkvmth
JunD      ikaerkrlrNiAaskcRkrk l erisr l eekvkt l ksqnte Lastasl l reqvaq l kqkvlsh

GCN4      dpaalkrarNteAarrsBark l qrmkq l edkvee l lsknyh Lenevar l kklvger*
C/EBP     neyrvrrerNniAvrksBdkakqrnve t qqkvle l tsdndr l rkrveq l sreldt l rgifrql
CREB      rkrevrlmkNreAarecRrkkkeyvkc l enrvav l enqnkt Lieelka l kdlychksd*
CRE-BP1   ekrrkflerNraAasrqBqktrkvwvqs l ekkaed l sslngq l qsevtl l rnevaq l kqlllah..
```

Figure 2-4 Configuration of the bZip region in members of the Fos and Jun families and several related proteins showing structural conservation. Amino acids identical in all proteins are written in capital letters. Basic amino acids and the conserved leucines and histidines in every seventh position are underlined. Two stretches of amino acids containing predominantly basic amino acids (BR-I and BR-II) are clearly discernible in the N-terminal part, separated by a short alanine-rich segment starting with an asparagine residue. The spacing between the basic region and the leucine zipper is identical in all bZip proteins. Amino acids between the leucines are conserved solely among members of the same family. The amino acid sequences are listed for c-Fos (Van Beveren et al., 1983), FosB (Zerial et al., 1989), Fra-1 (Cohen and Curran, 1988), Fra-2 (Nishina et al., 1990), c-Jun (Ryseck et al., 1988), JunB (Ryder et al., 1988), JunD (Hirai et al., 1989; Ryder et al., 1989), GCN4 (Vogt et al., 1987), C/EBP (Landschulz et al., 1988a), CREB (Hoeffler et al., 1988), and CRE-BP1 (Maekawa et al., 1989).

the mutant protein missing this potential regulatory site exhibits a 10-fold increased affinity for the TRE. *In vitro* binding of the mutant is no longer affected by oxidizing agents. Moreover, this mutant also shows an increased transforming and colony forming potential when assayed on chicken embryo fibroblasts (Okuno et al., 1993).

In agreement with the bipartite structure of the DNA binding domain in all bZip proteins is the alternative use of the basic residues therein as a nuclear targeting sequence. The two clusters of basic amino acids are essential for import of the protein into the nucleus (Dingwall and Laskey, 1991). Such clusters are shared by a large number of nuclear proteins of different origin and function. Mutational analysis of the nucleoplasmin and N1 proteins from *Xenopus* that bear similar motifs, has revealed the necessity of at least 2 basic amino acids in each of the two basic regions separated by a variable spacer region of approximately 10 amino acids (Dingwall and Laskey, 1991). The first experimental evidence for the relevance of this motif in the nuclear translocation of Fos came from substitution experiments in which the basic region had been fused to the chicken pyruvate kinase, thereby enabling this fusion protein to be translocated to the nucleus (Tratner and Verma, 1991). Thus it seems likely that the α-helical and bipartite configuration of the basic region in all Fos proteins does indeed serve a dual function: the direct interaction with one half of the palindromic TRE and, in the absence of DNA, the interaction with receptor molecules involved in the nuclear translocation process.

INTERACTION WITH JUN PROTEINS:
INVOLVEMENT OF LEUCINE ZIPPERS

DNA binding not only is controlled by amino acids in the basic region but also requires the formation of specific heterodimeric complexes. Such complexes exist as either low-affinity homodimers among Jun family members or high-affinity, stable heterodimers between members of the Fos and Jun families; Fos proteins on their own, however, are unable to form homodimers. It has been shown that both Fos and Jun can also form heterodimers with selected members of the ATF/CREB family, thereby enlarging the "AP-1 superfamily" (Karin and Smeal, 1992). Thus, a multitude of homo- and heterodimeric protein complexes can be generated from a limited number of Fos and Jun monomers all of which exhibit similar affinities for the TRE *in vitro* (Vinson et al., 1989; see also Chapter 1). Once formed, these complexes are quite stable and cannot be disrupted even under relatively stringent conditions (Curran and Teich, 1982; Curran et al., 1985).

It is now generally accepted that both Fos and Jun interact via a specific short α-helical segment present in both proteins that is directly adjacent to the DNA binding region and which contains regularly arrayed leucines as the most prominent characteristic (leucine zipper) (Landschulz et al., 1988b; see above and Figure 2-4). A schematic view showing the configuration of such an interaction between Fos and Jun is represented in Figure 2-5 as a "helical wheel." According to this model, both leucine zippers are thought to align to each other in the form of a coiled coil, thereby forming a common interactive surface (O'Shea et al., 1989a, 1989b). Major predictions concerning the stability and specificity of the specific protein dimers formed in this way have been confirmed by experimental data mainly based on a systematic mutational analysis of both leucine and nonleucine residues in Fos and Jun. These experimental findings demonstrate the clear necessity of a 4-3 repeat of hydrophobic residues, including the leucines that provide the stability of the complexes (Schuermann et al., 1991a; positions *a* and *d*, according to O'Shea et al., 1989b; see also Figure 2-5 and Chapter 1, this volume). On the other hand, charged amino acids in the laterally aligned sequences largely determine the specificity of the Fos/Jun interaction, thus preventing the formation of Fos/Fos homodimers, but allowing Jun/Jun homodimers and, preferentially, Fos/Jun heterodimers to form (positions *e* and *g* in Figure 2-5) (Schuermann et al., 1991a; Smeal et al., 1989). In agreement with their functional

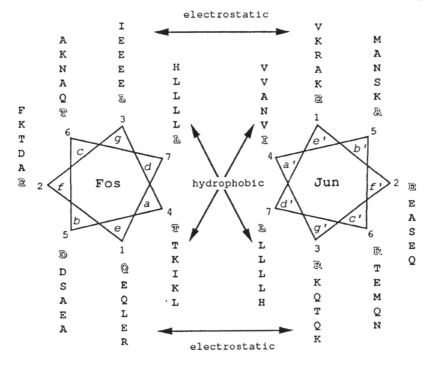

electrostatic

Figure 2-5 Helical wheel model of the leucine zipper in Fos and Jun, according to O'Shea et al. (1989a, 1989b). Amino acid sequences are listed *N*- to *C*-terminal, with the first *N*-terminal positions outlined. The helical positions are indicated by lower case letter code. Arrows show hydrophobic interactions between *d* and *d′* (leucines) and *a* and *a′*. Electrostatic interactions are supposed to form between residues in positions *g* and *e′* and *e* and *g′*.

implication, these residues are also the ones most conserved within each protein family (see Figure 2-4). Finally, evidence that leucine zippers are the major determinants of dimer formation comes from domain swap experiments. Transferring the leucine zipper of Jun or GCN4 into Fos, for example, allows the resulting protein to form homodimers and to bind specifically to a TRE-containing oligonucleotide (Neuberg et al., 1989a; Kouzarides and Ziff, 1989; Sellers and Struhl, 1989).

Apart from confirming structural predictions of the Fos/Jun protein interactions, leucine zipper mutants have been valuable in dissecting the molecular mechanisms through which Fos and Jun exert their transregulatory function. Only through the use of these mutants could it be shown that the interaction of Fos and Jun is a prerequisite for sequence-specific DNA binding, that both Fos and Jun are able to directly interact with the TRE, and that the formation of stable Fos/Jun complexes is fundamental for many protein functions *in vivo* (described in more detail in the following chapters).

DOMAINS INVOLVED IN TRANSCRIPTIONAL ACTIVATION

As for Jun, the Fos protein transcription factors contain specific domains required for the activation of transcription. A Jun mutant lacking its own transactivation domain becomes transcriptionally competent by heterodimerization with Fos (Angel et al., 1989). Moreover, when fused to the DNA binding domain of the yeast transcription factor GAL4, the

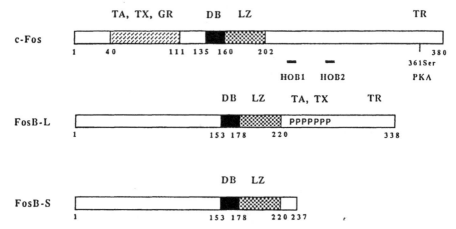

Figure 2-6 Modular composition of the Fos and Fos B protein. TA, area involved in transcriptional activation; TX, sequences required for optimal transformation capacity; GR, region necessary for the repression of glucocorticoid receptor activity; TR, transrepression function; HOB1, HOB2, homology boxes 1, 2; PKA, protein kinase A.

resulting GAL4-Fos hybrid protein is a potent activator of reporter plasmids containing GAL4 binding sites (Schuermann, unpublished results; Gebel, et al., unpublished results). Deletion of a large part of the C-terminus, however, completely abolishes transactivation by GAL4-Fos (Gebel, et al., unpublished results), demonstrating that Fos has transactivation abilities. Mapping of transactivation domains in the context of the wild-type, nonfused c-Fos protein has, however, proven to be more difficult. The potential to activate TRE-dependent transcription remains nearly unchanged when large parts of the *N*- or *C*-terminus are deleted. Most of this may be due to the fact that Fos and Jun are cooperative in this respect, so that loss of one particular domain can easily be compensated for by others existing in either protein (Abate et al., 1991). There is, however, evidence that the *N*-terminal half in Fos harbors at least two domains important for its ability to activate TRE-dependent transcription: one between amino acids 40 and 110, with no prevalence of particular residues typical for transactivation domains (Lucibello et al., 1991), and the other directly adjacent to the basic region (amino acids 116–139), which is predominantly acidic in character and conserved in all four Fos proteins (Abate et al., 1991; Lucibello et al., 1991) (see Figure 2-6). In addition, two motifs, referred to as HOB1 and HOB2, have recently been identified in the Fos *C*-terminus (amino acids 226–236 and 267–276) (Sutherland et al., 1992). Both are highly homologous to sequences in the Jun A1 activation domain. Although transactivation triggered by each of the two domains remains low, together they are efficient transactivators, thus pointing to a cooperative function in transcriptional activation. HOB1 and HOB2 motifs are conserved among members of the Jun family and are located within two regions previously identified to be required for transactivation (domains II and III in Angel et al., 1989). HOB2 is also found in C/EBP. Except for the acidic region, however, all activating regions mapped in Fos so far are unique to this protein, suggesting that gene regulation by the Fos-related proteins most likely also involves other mechanisms that remain to be identified. The same applies to the FBR-Fos protein, which lacks HOB1 and HOB2.

In FosB, an activation domain rich in prolines is represented by a short segment between amino acids 257 and 274 that is similar to the activation domains in one class of transcription factors (Mermod et al., 1989). *fos*B constructs lacking this domain exhibit a drastically reduced transcriptional and transforming potential (Wisdom et al., 1992). Furthermore, the sequences that constitute this region (amino acids 226–284) can restore the transcription activation potential of a truncated GAL4 protein, containing only the DNA binding domain, to about 50% of the rate yielded with the parental FosB protein in this place (Wisdom et al., 1992). Interestingly, the naturally occurring truncated form, FosB-S, does not contain this domain, which explains its low transcriptional and transforming potential. FosB-S, however, is not totally deficient in transcriptional activation. On a multimeric TRE reporter construct it can still stimulate transcription to about half the rate exhibited by FosB (Dobrzanski et al., 1991; Schuermann et al., 1993). This suggests that there is at least one additional activator domain residing in the N-terminal part, the contribution of which remains to be determined.

CONCLUSIONS

Molecular analysis of the Fos and Jun proteins has yielded a large body of new information concerning the structural requirements of the components of the AP-1 transcription factor. This analysis also has been a paradigm for the possibility to predict molecular functions from hypothetical assumptions on structural comparisons. The results obtained with respect to complex formation via leucine zippers and specific DNA binding mediated by a basic helical motif in Fos and Jun have had general implications for our understanding of the action of transcription factors. Our view on the Fos/Jun–DNA interaction alone, however, cannot explain why there is such a multitude of different Fos/Jun, Jun/Jun, and Jun/ATF complexes in one cell at the same time, all of which have similar *in vitro* properties. We therefore have to devise other mechanisms to explain some of the differences observed in the regulation of gene expression. At the molecular level, we can envisage several possibilities: (1) subtle differences in or around the DNA contact site, (2) additional factors in AP-1-mediated gene regulation that interact specifically with distinct members, and (3) differential responses to protein phosphorylation as a consequence of mitogen stimulation. Analogous to the structure–function analyses described above, we need to focus more on the structurally dissimilar regions within the Fos proteins that could participate in such specific interactions. Most important, however, is the need to find target genes for these AP-1 factors that could serve as a test model to analyze the subtle differences in transcriptional response. The years to come may, therefore, hold many challenging discoveries.

REFERENCES

Abate, C., Luk, D., Gentz, R., Rauscher, F. J., III, and Curran, T., Expression and purification of the leucine zipper and DNA binding domains of Fos and Jun: Both Fos and Jun contact DNA directly, *Proc. Natl. Acad. Sci. U.S.A.*, 87, 1032, 1990a.

Abate, C., Patel, L., Rauscher, F. J., III, and Curran, T., Redox regulation of Fos and Jun DNA-binding activity *in vitro*, *Science*, 249, 1157, 1990b.

Abate, C., Luk, D., and Curran, T., Transcriptional regulation by Fos and Jun *in vitro*: Interaction among multiple activator and regulatory domains, *Mol. Cell. Biol.*, 11, 3624, 1991.

Angel, P. and Karin, M., The role of Jun, Fos and the AP-1 complex in cell-proliferation and transformation, *Biochim. Biophys. Acta*, 1072, 129, 1991.

Angel, P., Imagawa, M., Chiu, R., Stein, B., Imbra, R. J., Rahmsdorf, H. J., Jonat, C., Herrlich, P., and Karin, M., Phorbol ester-inducible genes contain a common cis element recognized by a TPA-modulated trans-acting factor, *Cell*, 49, 729, 1987.

Angel, P., Allegretto, E. A., Okino, S. T., Hattori, K., Boyle, W. J., Hunter, T., and Karin, M., Oncogene *jun* encodes a sequence-specific trans-activator similar to AP-1, *Nature*, 332, 166, 1988.

Angel, P., Smeal, T., Meek, J., and Karin, M., Jun and v-Jun contain multiple regions that participate in transcriptional activation in an interdependent manner. *New Biol.*, 1, 35, 1989.

Barber, J. R. and Verma, I. M., Modification of Fos proteins: Phosphorylation of c-Fos, but not v-Fos, is stimulated by 12-tetradecanoyl-phorbol-13-acetate and serum. *Mol. Cell. Biol.*, 7, 2201, 1987.

Barker, P. E., Rabin, M., Watson, M., Breg, W. R., Ruddle, F. H., and Verma, I. M., Human c-fos oncogene mapped within chromosomal region 14q21 – q31, *Proc. Natl. Acad. Sci. U.S.A.*, 81, 5826, 1984.

Bohmann, D., Bos, T. J., Admon, A., Nishimura, T., Vogt, P., and Tjian, R., Human proto-oncogene c-*jun* encodes a DNA binding protein with structural and functional properties of transcription factor AP-1, *Science*, 38, 1386, 1987.

Boyle, W. J., Smeal, T., Defize, L. H. K., Angel, P., Woodgett, J. R., Karin, M., and Hunter, T., Activation of protein kinase C decreases phosphorylation of c-Jun at sites that negatively regulate its DNA-binding activity, *Cell*, 64, 573, 1991.

Chiu, R., Boyle, W. J., Meek, J., Smeal, T., Hunter, T., and Karin, M., The c-Fos protein interacts with c-Jun/AP-1 to stimulate transcription of AP-1 responsive genes. *Cell*, 54, 541, 1988.

Cohen, D. R. and Curran, T., *fra*-1: A serum-inducible, cellular immediate-early gene that encodes a Fos-related antigen. *Mol. Cell. Biol.*, 8, 2063, 1988.

Curran, T. and Teich, N. M., Identification of a 39,000 dalton protein in cells transformed by FBJ murine osteosarcoma virus, *Virology*, 116, 221, 1982a.

Curran, T. and Teich, N. M., Candidate product of the FBJ murine osteosarcoma virus oncogene: Characterization of a 55,000 kD phosphoprotein, *J. Virol.*, 42, 114, 1982b.

Curran, T., Van Beveren, C., Ling, N., and Verma, I. M., Viral and cellular fos proteins are complexed with a 39,000-dalton cellular protein, *Mol. Cell. Biol.*, 5, 167, 1985.

Curran, T. and Franza, B. R., Jr., Fos and jun: The AP-1 connection, *Cell*, 55, 395, 1988.

D'Eustachio, P., A genetic map of mouse chromosome 12 composed of polymorphic DNA fragments. *J. Exp. Med.*, 160, 827, 1984.

Dingwall, C. and Laskey, R. A., Nuclear targeting sequences — a consensus? *Trends Biochem. Sci.*, 16, 478, 1991.

Distel, R. J., Ro, H.-S., Rosen, B. S., Groves, D. L., and Spiegelman, B. M., Nucleoprotein complexes that regulate gene expression in adipocyte differentiation: Direct participation of c-Fos, *Cell*, 49, 835, 1987.

Dobrzanski, P., Noguchi, T., Kovary, K., Rizzo, C. A., Lazo, P. S., and Bravo, R., Both products of the *fos*B gene, FosB and its short form, FosB/SF, are transcriptional activators in fibroblasts, *Mol. Cell. Biol.*, 11, 5470, 1991.

Dony, C. and Gruss, P., Proto-oncogene c-*fos* expression in growth regions of fetal bone and mesodermal web tissue, *Nature*, 328, 711, 1987.

Finkel, M. P., Biskis, B. O., and Jinkins, P. B., Virus induction of osteosarcomas in mice, *Science*, 151, 698, 1966.

Finkel, M. P., Reilly, C. A., and Biskis, C. O., Viral etiology of bone cancer, *Front. Radiat. Ther. Oncol.*, 10, 28, 1975.

Franza, B. R., Jr., Sambucetti, L. C., Cohen, D. R., and Curran, T., Analysis of *fos* protein complexes and *fos*-related antigens by high-resolution two-dimensional gel electrophoresis, *Oncogene*, 1, 213, 1987.

Fujiwara, K. T., Ashida, K., Nishina, H., Iba, H., Miyajima, N., Nishizawa, M., and Kawai, S., The chicken c-*fos* gene: Cloning and nucleotide sequence analysis, *J. Virol.*, 61, 4012, 1987.

Gentz. R.. Rauscher. F. J., III. Abate. C., and Curran. T., Parallel association of Fos and Jun leucine zippers juxtaposes DNA binding domains, *Science*, 243, 1695, 1989.

Gius. D.. Cao. X.. Rauscher. F. J., III. Cohen. D. R.. Curran. T.. and Sukhatme. V. P., Transcriptional activation and repression by Fos are independent functions: The C terminus represses immediate-early gene expression via CArG elements, *Mol. Cell. Biol.*, 10. 4243. 1990.

Gizang-Ginsberg. E. and Ziff. E. B., Nerve growth factor regulates tyrosine hydroxylase gene transcription through a nucleoprotein complex that contains c-Fos. *Genes Dev.*, 4. 477. 1990.

Greenberg. M. E. and Ziff. E. B.. Stimulation of 3T3 cells induces transcription of the c-*fos* proto-oncogene. *Nature*, 311. 433. 1984.

Hai, T. and Curran. T.. Cross-family dimerization of transcription factors Fos/Jun and ATF/CREB alters DNA binding specificity. *Proc. Natl. Acad. Sci. U.S.A.*, 88. 3720, 1991.

Halazonetis. T. D.. Georgopoulos. K.. Greenberg. M. E., and Leder. P.. c-Jun dimerizes with itself and with c-Fos. forming complexes of different DNA binding affinities. *Cell*, 55. 917. 1988.

Herrlich. P. and Ponta. H.. Nuclear oncogenes convert extracellular stimuli in the genetic program. *Trends Genet.*, 5. 112, 1989.

Hirai. S.-I.. Ryseck. R.-P.. Mechta. F.. Bravo. R., and Yaniv. M.. Characterisation of *jun*D: A new member of the *jun* proto-oncogene family. *EMBO J.*, 8. 1433. 1989.

Hoeffler. J. P.. Meyer. T. E.. Yun. Y.. Jameson. J. L., and Habener. J. F.,'Cyclic AMP-responsive DNA-binding protein: Structure based on a cloned placental cDNA. *Science*, 242. 1430. 1988.

Jenuwein. T. and Müller. R.. Structure–function analysis of Fos protein: A single amino acid change activates the immortalizing potential of v-*fos*, *Cell*, 48. 647. 1987.

Karin. M. and Smeal. T.. Control of transcription factors by signal transduction pathways: The beginning of the end. *Trends Biol. Sci.*, 17. 418. 1992.

Kerppola. T. K. and Curran. T.. Transcription factor interactions: Basics on zippers. *Curr. Opinion Struct. Biol.*, 1. 71. 1991.

König. H.. Ponta. H.. Rahmsdorf. U.. Büscher. M.. Schönthal. A.. Rahmsdorf. H. J.. and Herrlich. P.. Autoregulation of *fos:* The dyad symmetry element as the major target of repression. *EMBO J.*, 8. 2559. 1989.

Kouzarides. T. and Ziff. E.. The role of the leucine zipper in the Fos–Jun interaction. *Nature*, 336. 646. 1988.

Kouzarides. T. and Ziff. E.. Leucine zippers of Fos, Jun and GCN4 dictate dimerization specificity and thereby control DNA binding. *Nature*, 341. 568, 1989.

Kruijer. W.. Cooper. J. S.. Hunter. T.. and Verma, I.. Platelet-derived growth factor induces rapid but transient expression of the c-*fos* gene and protein. *Nature*, 312. 711. 1984.

Lamb. N. J. C.. Fernandez. A.. Tourkine. N.. Jeanteur. P.. and Blanchard. J.-M.. Demonstration in living cells of an intragenic negative regulatory element within the rodent c-*fos* gene. *Cell*, 61. 485. 1990.

Landschulz. W. H.. Johnson. P. F.. Adashi. E. Y.. Graves. B. J.. and McKnight. S. L.. Isolation of a recombinant copy of the gene encoding C/EBP. *Genes Dev.*, 2. 786. 1988a.

Landschulz. W. H.. Johnson. P. F.. and McKnight. S. L.. The leucine zipper: A hypothetical structure common to a new class of DNA binding proteins, *Science*, 240, 1759, 1988b.

Lazo. P. S.. Dorfman. K.. Noguchi. T.. Mattéi. M.-G.. and Bravo. R.. Structure and mapping of the *fos*B gene. FosB downregulates the activity of the *fos*B promoter, *Nucleic Acids Res.*, 20. 343. 1992.

Lee. W.. Mitchell. P.. and Tjian. R.. Purified transcription factor AP-1 interacts with TPA-inducible enhancer elements, *Cell*, 49. 741. 1987.

Lucibello, F. C., Lowag, C., Neuberg, M., and Müller. R.. *Trans*-repression of the c-*fos* promoter: A novel mechanism of Fos-mediated trans-regulation. *Cell, 59, 999, 1989.

Lucibello, F. C., Slater, E. P., Jooss, K. U., Beato, M., and Müller, R.. Mutational transrepression of Fos and the glucocorticoid receptor: Involvement of a functional domain in Fos which is absent in FosB. *EMBO J., 9, 2827, 1990.*

Lucibello, F. C. and Müller R.. Proto-oncogenes encoding transcriptional regulators: Unraveling the mechanisms of oncogenic conversion. *Crit. Rev. Oncogen., 2, 259, 1991.*

Lucibello, F. C., Neuberg, M., Jenuwein, T., and Müller, R.. Multiple regions of v-Fos protein involved in the activation of AP-1 dependent transcription: Is trans-activation crucial for transformation? *New Biol., 3, 671, 1991.*

Maekawa, T., Sakura, H., Kanei-Ishii, C., Sudo, T., Yoshimura, T., Fujisawa, J.-I., Yoshida, M., and Ishii, S., Leucine zipper structure of the protein CRE-BP1 binding to the cyclic AMP response element in brain. *EMBO J., 8, 2023, 1989.*

Matsui, M., Tokuhara, M., Konuma, Y. N. N., and Ishizaki, R.. Isolation of human *fos*-related genes and their expression during monocyte-macrophage differentiation. *Oncogene, 5, 249, 1990.*

Meijlink, F., Curran, T., Miller, A. D., and Verma, I. M.. Removal of a 67-base-pair sequence in the noncoding region of protooncogene *fos* converts it to a transforming gene. *Proc. Natl. Acad. Sci. U.S.A., 82, 4987, 1985.*

Mermod, N., O'Neill, E. A., Kelly, T. J., and Tjian, R.. The proline-rich transcriptional activator of CTF/NF-1 is distinct from the replication and DNA binding domain. *Cell, 58, 741, 1989.*

Mölders, H., Jenuwein, T., Adamkiewicz, J., and Müller, R.. Isolation and structural analysis of a biologically active chicken c-*fos* cDNA: Identification of evolutionarily conserved domains in *fos* protein. *Oncogene, 1, 377, 1987.*

Morgan, J. I. and Curran, T., Stimulus-transcription coupling in neurons: Role of cellular immediate-early genes. *Trends Neurosci., 12, 459, 1989.*

Müller, R.. Cellular and viral *fos* genes: Structure, regulation of expression and biological properties of their encoded products. *Biochim. Biophys. Acta, 823, 207, 1986.*

Müller, R., Bravo, R., Burckhardt, J., and Curran, T., Induction of c-*fos* gene and protein by growth factors precedes activation of c-*myc*. *Nature, 312, 716, 1984.*

Müller, R., Bravo, R., Müller, D., Kurz, C., and Renz, M.. Different types of modification in c-*fos* and its associated protein p39: Modulation of DNA binding by phosphorylation. *Oncogene Res., 2, 19, 1987.*

Mumberg, D., Lucibello, F. C., Schuermann, M., and Müller, R.. Alternative splicing of *fos*B transcripts results in differentially expressed mRNAs encoding functionally antagonistic proteins. *Genes Dev., 5, 1212, 1991.*

Nakabeppu, Y. and Nathans, D.. A naturally occuring truncated form of FosB that inhibits Fos/Jun transcriptional activity. *Cell, 64, 751, 1991.*

Nakabeppu, Y., Ryder, K., and Nathans, D.. DNA binding activity of three murine Jun proteins: Stimulation by Fos. *Cell, 55, 907, 1988.*

Neuberg, M., Adamkiewicz, J., Hunter, J. B., and Müller, R.. A Fos protein containing the Jun leucine zipper forms a homodimer which binds to the AP-1 binding site. *Nature, 341, 243, 1989a.*

Neuberg, M., Schuermann, M., Hunter, J. B., and Müller, R., Two functionally different regions in Fos are required for the sequence-specific DNA interaction of the Fos/Jun protein complex. *Nature, 338, 589, 1989b.*

Neuberg, M., Schuermann, M., and Müller, R., Mutagenesis of the DNA contact site in Fos protein: Compatibility with the scissors grip model and requirement for transformation. *Oncogene, 6, 1325, 1991.*

Nishina, H., Sato, H., Suzuki, T., Sato, M., and Iba, H., Isolation and characterization of *fra-2*, an additional member of the *fos* gene family, *Proc. Natl. Acad. Sci. U.S.A.*, 87, 3619, 1990.

Ofir, R., Dwarki, V. J., Rashid, D., and Verma, I. M., Phosphorylation of the C-terminus of Fos-protein is required for transcriptional transrepression of the c-*fos* promoter, *Nature*, 348, 80, 1990.

Okuno, H., Akahori, A., Sato, H., Xanthoudakis, S., Curran, T., and Iba, H., Escape from redox regulation enhances the transforming activity of Fos, *Oncogene*, 8, 695, 1993.

O'Neil, K. T., Hoess, R. H., and DeGrado, W. F., Design of DNA-binding peptides based on the leucine zipper motif, *Science*, 249, 774, 1990.

O'Shea, E. K., Rutkowski, R., and Kim, P. S., Evidence that the leucine zipper is a coiled coil, *Science*, 243, 538, 1989a.

O'Shea, E. K., Rutkowski, R., Stafford, W. F., III, and Kim, P. S., Preferential heterodimer formation by isolated leucine zippers from Fos and Jun, *Science*, 245, 646, 1989b.

Perkins, K. K., Admon, A., Patel, N., and Tjian, R., The Drosophila Fos-related AP-1 protein is a developmentally regulated transcription factor, *Genes Dev.*, 4, 822, 1990.

Ransone, L. J., Visvader, J., Sassone-Corsi, P., and Verma, I. M., Fos–Jun interaction: Mutational analysis of the leucine zipper domain of both proteins, *Genes Dev.*, 3, 770, 1989.

Rauscher, F. J., III, Cohen, D. R., Curran, T., Bos, T. J., Vogt, P. K., Bohmann, D., Tjian, R., and Franza, B. R., Jr., Fos-associated protein p39 is the product of the *jun* proto-oncogene, *Science*, 240, 1010, 1988a.

Rauscher, F. J., III, Voulalas, P. J., Franza, B. R., Jr., and Curran, T., Fos and Jun bind cooperatively to the AP-1 site: Reconstruction *in vitro*, *Genes Dev.*, 2, 1687, 1988b.

Redemann-Fibi, B., Schuermann, M., and Müller, R., Stage- and tissue-specific expression of *fos*B during mouse development, *Differentiation*, 46, 43, 1991.

Renz, M., Verrier, B., Kurz, C., and Müller, R., Chromatin association and DNA binding properties of the c-*fos* proto-oncogene product, *Nucleic Acids Res.*, 15, 277, 1987.

Roux, P., Blanchard, J.-M., Fernandez, A., Lamb, N., Jeanteur, P., and Piechaczyk, M., Nuclear localization of c-Fos, but not v-Fos proteins, is controlled by extracellular signals, *Cell*, 63, 341, 1990.

Ryder, K., Lau, L. F., and Nathans, D., A gene activated by growth factors is related to the oncogene v-*jun*, *Proc. Natl. Acad. Sci. U.S.A.*, 85, 1487, 1988.

Ryder, K., Lanahan, A., Perez-Albuerne, E., and Nathans, D., Jun D: A third member of the *jun* gene family, *Proc. Natl. Acad. Sci. U.S.A.*, 86, 1500, 1989.

Ryseck, R.-P., Hirai, S.-I., Yaniv, M., and Bravo, R., Transcriptional activation of c-*jun* during G_0/G_1 transition in mouse fibroblasts, *Nature*, 334, 535, 1988.

Sambucetti, L. C. and Curran, T., The Fos protein complex is associated with DNA in isolated nuclei and binds to DNA cellulose, *Science*, 234, 1417, 1986.

Sassone-Corsi, P., Lamph, W. W., Kamps, M., and Verma, I. M., Fos-associated cellular p39 is related to nuclear transcription factor AP-1, *Cell*, 54, 553, 1988a.

Sassone-Corsi, P., Sisson, J. C., and Verma, I. M., Transcription autoregulation of the proto-oncogene *fos*, *Nature*, 334, 314, 1988b.

Sassone-Corsi, P., Ransone, L. J., Lamph, W. W., and Verma, I. M., Direct interaction between Fos and Jun nuclear oncoproteins: Role of the "leucine zipper" domain, *Nature*, 336, 692, 1988c.

Schönthal, A., Büscher, M., Angel, P., Rahmsdorf, H. J., Ponta, H., Hattori, K., Chiu, R., Karin, M., and Herrlich, P., The Fos and Jun/AP-1 proteins are involved in the downregulation of Fos transcription, *Oncogene*, 4, 629, 1989.

Schuermann, M., Neuberg, M., Hunter, J. B., Jenuwein, T., Ryseck, R.-P., Bravo, R., and Müller, R., The leucine repeat motif in Fos protein mediates complex formation with Jun/AP-1 and is required for transformation, *Cell*, 56, 507, 1989.

Schuermann, M., Hunter, J. B., Hennig, G., and Müller, R., Non-leucine residues in the leucine repeats of Fos and Jun contribute to the stability and determine the specificity of dimerization, *Nucleic Acids Res.*, 19, 739, 1991a.

Schuermann, M., Jooss, K., and Müller, R., *fos*B is a transforming gene encoding a transcriptional activator, *Oncogene*, 6, 567, 1991b.

Schuermann, M., Hennig, G., and Müller, R., Transcriptional activation and transformation by chimaeric Fos-estrogen receptor proteins: Altered properties as a consequence of gene fusion, *Oncogene*, 8, 2781, 1993.

Sellers, J. W. and Struhl, K., Changing Fos oncoprotein to a Jun-independent DNA-binding protein with GCN4 dimerization specificity by swapping leucine zippers, *Nature*, 341, 74, 1989.

Shuman, J. D., Vinson, C. R., and McKnight, S. L., Evidence of changes in protease sensitivity and subunit exchange rate on DNA binding by C/EBP, *Science*, 249, 771, 1990.

Shyu, A.-B., Greenberg, M. E., and Belasco, J. G., The c-*fos* transcript is targeted for rapid decay by two distinct mRNA degradation pathways, *Genes Dev.*, 3, 60, 1989.

Smeal, T., Angel, P., Meek, J., and Karin, M., Different requirements for formation of Jun:Jun and Jun:Fos complexes, *Genes Dev.*, 3, 2091, 1989.

Sutherland, J. A., Cook, A., Bannister, A. J., and Kouzarides, T., Conserved motifs in Fos and Jun define a new class of activating domain, *Genes Dev.*, 6, 1810, 1992.

Suzuki, T., Okuno, H., Yoshida, T., Endo, T., Nishina, H., and Iba, H., Difference in transcriptional regulatory function between c-Fos and Fra-2, *Nucleic Acids Res.*, 19, 5537, 1991.

Talanian, R. V., McKnight, C. J., and Kim, P. S., Sequence-specific DNA binding by a short peptide dimer, *Science*, 249, 769, 1990.

Tratner, I. and Verma, I. M., Identification of a nuclear targeting sequence in the Fos protein, *Oncogene*, 6, 2049, 1991.

Tratner, I., Ofir, R., and Verma, I. M., Alteration of a cyclic AMP-dependent protein kinase phosphorylation site in the c-Fos protein augments its transforming potential, *Mol. Cell. Biol.*, 12, 998, 1992.

Turner, R. V. and Tjian, R., Leucine repeats and an adjacent DNA binding domain mediate the formation of functional cFos-cJun heterodimers, *Science*, 243, 1689, 1989.

Van Beveren, C., van Straaten, F., Curran, T., Müller, R., and Verma, I. M., Analysis of FBJ-MuSV provirus and c-*fos* (mouse) gene reveals that viral and cellular *fos* gene products have different carboxy termini, *Cell*, 32, 1241, 1983.

Van Beveren, C., Enami, S., Curran, T., and Verma, I. M., FBR murine osteosarcoma virus. II. Nucleotide sequence of the provirus reveals that the genome contains sequences acquired from two cellular genes, *Virology*, 135, 229, 1984.

Van Straaten, F., Müller, R., Curran, T., Van Beveren, C., and Verma, I. M., Complete nucleotide sequence of a human c-*onc* gene: Deduced amino acid sequence of the human c-*fos* protein, *Proc. Natl. Acad. Sci. U.S.A.*, 80, 3183, 1983.

Varmus, H. E., Form and function of retroviral proviruses, *Science*, 216, 812, 1982.

Vinson, C. R., Sigler, P. B., and McKnight, S. L., Scissors-grip model for DNA recognition by a family of leucine zipper proteins, *Science*, 246, 911, 1989.

Vogt, P. K., Bos, T. J., and Doolittle, R. F., Homology between the DNA-binding domain of GCN4 regulatory protein of yeast and the carboxy-terminal region of a protein coded for by the oncogene *jun*, *Proc. Natl. Acad. Sci. U.S.A.*, 84, 3316, 1987.

Wisdom, R., Yen, J., Rashid, D., and Verma, I. M., Transformation by FosB requires a trans-activation domain missing in FosB2 that can be substituted by heterologous activation domains, *Genes Dev.*, 6, 667, 1992.

Xanthoudakis, S. and Curran, T., Identification and characterization of Ref-1, a nuclear protein that facilitates AP-1 DNA-binding activity, *EMBO J.*, 11, 653, 1992.

Yen, J., Wisdom, R., Tratner, I., and Verma, I. M., An alternative spliced form of FosB is a negative regulator of transcriptional activation and transformation by Fos proteins, *Proc. Natl. Acad. Sci. U.S.A.*, 88, 5077, 1991.

Zerial, M., Toschi, L., Ryseck, R.-P., Schuermann, M., Müller, R., and Bravo, R., The product of a novel growth factor activated gene, *fos*B, interacts with Jun proteins enhancing their DNA binding activity, *EMBO J.*, 8, 805, 1989.

Section II

REGULATION OF
AP-1 ACTIVITY AND SYNTHESIS

Based on the high degree of sequence homology between cJun, JunB, and JunD as well as between cFos, FosB, Fra-1, and Fra-2, one might assume that the AP-1 subunits are encoded by genes that were generated by gene duplication and that map as a cluster to one location. In fact, the genomic organization (intron–exon structure) of the different fos genes is highly conserved (see Chapter 2) and all the Jun proteins were found to be encoded by single exons (see Chapter 9). However, the fos loci have been localized on different chromosomes. Of the jun genes, jun-B and jun-D are positioned at the same chromosome (chromosome 8, subregion C) but are not closely linked, while c-jun is on chromosome 4. That the various Jun and Fos proteins greatly differ in their ability to activate transcription also strongly suggests that the members of the Jun and Fos (and CREB/ATF) protein families are not simply the result of evolutionary safeguard and redundancy. Rather, specific features and subtle differences among the various members of the bZip proteins and of their dimers are likely to be engaged in the transcriptional control of specific target genes. The activity of any one of the Jun and Fos proteins, therefore, has to be tightly controlled to establish the appropriate overall AP-1 activity.

For example, AP-1 has been identified by its function to regulate gene expression in response to growth factors, cytokines, tumor promoters, and carcinogens. Work from different laboratories over the past few years has established the existence of multiple mechanisms by which AP-1 activity is controlled to facilitate proper control of gene expression. These mechanisms will be discussed in the following chapters.

As the earliest measurable change of expression after a growth factor stimulus (immediate early response), the activity of preexisting AP-1 is altered by posttranslational modification, including glycosylation, phosphorylation, oxidization, and ubiquitination (see Chapters 5 and 6, this volume; D. Bohmann, personal communication). Alterations in phosphorylation of Fos and Jun seem to determine the DNA binding and transactivating functions and may also participate in regulating the interaction between Jun/Fos and other cellular proteins (Chapter 3). One of the most interesting examples studied concerns the interaction between Fos/Jun and one of several steroid hormone receptors (Chapter 4). Posttranslational modifications of Jun/Fos or interaction with other cellular proteins may also determine their cellular localization: stimulation of cells by extracellular signals seems to enhance nuclear uptake of Jun (Chapter 15) and Fos proteins (Chapter 7). The importance of nuclear uptake regulation is supported by the fact that the oncogenic counterparts of cFos and cJun — vFos and vJun — are always found in the nucleus.

Alterations in the activity of preexisting AP-1 seem to be required for subsequent transcriptional activation of immediate early (IE) genes, including c-jun (Chapter 9). While transcriptional activation of c-jun is likely to depend on "activated" cJun protein, transcriptional activation of the c-fos gene is mediated through the serum-responsive element (SRE) by the activities of SRF and p62-TCF (Chapter 8). Although the transcriptional activation of both c-jun and c-fos is only transient, lasting up to 3 hours poststimulation, enhanced synthesis of Fos and Jun protein leads to long-lasting induction of AP-1-dependent target genes, e.g., of the collagenase gene and other gene products (Chapter 10).

To function as a mediator between extracellular stimuli and the program of genes to be expressed, the activity of transcription factors like AP-1 must rapidly adjust to new microenvironmental conditions. Therefore, the signaling pathway leading to AP-1 activation as well as activated AP-1 itself must be deactivated to be susceptible for another round of stimulation. As with the activation process, multiple pathways to repress AP-1 activity have been identified. First, enhanced transcription returns to basal level within 1 hour. For c-fos, negative autoregulation is instrumental in transcriptional shut-off by interfering with the activity of SRF/p62-TCF by an as yet unknown mechanism (Chapter 8). Furthermore, c-jun is under negative regulation, possibly by cJun. The role of cJun in transcriptional repression of its own promoter is not yet understood. cJun is required for transcriptional activation of the c-jun promoter and transient transfection experiments have revealed positive autoregulation of truncated c-jun promoter constructs (Angel et al., 1988b, Cell, 55, 875). However, reduced c-jun promoter activity in cells overexpressing cJun has also been observed (Chapter 9 and references therein; Park and Ponta, unpublished results). It is possible that enhanced levels of Jun lead to subsequent synthesis or activation of a repressor acting on the c-jun promoter.

In addition to transcriptional repression, activated AP-1 protein has to be removed from the system. This could be achieved simply by reversion of the posttranslational modification (see Chapters 5 and 6) established during the activation process (e.g., by readjusting the equilibrium of a protein kinase–phosphatase modifying system), or by the rapid degradation of Jun and Fos proteins. In fact, the half-lives of cFos and cJun proteins were found to be 60 to 90 min. Another putative mechanism of repression may also play an important role: since the various jun and fos genes differ in transcriptional activation (Chapter 9), at certain points after induction, the number of AP-1 subunits with very weak transactivation function (e.g., JunB or JunD homodimers) may exceed that of active dimers. Inactive AP-1 proteins may then replace active ones by competitive binding to AP-1 sites. Transcriptionally inactive members may also sequester active ones by forming inactive heterodimers or by changing their sequence specificity.

In summary, regulation of AP-1 has been identified as a complex network of positive and negative pathways on the level of transcription as well as posttranslational modification that guarantees an efficient and fine-tuned modulation of AP-1-dependent gene regulation in response to environmental signals.

Chapter 3

Factors Modulating AP-1 Activity

Johan Auwerx and Paolo Sassone-Corsi

CONTENTS

INTRODUCTION

Understanding the regulatory networks linking signals generated at the cell surface to changes in transcriptional responses in the nucleus represents a major challenge of modern molecular biology. The products of several cellular proto-oncogenes appear to function in the transmission of inter- and intracellular information through several signal transduction pathways. The AP-1 family of transcription factors, exemplified by the prototype composed of the products of two nuclear proto-oncogenes, *fos* and *jun*, has been shown to play a pivotal role in diverse aspects of cell growth, differentiation, and development (Angel and Karin, 1991). By regulating the transcription of downstream genes, these transcription factors can be considered as the final link in the signal transduction cascade.

The c-Fos and c-Jun proto-oncoproteins are normally expressed at a low basal level in the cell but are rapidly and transiently induced by a variety of stimuli. These proteins belong to the class of basic leucine zipper (bZip) transcription factors (Landschulz et al., 1988; Busch and Sassone-Corsi, 1990; see also Chapters 1 and 2). The fact that components of the AP-1 transcription factor are involved as "third messengers" in several signal transduction pathways implies that their expression, DNA binding activity and transcriptional activity, is finely controlled. The tuning of AP-1 activity is necessary to allow a flexible regulation in response to different cellular stimuli. Alterations in AP-1 function can be brought about by transcriptional induction or by posttranslational modifications of both oncoproteins or their regulatory factors. This chapter focuses on these modifications, which take place in the absence of protein synthesis, and on auxiliary factors involved in modulating AP-1 function. The transcription factor activity can be affected on two levels: DNA binding and transcriptional activation.

EFFECTS ON DNA BINDING

Recently, several lines of evidence have indicated that posttranscriptional regulation plays a major role in determining Fos/Jun DNA binding. c-Jun contains two clusters of amino acids that are prone to phosphorylation. Phosphorylation of Jun homodimers at the cluster near the c-Jun DNA binding domain (residues 231, 239, 243, 249), by glycogen synthase kinase-3 (GSK-3) (Boyle et al., 1990), casein kinase I (CKI) (Auwerx and Sassone-Corsi, unpublished data), or casein kinase II (CKII) (Lin et al., 1992), appears

0-8493-4573-1/94/$0.00+$.50

42

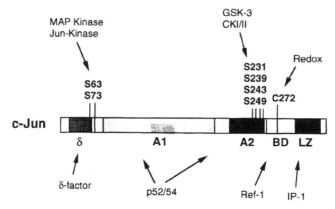

Figure 3-1 Scheme of the c-Jun protein structure and site of action of various agents. LZ, leucine zipper; BD, basic domain; A1 and A2, activation domains. The phosphoserines and the redox site (C272) are indicated. Factors that have been postulated to modulate Jun function are also shown. See the text for details and the original references (Abate et al., 1990; Boyle et al., 1990; Baichwal et al., 1990; Binétruy et al., 1991; Auwerx and Sassone-Corsi, 1991; Pulverer et al., 1991; Smeal et al., 1991; Lin et al., 1992; Oehler and Angel, 1992; de Groot et al., 1993; Nikolakaki et al., 1993; Oehler et al., 1993).

to decrease its DNA binding potential and, therefore, has an inhibitory effect on transcriptional activation. Exposure of cells to the tumor promoter TPA (12-O-tetradecanoyl-phorbol-13-acetate) causes dephosphorylation of Ser243, resulting in enhanced DNA binding (Boyle et al., 1990). Whether all the above-mentioned kinases play a role in phosphorylating the inhibitory sites of c-Jun *in vivo* awaits further analysis. However, GSK-3 and CKII appear to be good candidates. Indeed, it has been shown recently that GSK-3, and the analogous *Drosophila* protein kinase, *shaggy*, can reduce transcriptional activation elicited by the AP-1 complex (de Groot et al., 1993; Nikolakaki et al., 1993). It is noteworthy that in v-Jun the Ser243 is mutated to phenylalanine. This not only causes the loss of phosphorylation of Ser243 but also prevents phosphorylation of other sites, resulting in a molecule that is 10-fold more active relative to c-Jun. This mechanism has been suggested to contribute to the increased oncogenic potential of v-Jun (Boyle et al., 1990; Lin et al., 1992).

The serines in positions 63 and 73 in c-Jun are phosphoacceptor sites in response to stimulation by phorbol esters (see Figure 3-1). Recent results show that phosphorylation of these sites could be responsible for the increased activity of the AP-1 complex in response to TPA and transformation by some nonnuclear oncogenes, such as Ha-*ras* (Binétruy et al., 1991). This molecular mechanism is likely to be at the basis of the cooperation between cJun and Ha-ras proteins at both the transcriptional (Sassone-Corsi et al., 1989) and transformation (Schütte et al., 1989) levels. It seems that the kinase involved in phosphorylation of the 63 and 73 serines may be distinct from other already characterized kinases phosphorylating Jun *in vitro* (Pulverer et al., 1991), and it has been recently referred to as the Jun-kinase (M. Karin, personal communication).

In addition to phosphorylation, other posttranslational modifications are apparently involved in the regulation of AP-1 function. Reduction–oxidation (redox) has been shown to regulate DNA binding activity of several transcription factors, including Myb, ATF (Xanthoudakis et al., 1992), NF-κB (Toledano and Leonard, 1991), TFIIIC (Cromlish and Roeder, 1989), USF (Pognonec et al., 1992), and AP-1 *in vitro* (Abate et al., 1990). Redox regulation of AP-1 DNA binding activity is mediated by a conserved cysteine residue (Cys272 of mouse c-Jun) in the c-Jun DNA binding domain (Abate et al., 1990). Using the DNA binding domain of c-Fos or c-Jun, reduction of this residue stimulates AP-1 DNA binding, whereas oxidation or chemical modification inhibits DNA binding (see Figure 3-1). This cysteine residue, which is mutated to a serine in v-Jun, is highly conserved in the AP-1 family of transcription factors. When Cys154 of c-Fos or Cys272 of c-Jun are substituted by a serine residue, redox control of DNA binding is lost, resulting in increased DNA binding. While this is true for *in vitro* binding studies using a truncated version of Jun spanning only the minimal DNA binding domain, substitution of the Cys residue leads to a complete loss of DNA binding *in vitro* and transcriptional activation *in vivo* (Oehler et al., 1993). However, redox-independent DNA binding *in vitro* and transcriptional activation is regained on additional mutation of Ser243 (Oehler et al., 1993), which was found to be the phosphorylation site responsible for a loss of the DNA binding activity of Jun (Boyle et al., 1990; Lin et al., 1992). These amino acid substitutions are the only ones found in the DNA binding domain of v-Jun (when compared to c-Jun), suggesting that v-Jun might have acquired its oncogenic potential by a loss of negative regulation of its DNA binding activity by phosphorylation and oxidation. Redox stimulation of c-Fos and c-Jun DNA binding activity is accomplished by a ubiquitous 37-kDa nuclear protein, Ref-1, that has been recently purified and cloned (Xanthoudakis et al., 1992). This protein stimulates the DNA binding activity of not only the AP-1 transcription factor family but also other transcription factors, including NF-κB, Myb, and members of the ATF family. The Ref-1 protein is bifunctional and possesses apurinic/apyrimidinic endonuclease DNA repair activity. Therefore, a possible link between transcriptional regulation, oxidative signaling, and DNA repair has been suggested (Xanthoudakis et al., 1992).

Another way to influence DNA binding is the regulation of cytoplasmic sequestration of transcription factors. The nuclear/cytoplasmic compartmentation in eucaryotes allows for easy and fast control of gene expression. This control mechanism for transcription factor activity has been demonstrated for several transcription factors, including SWI5 and the *rel*-related family of transcription factors (Hunter and Karin, 1992). AP-1 activity also appears to be modulated by cellular translocation. Indeed, Roux et al. (1990; see also Chapter 7, this volume) showed that the c-Fos protein was anchored in the cytoplasm when cells were maintained in the absence of serum. Serum stimulation, as well as treatment of cells with dBcAMP or cycloheximide, induced an active nuclear transloca-tion of c-Fos. This regulated nuclear transport was specific for the c-Fos protein since the v-Fos protein was constitutively present in the nucleus. These investigators, therefore, speculated that this constitutive nuclear localization of v-Fos relative to c-Fos might contribute to its increased tumorigenic potential.

DNA binding can also be altered by interaction of AP-1 with other proteins. Cross-family dimerization between bZip transcription factors of the Fos–Jun and ATF groups can also alter DNA binding specificity (Ivashkiv et al., 1990; Hai and Curran, 1991). For instance, heterodimerization of c-Jun with ATF-2 changes its specificity of binding from an AP-1 site to a cAMP-responsive element (Ivashkiv et al., 1990). Alternatively, AP-1 function can be blocked by repressor factors to which it does not dimerize; this is true for CREM (Foulkes et al., 1991), which is able to bind to an AP-1 site and thus block Jun-mediated transactivation via occupation of the regulatory site (Masquilier and Sassone-Corsi, 1992).

Protein–protein interactions with AP-1, however, are not limited to bZip proteins. In fact, c-Jun has been shown to interact with several members of the steroid hormone receptor gene family (Jonat et al., 1990; Yang-Yen et al., 1990; Schüle et al., 1990; Gaub et al., 1990). Most information available is on the interaction of AP-1 and the glucocorticoid hormone receptor. Overexpression of Jun prevents glucocorticoid-induced activation of genes carrying a glucocorticoid response element, while, conversely, glucocorticoid hormone receptor is a potent inhibitor of AP-1-mediated gene activation. *In vitro*, this mutual inhibition is the result of a decrease in DNA binding to their respective response elements, due to the fact that steroid hormone receptors form a nonproductive complex with c-Jun (Yang-Yen et al., 1990; Schüle et al., 1990), while *in vivo*, repression of DNA binding is not instrumental in inhibiting AP-1-dependent collagenase expression (König et al., 1992). Direct interaction between the two proteins is mediated via the DNA binding domain of c-Jun and the ligand and DNA binding domain of the glucocorticoid receptor. Similar data are also available for the interaction of AP-1 with the retinoic acid receptor (Schüle et al., 1991). Interestingly, the oncogenic v-ErbA protein can abrogate the inactivation of AP-1 by the retinoic acid receptor (Desbois et al., 1991).

Another kind of heterologous interaction has been described for the physical association between c-Jun and MyoD (Bengal et al., 1992). MyoD is a member of the helix–loop–helix group of proteins, which acts as a key regulator in the myogenic differentiation process and is capable of activating many muscle-specific genes. The interaction between c-Jun and MyoD results in functional antagonism at the transcriptional level and could be the mechanism underlying the Jun-mediated inhibition of myogenesis.

The existence of another protein controlling AP-1 DNA binding activity has been described (Auwerx and Sassone-Corsi, 1991). The protein, termed IP-1, is present in both the cytoplasm and nucleus of cells and reduces AP-1 complex formation with DNA in a rapid and phosphorylation-dependent fashion. The IP-1 protein appears to be highly unstable. It is regulated by phosphorylation, and only in its nonphosphorylated form exerts an inhibitory activity on AP-1 DNA binding. Initial purification attempts showed that the protein had a molecular weight of approximately 43 kDa. IP-1 is subject to complex regulation. In fact, IP-1 activity was shown to be modulated after activation of several signal transduction pathways, including PKC, PKA, and Ca^{2+}/calmodulin-dependent kinase pathways, as well as after serum stimulation of cells. These data, obtained after *in vivo* stimulation of cells with various agents are in good agreement with previous data showing that: (1) IP-1 is inactivated via phosphorylation by protein kinase A *in vitro* (Auwerx and Sassone-Corsi, 1991) and (2) AP-1 function is activated by protein kinase A, although c-Jun is not phosphorylated by this kinase (de Groot and Sassone-Corsi, 1992). Furthermore, induced differentiation of cultured cells appeared to influence IP-1 activity, since treatment of P19 EC cells caused enhanced AP-1 DNA binding and a correlated reduction of IP-1 activity (Auwerx and Sassone-Corsi, 1992). Further characterization of IP-1 awaits purification and isolation.

EFFECTS ON TRANSACTIVATION POTENTIAL

Posttranslational modifications, such as phosphorylation, not only have an affect on DNA binding activity but also can substantially modify transregulation elicited by transcription factors. For instance, phosphorylation of the sites clustered within the *N*-terminal activation domain (Ser63–73) of c-Jun enhances transactivation potential without having an effect on DNA binding (Pulverer et al., 1991; Smeal et al., 1991). As discussed above, the nature of the kinase that phosphorylates these sites is not clearly established; MAP/Erk kinases have been involved (Pulverer et al., 1991), but others suggest the presence of a Jun-specific kinase (M. Karin, personal communication). The activity of this kinase is enhanced by transforming oncoproteins such as Ha-ras (Binétruy et al., 1991; Smeal et al., 1991). In some cells activation of PKC by TPA also leads to stimulation of these sites (Pulverer et al., 1991; see Figure 3-1). c-Fos function also is regulated by phosphorylation; c-Fos was shown to negatively autoregulate its own expression by interacting with factors binding to c-*fos* regulatory sequences (Sassone-Corsi et al., 1988; see also Chapter 6). The *C*-terminal region of c-Fos, but not v-Fos (Barber and Verma, 1987), contains three Ser clusters (362–364, 368–369, 371–374) that are implicated in the transrepression (Lucibello et al., 1990; Ofir et al., 1990; Gius et al., 1990). Mutation of these Ser residues to Ala results in hypophosphorylation, causing a loss in transrepression but not in the transactivation potential (Ofir et al., 1990). The kinases involved in phosphorylation are unknown, but PKA (Tratner et al., 1992) has been shown *in vitro* to phosphorylate residues in this region of the protein.

Various protein–protein interactions can modify transcriptional activation by AP-1, which is active as a dimer. Changing the composition of the dimers can result in AP-1 proteins with a dramatically different transactivation potential. A clear example can be found in the *fos*B gene products. Alternative splicing gives rise to two different products: FosB-L, containing 338 amino acids and FosB-S, which contains only the NH_2-terminal 237 amino acids (Dobrzanski et al., 1991; Mumberg et al., 1991; Yen et al., 1991; Nakabeppu and Nathans, 1991). Although both proteins contain the conserved bZip motif and thus can heterodimerize with Jun and bind DNA in a sequence-specific manner, only the large form of FosB can transform fibroblasts efficiently. This difference in transformation capacity could be correlated to differences in transactivating properties of the two FosB forms, since the shorter form lacks an activation domain present in the longer form (Nakabeppu and Nathans, 1991; Wisdom et al., 1992). Another interesting situation has been described for two Jun proteins. JunB is a less potent transactivator than c-Jun and needs cooperative interaction to transactivate (Chiu et al., 1989; Schütte et al., 1989). This decreased transactivation potential has been correlated to the diminished transformation potential of JunB relative to c-Jun (Schütte et al., 1989).

Baichwal and Tjian (1990) have described the regulation of Jun activity by a cell-specific factor. In some cell lines, transactivation by chimeric Jun proteins, having the δ, A1, and A2 domains of Jun and a DNA binding and dimerization domain from a different transcription factor, was enhanced by transfection of c-Jun, suggesting that a factor in these cells was inhibiting transcriptional activation by the Jun domains in the chimera. The excess Jun deregulated transactivation of the chimera by competing for an inhibitory factor and titrating it. The region responsible for interaction with this cellular factor was localized to the δ and ε (a part of the A1 activation region) domains of Jun. v-Jun lacks the δ region, and chimeras with the amino-terminal half of v-Jun are not repressed to the same extent as chimeras with the amino-terminal of c-Jun. The fact that removal of the δ region from c-Jun enhances its transforming activity (Bos et al., 1990), together with the notion that part of the increase in AP-1 activity after expression of the *ras* and *src* oncogenes is caused by downregulation of the δ regulator (Baichwal et al., 1991), underscores the importance of this putative negative regulatory protein. By which mechanism this negative regulatory protein controls the activation of Jun is unclear. The negative regulator could be inhibiting or reversing inducible activation of the A1 activation region. Interestingly, phosphorylations occurring in or near the A1 region have also been described in c-Jun (Pulverer et al., 1991; Smeal et al., 1991). Another possible scenario would be that the inhibitor binds to a site in or near a fully active transcriptional activation domain to repress it (see Figure 3-1). This theory would mean that activation of the Jun molecule requires titration of the inhibitor, resulting in relief of repression (Baichwal and Tjian, 1990). A detailed biochemical characterization is required to better understand the function of this putative inhibitory protein.

More recently, Oehler and Angel (1992) have described the presence of intermediary factors modulating the activator function of c-Jun. These authors were able to find physical interaction between the highly negatively charged domains of Jun and three proteins of 52, 53, and 54 kDa (p52/54). The p52/54 intermediary factor is limiting in the cell and is postulated to be required for efficient interaction of Jun with the other components of the transcriptional machinery.

In conclusion, it is clear that AP-1 is able to interact with a multitude of additional regulatory proteins. How these cofactors are able to modulate AP-1 activity or its phosphorylation state is still unclear. Additional studies are required to unravel the molecular architecture and the physiological functions of the AP-1 complex.

ACKNOWLEDGMENTS

Communication of unpublished work by M. Karin and P. Pognonec is kindly acknowledged. Part of the work reported was supported by an ATIPE grant to J. A. Work in the laboratory of P. Sassone-Corsi is supported by CNRS, INSERM, ARC, and Rhône-Poulenc-Rorer.

REFERENCES

Abate, C., Patel, L., Rauscher, F. J., and Curran, T., Redox regulation of Fos and Jun DNA binding activity *in vitro*, *Science*, 249, 1157, 1990.

Angel, P. and Karin, M., The role of Fos, Jun and AP-1 complex in cell-proliferation and transformation, *Biochim. Biophys. Acta.*, 1072, 129, 1991.

Auwerx, J. and Sassone-Corsi, P., IP-1: A dominant inhibitor of Fos/Jun whose activity is modulated by phosphorylation, *Cell*, 64, 983, 1991.

Auwerx, J. and Sassone-Corsi, P., AP-1 (Fos-Jun) regulation by IP-1: Effect of signal transduction pathways and cell growth, *Oncogene*, 7, 2271, 1992.

Baichwal, V. R. and Tjian, R., Control of c-Jun activity by interaction of a cell-specific inhibitor with regulatory domain d: Differences between v- and c-Jun, *Cell*, 63, 815, 1990.

Baichwal, V. R., Park, A., and Tjian, R., v-Src and EJ Ras alleviate repression of c-Jun by a cell-specific inhibitor, *Nature*, 352, 165, 1991.

Barber, J. R. and Verma, I. M., Modification of Fos proteins: Phosphorylation of c-fos but not v-fos is stimulated by TPA and serum, *Mol. Cell. Biol.*, 7, 2201, 1987.

Bengal, E., Ransone, L., Scharfman, R., Dwarki, V. J., Tapscott, S. J., Weintraub, H., and Verma, I. M., Functional antagonism between c-Jun and MyoD proteins: A direct physical association, *Cell*, 68, 507, 1992.

Binétruy, B., Smeal T., and Karin, M., Ha-Ras augments c-Jun activity and stimulates phosphorylation of its activation domain, *Nature*, 351, 122, 1991.

Bos, T. J., Monteclaro, F. S., Mitsunubo, F., Ball, A. R., Chang, C. H. W., Nishimura, T., and Vogt, P. K., Efficient transformation of chicken embryo fibroblasts by c-Jun requires structural modification in coding and noncoding sequences, *Genes Dev.*, 4, 1677, 1990.

Boyle, W. J., Smeal, T., Defize, L. H. K., Angel, P., Woodget, J. R., Karin, M., and Hunter, T., Activation of protein kinase C decreases phosphorylation of c-Jun at sites that negatively regulate its DNA-binding, *Cell*, 64, 573, 1990.

Busch, S. J. and Sassone-Corsi, P., Dimers, leucine-zippers and DNA-binding domains, *Trends Genet.*, 6, 36, 1990.

Chiu, R., Angel, P., and Karin, M., Jun-B differs in its biological properties from and is a negative regulator of c-Jun, *Cell*, 59, 979, 1989.

Cromlish, J. A. and Roeder, R. G., Human transcription factor IIIC (TFIIIC): Purification, polypeptide structure, and involvement of thiol groups in specific DNA binding, *J. Biol. Chem.*, 264, 18100, 1989.

Foulkes, N. S., Borrelli, E., and Sassone-Corsi, P., CREM gene: Use of alternative DNA binding domains generates multiple antagonists of cAMP-induced transcription, *Cell*, 64, 739, 1991.

de Groot, R. and Sassone-Corsi, P., Activation of Jun/AP-1 by protein kinase A. *Oncogene,* 7, 2281, 1992.

de Groot, R., Auwerx, J., Bourouis, M., and Sassone-Corsi, P., Negative regulation of Jun/AP-1 by glycogen synthase kinase 3 and the Drosophila kinase *Shaggy, Oncogene,* 8, 841, 1993.

Desbois, C., Aubert, D., Legrand, C., Pain, B., and Samarut, J., A novel mechanism of action for v-ErbA: Abrogation of the inactivation of transcription factor AP-1 by retinoic acid and thyroid hormone receptors, *Cell,* 67, 731, 1991.

Dobrzanski, P., Noguchi, T., Kovari, K., Rizzo, C. A., Lazo, P. S., and Bravo, R., Both products of the fosB gene, FosB and its short form, FosB/SF, are transcriptional activators in fibroblasts, *Mol. Cell. Biol.,* 11, 5470, 1991.

Gaub, M. P., Bellard, M., Scheuer, I., Chambon, P., and Sassone-Corsi, P., Activation of the ovalbumin gene by the estrogen receptor involves the Fos-Jun complex, *Cell,* 63, 1267, 1990.

Gius, D., Cao, X., Rauscher, F. J., Cohen, D. R., and Curran, T., Transcriptional activation and repression by Fos are independent functions: The C terminus represses immediate-early gene expression via CArG elements, *Mol. Cell. Biol.,* 10, 4243, 1990.

Hai, T. and Curran, T., Cross-family dimerization of transcription factors Fos/Jun and ATF/CREB alters DNA binding specificity, *Proc. Natl. Acad. Sci. U.S.A.,* 88, 3720, 1991.

Hunter, T. and Karin, M., The regulation of transcription by phosphorylation, *Cell,* 70, 375, 1992.

Ivashkiv, L. B., Liou, H. C., Kara, C. J., Lamph, W. W., Verma, I. M., and Glimcher, L. H., mXBP/CRE-BP2 and c-Jun form a complex which binds to the cAMP, but not to the 12-O-tetradecanoyl phorbol-13-acetate response element, *Mol. Cell. Biol.,* 10, 1609, 1990.

Jonat, C., Rahmsdorf, H. J., Park, K. K., Cato, A. C. B., Gebel, S., Ponta, H., and Herrlich, P., Antitumor promotion and antiinflammation: Down-modulation of AP-1 (Fos/Jun) activity by glucocorticoid Hormone, *Cell,* 62, 1189, 1990.

König, H., Ponta, H., Rahmsdorf, H. J., and Herrlich, P., Interference between pathway-specific transcription factors: Glucocorticoids antagonize phorbol ester-induced AP-1 activity without altering AP-1 site occupation in vivo, *EMBO J.,* 11, 2241, 1992.

Landschulz, W. H., Johnson, P. F., and McKnight, S. L., The leucine zipper: A hypothetical structure common to a new class of DNA binding proteins, *Science,* 240, 1759, 1988.

Lin, A., Frost, J., Deng, T., Smeal, T., Al-Alawi, N., Kikkawa, U., Hunter, T., Brenner, D., and Karin, M., Casein kinase II is a negative regulator of c-Jun DNA binding and AP-1 activity, *Cell,* 70, 777, 1992.

Lucibello, F. C., Lowag, C., Neuberg, M., and Müller, R., Trans-repression of the mouse c-*fos* promoter: a novel mechanism of Fos-mediated trans-regulation, *Cell,* 59, 999, 1990.

Masquilier, D. and Sassone-Corsi, P., Transcriptional cross-talk: Nuclear factors CREM and CREB bind to AP-1 sites and inhibit activation by Jun, *J. Biol. Chem.,* 267, 22460, 1992.

Mumberg, D., Lucibello, F. C., Schuermann, M., and Müller, R., Alternative splicing of fosB transcripts results in differentially expressed mRNAs encoding functionally antagonistic proteins, *Genes Dev.,* 5, 1212, 1991.

Nakabeppu, Y. and Nathans, D., A naturally occurring truncated form of FosB that inhibits Fos/Jun transcriptional activity, *Cell,* 64, 751, 1991.

Nikolakaki, E., Coffer, P. J., Hemelsoet, R., Woodgett, J. R., and Defize, L. H. K., Glycogen synthase kinase-3 phosphorylates Jun-family members *in vitro* and negatively regulates their transactivating function in intact cells, *Oncogene,* 8, 883, 1993.

Oehler, T. and Angel, P., A common intermediary factor (p52/54) recognizing "acidic-blob"-type domains is required for transcriptional activation by the Jun proteins, *Mol. Cell. Biol.*, 12, 5508, 1992.

Oehler, T., Pintzas, A., Stumm, S., Darling, A., Gillespie, D., and Angel, P. Mutation of a phosphorylation site in the DNA binding domain is required for redox-independent transactivation of AP-1 dependent genes by v-Jun, *Oncogene*, 8, 1141, 1993.

Ofir, R., Dwarki, V. J., Rashid, D., and Verma, I. M., Phosphorylation of the C terminus of Fos protein is required for transcriptional transrepression of the c-Fos promoter, *Nature*, 348, 80, 1990.

Pognonec, P., Kato, H., and Roeder, R. G., The helix-loop-helix/leucine repeat transcription factor USF can be functionally regulated in a redox dependent manner, *J. Biol. Chem.*, 267, 24563, 1992.

Pulverer, B. J., Kyriakis, J. M., Avruch, J., Nikolakaki, E., and Woodgett, J. R., Phosphorylation of c-Jun mediated by MAP kinases, *Nature*, 353, 670, 1991.

Ransone, L. J. and Verma, I. M., Nuclear proto-oncogenes Fos and Jun, *Annu. Rev. Cell. Biol.*, 6, 539, 1990.

Roux, P., Blanchard, J. M., Fernandez, A., Lamb, N., Janteur, P., and Piechaczyk, M., Nuclear localization of c-Fos, but not v-Fos proteins is controlled by extracellular signals, *Cell*, 63, 341, 1990.

Sassone-Corsi, P., Sisson, J. C., and Verma, I. M., Transcriptional autoregulation of the proto-onogene *fos*, *Nature*, 334, 314, 1988.

Sassone-Corsi, P., Der, C. J., and Verma, I. M., Ras-induced neuronal differentiation of PC12 cells: Possible involvement of *fos* and *jun*, *Mol. Cell. Biol.*, 9, 3174, 1989.

Schüle, R., Rangarajan, P., Kliewer, S., Ransone, L. J., Bolado, J.,Yang, N., Verma, I. M., and Evans, R. M., Functional antagonism between oncoprotein c-Jun and the glucocorticoid receptor, *Cell*, 62, 1217, 1990.

Schüle, R., Rangarajan, P., Yang, N., Kliewer, S., Ransone, L. J., Bolado, J., Verma, I. M., and Evans, R. M., Retinoic acid is a negative regulator of AP-1 responsive genes, *Proc. Natl. Acad. Sci. U.S.A.*, 88, 6092, 1991.

Schütte, J., Viallet, J., Nau, M., Segal, S., Fedorko, J., and Minna, J., Jun-B inhibits and c-Fos stimulates the transforming activities of c-Jun, *Cell*, 59, 987, 1989.

Smeal, T., Binétruy, B., Mercola, D. A., Birrer, M., and Karin, M., Oncogenic and transcriptional cooperation with H-Ras requires phosphorylation of c-Jun on serines 63-73, *Nature*, 354, 494, 1991.

Toledano, M. B. and Leonard, W. J., Modulation of transactivation of transcription factor NF-kB binding activity by oxidation reduction in vitro, *Proc. Natl. Acad. Sci. U.S.A.*, 88, 4328, 1991.

Tratner, I., Ofir, R., and Verma, I. M., Alterations of a cyclic AMP-dependent protein kinase phosphorylation site in c-Fos protein augments its transforming potential, *Mol. Cell. Biol.*, 12, 998, 1992.

Wisdom, R., Yen, J., Rashid, D., and Verma, I. M., Transformation of FosB requires a transactivation domain missing in FosB2 that can be substituted by heterologous activation domains, *Genes Dev.* 6, 667, 1992.

Xanthoudakis, S., Miao, G., Wang, F., Pan, Y. C. E., and Curran, T., Redox activation of Fos-Jun DNA binding activity is mediated by a DNA repair enzyme, *EMBO J.*, 11, 3323, 1992.

Yang-Yen, H. F., Chambard, J. C., Sun, Y. L., Smeal, T., Schmidt, T. J., Drouin, J., and Karin, M., Transcriptional interference between c-Jun and the glucocorticoid receptor: Mutual inhibition of DNA binding due to direct protein–protein interaction, *Cell*, 62, 1205, 1990.

Yen, J. R., Wisdom, R., Tratner, I., and Verma, I. M., An alternative spliced form of FosB is a negative regulator of transcriptional activation and transformation by Fos proteins, *Proc. Natl. Acad. Sci. U.S.A.*, 88, 5077, 1991.

Chapter 4

Steroid Receptors–AP-1 Interaction: Cross-Coupling of Signal Transduction Pathways

Andrew C. B. Cato

CONTENTS

INTRODUCTION

Various decisions in development and cellular proliferation occur through changes in gene expression. Of particular importance to the development of a multicellular organism is the need for the individual cells to reprogram transcriptional activities in response to stimuli from the environment or neighboring cells. This reprogramming could be in the direction of either differentiation or proliferation.

At the disposal of such reprogramming actions are intracellular and membrane bound receptors that intercept various stimuli and translate them into changes in gene activity. When a given cell is exposed to a large number of physiological cues, the information received must be brought to some form of coherency to avoid total chaos. In other words, the different signal transduction pathways need to talk to each other to convey their information in one direction at a time. This is achieved by a coordinate amplification or extinction of some of the signals intercepted. In this article, I will consider two main signal transduction pathways: the effects of steroid hormones and growth promoting substances on a cell. I will review recently published efforts to understand cross-talk between these two signaling pathways and discuss divergent findings to find out whether the reported inconsistencies are real or have been generated by the use of different experimental systems. Where common ground exists in the proposed interaction of the two pathways, I will try to expose it, with the hope of answering the question as to whether a given cell uses one mechanism or different mechanisms in establishing a rapprochement between the two signaling pathways.

STEROID HORMONES

Steroid hormones are naturally occurring cyclopentanophenanthrene molecules that regulate diverse processes such as homeostasis, reproduction, development, and differentiation. They function via binding to intracellular receptors that convey their action through interaction with discrete nucleotide sequences on regulatable regions of specific genes (Evans, 1988; Beato, 1989; Cato et al., 1992b). Steroid hormone receptors regulate their own activity through repression of the timely production of their ligands by derivatives of the central nervous system and through negative regulation of their synthesis (Keller-Wood and Dallman, 1984). Glucocorticoid and progesterone receptors (GR and PR), for example, both repress the RNA and protein levels of their receptors (Okret et al., 1986; Rosewicz et al., 1988; Wei et al., 1988; Read et al., 1988; Hoeck et al., 1989; Burnstein et al., 1990). The PR can also downregulate its own expression through protein–protein interaction with the estrogen receptor (ER), which is an activator of expression of the PR (Savouret et al., 1991). Different steroid receptors can also enhance each other's activity when their binding sites are close to each other, as has been reported for the ER and PR, or the ER and GR (Cato et al., 1988; Ankenbauer et al., 1988).

GROWTH FACTORS

Growth factors are polypeptides that, as their names imply, affect the growth of cells. Peptid hormones and growth factors bind to specific receptors on the outer face of the cellular membrane and transduce signals that activate enzymes on the internal face of the cell membrane. These signals lead to the production or activation of other proteins. One of the key gene products synthesized in response to growth factor stimulation belongs to the family of transcription factor AP-1. This family consists of a number of polypeptides, the prototypes of which are c-Jun and c-Fos. Other members in the family are JunB, JunD, FosB, Fra1, and, most likely, Fra2 (Angel and Karin, 1991). These proteins function by binding to discrete regulatable elements of genes as homo- or heterodimers. Depending on the composition of the dimer species, different elements are preferentially recognized: for example, c-Jun/c-Jun homodimers and c-Jun/c-Fos heterodimers bind to the AP-1 site represented by the consensus T G/T A C/G TCA. On the other hand, a homologous sequence such as T G/T ACGTCA is recognized by a c-Jun/ATF2 heterodimer, or homodimers of ATF2, or heterodimers of ATF/CREB (Benbrook and Jones, 1990; Ivashkiv et al., 1990; Hai and Curran, 1991).

INTERACTION OF STEROID AND GROWTH-PROMOTING PATHWAYS

Various lines of evidence suggest that the steroid receptor pathway interacts with the signal transduction pathway for growth-promoting factors. Such interaction could have negative or positive repercussions. For example, while epidermal growth factor (EGF) in the presence of estradiol and progesterone stimulates growth of the mammary gland during pregnancy, it also inhibits premature differentiation induced by glucocorticoids, insulin and prolactin (Tonelli and Sorof, 1980; Taketani and Oka, 1983a, 1983b). In human prostate epithelium, in the mouse keratinocyte (MK) cell line, and in rabbit cornea, glucocorticoids oppose EGF-induced growth (Chaproniere and Webber, 1985; Woost et al., 1985; Zendegui et al., 1988), but in rat mammary epithelium and submandibular gland and human amniotic cells, glucocorticoids enhance EGF-induced cell growth and postanoid synthesis (Salomon et al., 1981; Redman et al., 1988; Mitchell and Tjian, 1989). Such complex activity of steroid and growth promoting substances can be studied by simplifying the complex physiological processes to the level of activity regulation of single genes. These simplified studies have provided different mechanistic models that broadly explain the changes in gene activity brought about by the interplay of steroid receptors and AP-1. The models can be categorized as superimposition, tethering, and mutual inhibition.

SUPERIMPOSITION

In this model, the binding site for AP-1 overlaps or is juxtaposed to the binding site of the GR. Examples of this type of arrangement of regulatory elements can be found in the rat α-fetoprotein and mouse proliferin genes (Diamond et al., 1989; Mordacq and Linzer, 1989; Zhang et al., 1991). Another example is found in the osteocalcin gene, where a vitD3 binding site overlaps an AP-1 recognition site (Schüle et al., 1990b).

In the proliferin gene, where the phorbol ester TPA induces expression while glucocorticoids downregulate this activity, the GR binding site is just upstream of an AP-1 site, which suggests that the GR and AP-1 might compete for the same binding sites (Mordacq and Linzer, 1989; Diamond et al., 1990). In transfection experiments in HeLa cells, with a chimeric construct consisting of a minimal promoter to which an oligonucleotide containing the AP-1 and GR binding sites had been linked, glucocorticoids did not repress but rather activated expression (Diamond et al., 1990). This increased expression occurred in HeLa cells when the intracellular concentration of c-Jun was elevated (Diamond et al., 1990). Similar experiments performed with a minimal promoter containing the α-fetoprotein regulatory element confirmed that the GR stimulates the action of c-Jun homodimers (Zhang et al., 1991). However, in both systems a hormone-dependent repression occurred when the intracellular concentrations of c-Jun and c-Fos were elevated (Diamond et al., 1990; Zhang et al., 1991). These results, particularly with the proliferin construct, led to the concept of "composite response elements" (CREs), whereby the steroid receptor in a particular cell line can mediate positive or negative regulation depending on the relative concentration of nonreceptor protein in the cell.

DNase I footprinting studies confirmed that the GR binds to the glucocorticoid response element (GRE) in the proliferin gene construct (Diamond et al., 1990). Consistent with this finding, transfection of receptor mutants that lack DNA binding activity failed both to repress the action of c-Jun and c-Fos and to activate the activity of the c-Jun homodimer (Diamond et al., 1990). These results led to the conclusion that the DNA binding activity of the GR is essential for hormone regulation at the proliferin (plfG) chimeric gene promoter. In addition to the contribution of the DNA binding activity of the GR, Pearce and Yamamoto (1993) used glucocorticoid–mineralocorticoid receptor (GR–MR) chimeras to show that the N-terminus of the GR is necessary for repressing the action of c-Fos and c-Jun at the plfG chimeric promoter. They found that the MR, as opposed to the GR, does not repress the action of c-Fos and c-Jun (Pearce and Yamamoto, 1993). Replacement of the N-terminal region of the MR by the corresponding region of the GR produced a chimeric GR–MR that repressed c-Fos/c-Jun expression of the plfG–CAT construct. Through the use of other chimeras they delineated a region between 105 and 440 of the rat GR as important for the repression (Pearce and Yamamoto, 1993).

On the c-Jun protein, the basic zipper region was shown to be responsible for specifying enhancement by c-Jun homodimers and repression by c-Jun–c-Fos heterodimers (Miner and Yamamoto, 1992). Such experiments on the delineation of regions on AP-1 or the GR responsible for the regulatory action at the plfG–CAT composite element presupposes that these factors interact physically with each other. Receptor-specific monoclonal antibody coimmunoprecipitates either *in vitro* translated c-Jun or c-Jun/c-Fos heterodimers from reticulocyte extracts mixed with a purified receptor preparation (Miner and Yamamoto, 1992). AP-1 immunoprecipitation was receptor dependent but c-Fos alone was not found to be associated with the GR (Miner and Yamamoto, 1992). This indicates that the receptor interacts with only the dimerized form of AP-1 (c-Fos cannot form dimers), but whether the receptor contacts c-Fos in the receptor c-Jun/c-Fos complex is not known.

TETHERING

In contrast to the reported regulation at the CRE of the plfG construct, we and other investigators (Jonat et al., 1990; Schüle et al., 1990b; Yang-Yen et al., 1990) have reported another mechanism through which GR interacts with AP-1. The experiments that led to the mechanism were performed on the human collagenase I gene or a minimal promoter construct containing a multimerized AP-1 site derived from this gene. The AP-1 binding site of the collagenase gene mediates increased expression by phorbol ester. The same site also mediates negative regulation of expression of the collagenase gene by glucocorticoid, although there is no binding site for the receptor overlapping the AP-1 site nor does the receptor bind to the AP-1 site. Inhibition of the phorbol ester-induced expression of collagenase by glucocorticoids does not require protein synthesis, indicating that de novo protein synthesis is not necessary (Jonat et al., 1990). Cotransfection of a reporter gene consisting of the collagenase AP-1 site and the GR established that the receptor was required for the inhibition (Jonat et al., 1990; Schüle et al., 1990b; Yang-Yen et al., 1990). One key experiment that provided insight into the mechanism of this type of inhibition of AP-1 action was reported by König et al. (1992), who showed in *in vivo* footprinting analyses that occupancy of the AP-1 binding site on the collagenase gene was not altered under the same conditions that inhibit AP-1 action. This result implies that the GR in the presence of hormone is tethered to the DNA-bound AP-1 at the collagenase regulatory element without itself binding to DNA. The evidence to support this model of protein–protein interaction is the alteration of factor binding to sequences, such as the PEA3 recognition sequence close to the AP-1 site in the collagenase gene, that may reflect steric hinderance at these sites as the GR is tethered to the AP-1 recognition site (König et al., 1992).

Experiments to determine the sequences on the GR responsible for the repression have produced discordant results. Jonat et al. (1990), reported that amino-terminal sequences are required for the negative action of the GR, whereas other investigators have shown that the DNA binding domain of the GR is important for repression (Lucibello et al., 1990; Schüle et al., 1990b). More recent studies have shown that the repressing activity of the receptor is generated by intramolecular contact of several regions of the receptor to generate a repressing surface (Heck et al., 1994). The experiments that contributed to this result were performed using chimeric GR–MR constructs. As in the case of the proliferin gene, we have also observed that the MR does not inhibit AP-1 activity (Cato et al., 1992; Ponta et al., 1992). This allows chimeric GR–MR constructs to be used for determination of the regions of the GR that are important for repression. Depending on the combinations of MR and GR domains used or the hormone administered (dexamethasone or aldosterone), different regions of the GR were shown to be important for repression (Heck et al., 1994). For example, MMM (for the 3 domains of the GR), in the presence of its own ligand or of dexamethasone, did not repress the activity of the minimal promoter with multimerized collagenase AP-1 sequences. However, GMM, in the presence of dexamethasone, repressed, indicating the contribution of the amino-terminal domain of the GR to repression. A chimeric construct consisting of only the DNA and hormone binding domain of the MR (ΔMM) in the presence of dexamethasone did not repress, whereas a ΔGM construct did, albeit to a lower extent than did the GR (Heck et al., 1994). These results indicate a contribution of the DNA binding domain (DBD) of the GR.

In the DBD of the receptors, introduction and analyses of a series of amino acid exchanges showed that three mutants were unable to repress AP-1 activity, although they did bind DNA and transactivated (Heck et al., 1994). This indicates that transactivation and transrepression can be dissociated. Two of the three amino acids exchanged in the GR occur at equivalent positions in the PR, which would suggest that the PR is not a repressing protein. The PR, nevertheless, inhibited gene activity controlled by the collagenase AP-1 sequences, confirming our finding that not just a few amino acid sequences but rather a large surface of the GR or PR contribute to the repressing activity of these molecules. Further confirmation of this finding comes from experiments with steroid analogs in which we inhibited the transactivation function of the GR without affecting the ability of the receptor to inhibit AP-1 activity (Heck et al., 1994). An important question in all these studies is whether or not the GR interacts directly with AP-1. In experiments aimed to demonstrate this, *in vitro* translation of the GR, Fos, and Jun, followed by immunoprecipitation, showed that antibodies to c-Jun but not to c-Fos coprecipitate the receptor in the presence and absence of protein cross-linkers (Heck, unpublished results). The mutations in the DBD that destroyed the ability of the GR to repress can still be coprecipitated by c-Jun antibodies suggesting perhaps that the differences in interaction do not appear in the coprecipitation protocol. Alternatively, the receptor interaction with c-Jun alone may not form the basis for transrepressing activity of the receptor.

MUTUAL INHIBITION

Other studies on inhibition by the GR of AP-1 activity at the collagenase gene have provided results for yet another mechanism. In this mechanism AP-1 is thought to interact with the GR and both are thought to inhibit each other's DNA binding and transactivation functions. Studies from Yang-Yen et al. (1990) and Schüle et al. (1990a) have provided evidence for this model. They showed that incubation of a purified receptor with an excess of purified c-Jun inhibited the binding of the GR to a simple palindromic GRE. The reverse is also true, as incubation of c-Jun with excess GR reduced somewhat the binding of c-Jun to a simple, collagenase AP-1 binding site. Transfection experiments indicated that the receptor zinc finger DBD was essential for the GR repression of AP-1 function, but specific DNA binding may not be necessary for this repression. Substitution of the GR zinc finger domain with that of either retinoic acid or thyroid receptor produced chimeras that still retained their ability to repress (Schüle et al., 1990a). Overexpression of c-fos or c-jun inhibited GR-mediated activation of expression of constructs with simple GREs (Jonat et al., 1990; Yang-Yen et al., 1990).

DIFFERENCES AND SIMILARITIES IN THE DIFFERENT MODELS: PROLIFERIN VS. COLLAGENASE AP-1–GR REGULATING SYSTEM

At first glance it appears that major differences exist in the negative mode of action of the GR at the AP-1 sites of the proliferin gene and the collagenase AP-1 regulatory site. To determine whether these differences are real, I will first discuss the two major "apparent" differences of DNA binding and AP-1–GR interaction that are thought to distinguish the proliferin from the collagenase regulatory systems.

DNA BINDING

DNA binding by the GR has been reported necessary for repression at the proliferin regulatory element but not for collagenase AP-1 activity (Diamond et al., 1990; Jonat et al., 1990). Classical GREs are not present in the proliferin or α-fetoprotein genes, where DNA binding activity of the receptors has been shown to be responsible for repression of AP-1 activity (Diamond et al., 1990; Zhang et al., 1991). These genes contain the so-called negative GRE (nGRE). Recent experiments have shown that nGREs such as those found in the pro-opiomelanocortin gene bind three molecules of the GR rather than the dimeric GR molecules that bind to classical GRE (Drouin et al., 1993). It is possible that on an nGRE the initial binding by the receptor dimer serves only to tether the third receptor molecule to make the appropriate protein–protein interaction with AP-1. This inhibiting GR monomer may not need to bind DNA in the negative action of the receptor at AP-1 binding sites.

Receptor mutants that do not bind DNA, unfortunately, cannot be used to demonstrate the need for DNA binding activity for repression. Such mutants are functionally inactive and do not mediate repression even through the collagenase AP-1 sites, where DNA binding activity has been shown to be unimportant for repression. In such mutants the particular conformation of the GR needed for repressing is destroyed.

AP-1–GR INTERACTION

Gene expression at the proliferin chimeric construct is said to be enhanced or repressed by the GR depending on the composition of AP-1 present in the recipient cells used for the transfection (Diamond et al., 1990; Miner and Yamamoto, 1992). In contrast, various reports on AP-1–GR interaction through the collagenase gene AP-1 site have only indicated repression by the GR (Jonat et al., 1990; Schüle et al., 1990b; Yang-Yen et al., 1990). It must be kept in mind that a number of cell lines used for experiments on the AP-1–GR interaction may contain varying endogenous levels of c-fos, c-Jun, and related family members, which might complicate interpretation of the results obtained from the transrepression experiments. Ideally, a cell line with low or almost undetectable levels of AP-1 polypeptides into which different members could be introduced is useful for determining the functional interaction of the individual Fos/Jun family members, as either homodimers or heterodimers with the GR. Such experiments in murine teratocarcinoma F9 cells have confirmed the role of the hetero- and homodimer AP-1 species in their functional interaction with the GR. Even at the collagenase regulatory element, preliminary studies in F9 cells have shown a positive regulation by the GR in the presence of a c-Jun homodimer and a negative regulation in the presence of c-Jun–c-Fos heterodimers (Angel, unpublished results).

There have been several attempts to show that the GR interacts with AP-1 in coimmunoprecipitation experiments in the absence of DNA. While some have been able to show such interactions (Touray et al., 1991; Diamond et al., 1990; Yang-Yen et al., 1990; Miner and Yamamoto, 1992), others could not demonstrate it (Lucibello et al., 1990), indicating that the complexes are rather unstable and allow detection only under precisely controlled conditions. One successful experiment performed by Miner and Yamamoto (1992), showed that GR–c-Jun and GR–c-Jun–c-Fos, but not GR–cFos, complexes are formed. This indicates that the GR associates with AP-1 through c-Jun without disrupting the c-Jun–c-Fos dimer. Recent results from my laboratory confirm these findings (Heck, unpublished results). Yang-Yen et al. (1990a) have shown, however, that in their hands the GR can be cross-linked to either c-Jun or c-Fos. Touray et al. (1991), working in the absence of cross-linking agents, showed, in extracts of NIH3T3 cells that overexpress the GR, that c-Jun and c-Fos each form stable complexes with the GR. They further showed that even *in vitro* synthesized c-Fos and c-Jun associate with the GR. These apparent differences can be explained by the use of different antibodies, some of which may recognize epitopes that disrupt the interactions being investigated.

The role of Fos in interaction with the GR is, therefore, far from clear from the results of the coprecipitation studies. In fact, it may be that c-Fos on its own does not interfere with the action of the GR. The reported findings of c-Fos inhibition of GR response (Lucibello et al., 1990) could result from heterodimerization of the transfected c-Fos with endogenous c-Jun, as the HeLa cells used for these experiments contain large amounts of endogenous Jun (Angel, personal communication). A similar explanation can be provided for other reports in which Fos alone has been claimed to repress the response of the GR (Jonat et al., 1990).

In studies by Lucibello et al. (1990), N-terminal amino acids of c-Fos between positions 40 and 111 were claimed to be involved in repression of GR action. Miner and Yamamoto (1992), on the other hand, reported that four amino acids in the basic domain of c-Fos (positions 128–164) may constitute a signal (when c-Fos and c-Jun heterodimers are bound to the proliferin regulatory element) for repression by the GR. Thus, there seems to be some uncertainty concerning the regions of AP-1 involved in repression, and future work is required to clarify this.

DOMAINS OF THE GR

Various investigators have suggested that the DNA binding domain of the GR is involved in the repression of expression of chimeric genes containing the AP-1 binding sites from the collagenase gene (Jonat et al., 1990; Schüle et al., 1990a). Other reports have shown that deletion of the DBD still led to partial repression of AP-1 activity (Jonat et al., 1990), while yet another report demonstrated that chimeric constructs containing Gal4 DBD and hormone binding domain of the GR could still repress AP-1 activity (Shemshedini et al., 1991). The latter report implies that sequences in the hormone binding domain of the GR may also be important for repression. From studies with the proliferin chimeric gene constructs, N-terminal sequences of the GR have been implicated in the repression of AP-1 response (Pearce and Yamamoto, 1993). These varying reports may indicate that, as in transactivation, no one particular domain of the receptor is involved in repression, but that different segments of the receptor function together to provide the interacting surface for repression. Recent studies show that N-terminal, hormone binding, and DNA binding regions of the GR all participate in the repression of AP-1 activity (Heck et al., 1994). Taken together, these considerations do not provide any clear differences to distinguish the negative action of the GR on AP-1 response into two regulating systems of proliferin and collagenase genes. The GR may thus use only one particular mechanism for inhibiting AP-1 action independent of the gene studied.

CROSS-TALK BETWEEN STEROID RECEPTORS
AND OTHER TRANSCRIPTION FACTORS

The cross-talk between steroid hormone receptors and other transcription factors is not limited to the GR and AP-1 nor does it always result in negative regulation of gene expression. In addition to the GR, other members of the steroid/retinoic acid/thyroid family, such as estrogen, androgen, progesterone, retinoic acid, vitamin D3, and thyroid receptors, are involved in various forms of cross-talk. The list of other transcription factors that take part in the cross-talk is growing. At the moment they range from CREB, NF-1L6 (member of the basic zipper DNA binding protein), and NFAT to members of the POU homeodomain family (OTF1 and OTF2A) (Akerblom et al., 1988; Schüle et al., 1990b; Owen et al., 1990; Shemshedini et al., 1991; Doucas et al., 1991; Felli et al., 1991; Wieland et al., 1991; Northrop et al., 1993; Kutoh et al., 1992; Vacca et al., 1992; Nishio et al., 1993). Details of the mechanism of cross-talk with members of the steroid family have not been studied in these interactions to the same level as that of the AP-1–GR interaction. However, they are thought to require either DNA or non-DNA binding activities of the various steroid receptors.

The GR, for example, represses the action of both OTF1 and OTF2A in the absence of DNA binding (Wieland et al., 1991; Kutoh et al., 1992). The GR achieves this by associating with Oct1, and immunoprecipitation experiments have shown that the homeodomain of OTF1 is necessary for the interaction (Kutoh et al., 1992). A related homeoprotein, OTF2A, that diverges slightly in sequences from the homeodomain region of OTF1 is also functionally repressed by the GR, implying that the homeodomain region of the OTF2A protein may not interact with the GR. Competition of intermediary factors necessary for the OTF2A activity by the GR has been suggested as the mechanism underlying the negative action of the GR on OTF2A activity. As homeodomain proteins comprise a large family of transcription factors, it is likely that the interaction of the GR with other members in this family will continue. The same is true of the interaction of the GR and NFAT, which belongs to an extensive family of proteins with diverse functions (Wasylyk et al., 1993). The list of transcription factors involved in cross-talk with steroid hormone receptors is growing enormously, and this is a challenge for the resolution of complex biological processes at the level of alteration in transcriptional activity of genes. As more and more transcriptional factors are implicated in this action, the chance of further cross-coupling of signal transduction pathways involving steroid receptors being unravelled will be tremendously increased.

REFERENCES

Akerblom, I. E., Slater, E. P., Beato, M., Baxter, J. D., and Mellon, P. L., Negative regulation by glucocorticoids through interference with a cAMP responsive enhancer, Science, 241, 350, 1988.

Angel, P., and Karin, M., The role of Jun, Fos and the AP-1 complex in cell-proliferation and transformation, Biochim. Biophys. Acta, 1072, 129, 1991.

Ankenbauer, W., Strähle, U., and Schütz, G., Synergistic action of glucocorticoid and estradiol responsive elements, Proc. Natl. Acad. Sci. U.S.A., 85, 7526, 1988.

Beato, M., Gene regulation by steroid hormones, Cell, 56, 335, 1989.

Benbrook, D. M. and Jones, N. C., Heterodimer formation between CREB and JUN proteins, Oncogene, 5, 295, 1990.

Burnstein, K. L., Jewell, C. M., and Cidlowski, J. A., Human glucocorticoid receptor cDNA contains sequences sufficient for receptor down-regulation, J. Biol. Chem., 265, 7284, 1990.

Cato, A. C. B., Heitlinger, E., Ponta, H., Klein-Hitpass, L., Ryffel, G. U., Bailly, A., Rauch, C., and Milgrom, E., Estrogen and progesterone receptor-binding sites on the chicken vitellogenin II gene: Synergism of steroid hormone action. Mol. Cell. Biol., 8, 5323, 1988.

Cato, A. C. B., König, H., Ponta, H., and Herrlich, P., Steroids and growth promoting factors in the regulation of expression of genes and gene networks, J. Steroid Biochem. Molec. Biol., 43, 63, 1992a.

Cato, A. C. B., Ponta, H., and Herrlich, P., Regulation of gene expression by steroid hormones, Progr. Nucleic Acid Mol. Biol., 43, 1, 1992b.

Chaproniere, D. M. and Webber, M. M., Dexamethasone and retinyl acetate similarly inhibit and stimulate EGF- or insulin-induced proliferation of prostate epithelium, J. Cell Physiol., 122, 249, 1985.

Diamond, A. M., Derand, C. J., and Schwartz, J. L., Alterations in transformation efficiency by the ADPRT-inhibitor 3-aminobenzamide are oncogene specific, Carcinogenesis, 10, 383, 1989.

Diamond, M. I., Miner, J. N., Yoshinaga, S. K., and Yamamoto, K. R., Transcription factor interactions: Selectors of positive or negative regulation from a single DNA element, Science, 249, 1266, 1990.

Doucas, V., Spyrou, G., and Yaniv, M., Unregulated expression of c-Jun or c-Fos proteins but not Jun D inhibits oestrogen receptor activity in human breast cancer derived cells, EMBO J., 10, 2237, 1991.

Drouin, J., Trifiro, M. A., Plante, R. K., Nemer, M., Eriksson, P., and Wrange, Ö., Glucocorticoid receptor binding to a specific DNA sequence is required for hormone-dependent repression of pro-opiomelanocortin gene transcription, *Mol. Cell. Biol.*, 9, 5305, 1989.

Drouin, J., Sun, Y. L., Chamberland, M., Grauthier, Y., DeLian, A., Nemer, M., and Schmidt, T. J., Novel glucocorticoid receptor complex with DNA element of the hormone-repressed POMC gene, *EMBO J.*, 12, 145, 1993.

Evans, R. M., The steroid and thyroid hormone receptor superfamily, *Science*, 240, 889, 1988.

Felli, M. P., Vacca, A., Meco, D., Screpanti, I., Farina, A. R., Maroder, M., Martinotti, S., Petrangeli, E., Frati, L., and Gulino, A., Retinoic acid-induced down-regulation of the interleukin-2 promoter via cis-regulatory sequences containing an octamer motif, *Mol. Cell. Biol.*, 11, 4771, 1991.

Gaub, M.-P., Bellard, M., Scheuer, I., Chambon, P., and Sassone-Corsi, P., Activation of the ovalbumin gene by the estrogen receptor involves the Fos-Jun complex, *Cell*, 63, 1267, 1990.

Hai, T. and Curran, T., Cross-family dimerization of transcription factors Fos/Jun and ATF/CREB alters DNA binding specificity, *Proc. Natl. Acad. Sci. U.S.A.*, 88, 3720, 1991.

Heck, S., Kullmann, M., Gast, A., Ponta, H., Rahmsdorf, H. J., Herrlich, P., and Cato, A. C. B., A distict modulating domain in glucocorticoid receptor monomers in the repression of activity of the transcription factor AP-1, *EMBO J.*, in press.

Hoeck, W., Rusconi, S., and Groner, B., Down-regulation and phosphorylation of glucocorticoid receptors in cultured cells, *J. Biol. Chem.*, 264, 14396, 1989.

Ivashkiv, L. B., Liou, H.-C., Kara, C. J., Lamph, W. W., and Verma, I. M., mXBP/CRE-BP2 and c-Jun form a complex which binds to the cyclic AMP, but not to the 12-O-tetradecanoylphorbol-13-acetate, response element, *Mol. Cell. Biol.*, 10, 1609, 1990.

Jonat, C., Rahmsdorf, H. J., Park, K.-K., Cato, A. C. B., Gebel, S., Ponta, H., and Herrlich, P., Anti-tumor promotion and antiinflammation: Down-modulation of AP-1 (Fos/Jun) activity by glucocorticoid hormone, *Cell*, 62, 1189, 1990.

Keller-Wood, M. E. and Dallman, M. F., Cortisteroid inhibition of ACTH secretion, *Endocr. Rev.*, 5, 1, 1984.

König, H., Ponta, H., Rahmsdorf, H. J., and Herrlich, P., Interference between pathway-specific transcription factors: Glucocorticoids antagonize phorbol ester-induced AP-1 activity without altering AP-1 site occupation in vivo, *EMBO J.*, 11, 2241, 1992.

Kutoh, E., Strömstedt, P.-E., and Poellinger, L., Functional interference between the ubiquitous and constitutive octamer transcription factor 1 (OTF-1) and the glucocorticoid receptor by direct protein–protein interaction involving the homeo subdomain of OTF-1, *Mol. Cell. Biol.*, 12, 4960, 1992.

Lucibello, F. C., Slater, E. P., Jooss, K. U., Beato, M., and Müller, R., Mutual transrepression of Fos and the glucocorticoid receptor: Involvement of a functional domain in Fos which is absent in FosB, *EMBO J.*, 9, 2827, 1990.

Miner, J. N. and Yamamoto, K. R., The basic region of AP-1 specifies glucocorticoid receptor activity at a composite response element, *Genes Dev.*, 6, 2491, 1992.

Mitchell, P. J. and Tjian, R., Transcriptional regulation in mammalian cells by sequence-specific DNA binding proteins, *Science*, 245, 371, 1989.

Mordacq, J. C. and Linzer, D. I. H., Co-localization of elements required for phorbol ester stimulation and glucocorticoid repression of proliferin gene expression, *Genes Dev.*, 3, 760, 1989.

Nicholson, R. C., Mader, S., Nagpal, S., Leid, M., Rochette-Egly, C., and Chambon, P., Negative regulation of the rat stromelysin gene promoter by retinoic acid is mediated by an AP1 binding site, *EMBO J.*, 9, 4443, 1990.

Nishio, Y., Ishiki, H., Kishimoto, T., and Akira, S., A nuclear factor for interleukin-6-expression (NF-IL6) and the glucocorticoid receptor synergistically activate transcription of the rat α1-acid glycoprotein gene via direct protein-protein interaction, *Mol. Cell. Biol.*, 13, 1854, 1993.

Northrop, J. P., Crabtree, G. R., and Mattila, P. S., Negative regulation of interleukin 2 transcription by the glucocorticoid receptor, *J. Exp. Med.*, 175, 1235, 1993.

Okret, S., Prellinger, L., Dong, Y., and Gustafsson, J.-Å., Down-regulation of glucocorticoid receptor mRNA by glucocorticoid hormones and recognition by the receptor of a specific binding sequence within a receptor cDNA clone, *Proc. Natl. Acad. Sci. U.S.A.*, 83, 5899, 1986.

Owen, T. A., Bortell, R., Yocum, S. A., Smock, S. L., Zhang, M., Abate, C., Shalhoub, V., Aronin, N., Wright, K. L., van Wijnen, A. J., Stein, J. L., Curran, T., Lian, J. B., and Stein, G. S., Coordinate occupancy of AP-1 sites in the vitamin D-responsive and CCAAT box elements of Fos-Jun in osteocalcin gene: Model for phenotype suppression of transcription, *Proc. Natl. Acad. Sci. U.S.A.*, 87, 9990, 1990.

Pearce, D. and Yamamoto, K. R., Mineralocorticoid and glucocorticoid receptor activities distinguished by nonreceptor factors at a composite response element, *Science*, 259, 1161, 1993.

Ponta, H., Cato, A. C. B., and Herrlich, P., Interference of pathway specific transcription factors, *Biochim. Biophys. Acta*, 1129, 255, 1992.

Pratt, W. B., Jolly, D. J., Pratt, D. V., Hollenberg, S. M., Giguére, V., Cadepond, F. M., Schweizer-Groyer, G., Catelli, M.-G., Evans, R. M., and Baulieu, E.-E., A region in the steroid binding domain determines formation of the non-DNA-binding, 9 S glucocorticoid receptor complex, *J. Biol. Chem.*, 263, 267, 1988.

Read, D. L., Sneider, C. E., Müller, J. S., Green, G. L., and Katzenellenbogen, B. S., Ligand-modulated regulation of progesterone receptor messenger ribonucleic acid and protein in human breast cancer cell line, *Mol. Endocrinol.*, 2, 263, 1988.

Redman, R. S., Quissell, D. O., and Barzen, K. A., Effects of dexamethasone, epidermal growth factor, and retinoic acid on rat submandibular acinor-intercalated duct complexes in primary culture. *In Vitro Cell Dev. Biol.*, 24, 734, 1988.

Rosewicz, S., McDonald, A. R., Maddux, B. A., Goldfine, I. D., Miesfeld, R. L., and Logsdon, C. D., Mechanism of glucocorticoid receptor down-regulation by glucocorticoid, *J. Biol. Chem.*, 263, 2581, 1988.

Salomon, D.-S., Liotta, L. A., and Kidwell, W. R., Differential response to growth by rat mammary epithelium plated in different collagen substrata in serum-free medium, *Proc. Natl. Acad. Sci. U.S.A.*, 78, 382, 1981.

Savouret, J. F., Bailly, A., Misrahi, M., Rauch, C., Redenich, G., Chauchereau, A., and Milgrom, E., Characterization of the hormone responsive element involved in the regulation of the progesterone receptor gene, *EMBO J.*, 10, 1875, 1991.

Schüle, R., Rangarajan, P., Kliewer, S., Ransone, L. J., Bolado, J., Yang, N., Verma, I. M., and Evans, R. M., Functional antagonism between oncoprotein c-Jun and the glucocorticoid receptor, *Cell*, 62, 1217, 1990a.

Schüle, R., Umesono, K., Mangelsdorf, D. J., Bolado, J., Pike, J. W., and Evans, R. M., Jun-Fos and receptors for vitamins A and D recognize a common response element in the human osteocalcin gene. *Cell*, 61, 497, 1990b.

Shemshedini, L., Knauthe, R., Sassone-Corsi, P., Pornon, A., and Gronemeyer, H., Cell-specific inhibitory effects of Fos and Jun on transcription activation by nuclear receptors, *EMBO J.*, 10, 3839, 1991.

Taketani, Y. and Oka, T., Epidermal growth factor stimulates cell proliferation and inhibits functional differentiation of mouse mammary epithelial cells in culture. *Endocrinology*, 113, 871, 1983a.

Taketani, Y. and Oka, T., Possible physiological role of epidermal growth factor in the development of the mouse mammary gland during pregnancy, *FEBS Lett.*, 152, 256, 1983b.

Tonelli, Q. J. and Sorof, S., Epidermal growth factor requirement for development of culture mammary gland, *Nature*, 285, 250, 1980.

Touray, M., Ryan, F., Jaggi, R., and Martin, F., Characterisation of functional inhibition of glucocorticoid receptor by Fos/Jun, *Oncogene*, 6, 1227, 1991.

Vacca, A., Felli, M. P., Farina, A. R., Martinotti, S., Maroder, M., Screptani, I., Meco, D., Petrangeli, E., Frati, L., and Gulino, A., Glucocorticoid receptor-mediated suppression of the interleukin 2 gene expression through impairment of the cooperativity between nuclear factor of activated T cells and AP-1 enhancer elements, *J. Exp. Med.*, 175, 637, 1992.

Wasylyk, B., Hahn, L., and Giovane, A., The Ets family of transcription factors, *Eur. J. Biochem.*, 211, 7, 1993.

Wei, L. L., Krett, N. L., Francis, M. D., Gordon, D. F., Wood, W. M., O'Malley, B. W., and Horowitz, K. B., Multiple human progesterone receptor messenger ribonucleic acids and their autoregulation by progestin agonists and antagonists in breast cancer cells, *Mol. Endocrinol.*, 2, 62, 1988.

Wieland, S., Döbbeling, U., and Rusconi, S., Interference and synergism of glucocorticoid receptor and octamer factors, *EMBO J.*, 10, 2513, 1991.

Woost, P. G., Brightwell, J., Eiferman, R. A., and Schultz, G. S., Effect of growth factors with dexamethasone in healing of rabbit corneal stromal incisions, *Exp. Eye Res.*, 40, 47, 1985.

Yang-Yen, H.-F., Chambard, J.-C., Sun, Y.-L., Smeal, T., Schmidt, T. J., Drouin, J., and Karin, M., Transcriptional interference between c-Jun and the glucocorticoid receptor: Mutual inhibition of DNA binding due to direct protein-protein interaction, *Cell*, 62, 1205, 1990.

Zendegui, J. G., Inman, W. H., and Carpenter, G., Modulation of the mitogenic response of an epidermal growth factor-dependent keratinocyte cell line by dexamethasone, insulin and transforming growth factor-β, *J. Cell Physiol.*, 136, 257, 1988.

Zhang, X.-K., Dong, J.-M., and Chiu, J.-F., Regulation of α-fetoprotein gene expression by antagonism between AP-1 and the glucocorticoid receptor at their overlapping binding site, *J. Biol. Chem.*, 266, 8248, 1991.

Chapter 5

Regulation of c-Jun Activity by Phosphorylation

Christoph Sachsenmaier and Adriana Radler-Pohl

CONTENTS

INTRODUCTION

The transcription factor AP-1 (Fos/Jun) is a key regulator in converting a multitude of signals into genetic responses (Angel and Karin, 1991). It is, therefore, subjected to tight regulation, both at the level of synthesis of Jun and Fos and by posttranslational modification of preexisting proteins (e.g., phosphorylation, glycosylation, redox regulation, association with proteins, and degradation). In this chapter we will focus on phosphorylation/dephosphorylation of the c-Jun protein in response to extracellular signals and the protein kinases and phosphatases involved in this process. Several reports put the c-Jun protein at the receiving end of a kinase cascade that is initiated by different growth factors, cytokines, activated oncogenes, phorbol ester tumor promoters, and UV irradiation. These reactions take place immediately after cell stimulation and do not require new protein synthesis. Important insights have been gained into this mechanism regulating c-Jun's activity. Hence, it is discussed in several other chapters of this book.

In contrast to the Jun protein, much less is known about the relationship between posttranslational modification and transcriptional function of the Fos protein. That issue is discussed in Chapter 2.

c-JUN AS A PHOSPHOPROTEIN

Unstimulated mammalian cells contain low, but detectable, amounts of c-Jun protein. In this state, c-Jun is phosphorylated constitutively at serines and threonines close to its *C*-terminal DNA binding domain. Phosphorylation in this region markedly reduces DNA binding and transactivation ability of Jun *in vitro* (Boyle et al., 1991) and *in vivo* (Lin et al., 1992, Hagmeyer et al., 1993), most likely via electrostatic repulsion between the negative charges near the DNA binding domain of Jun and of the sugar–phosphate backbone of the DNA.

Four possible inhibitory sites (Thr231, Thr239, Ser243, and Ser249) become phosphorylated within this region when bacterially expressed, nonphosphorylated Jun is incubated *in vitro* with purified protein kinases (Boyle et al., 1991; Lin et al., 1992; the numbering corresponds to human c-Jun in Angel et al., 1988) (see Figure 5-1A). All four sites reside within one tryptic peptide (amino acids 227–253). The mono-, di- and triphosphorylated "phosphoisomers" of this peptide are resolved on a diagonal when separated by two-dimensional tryptic phosphopeptide mapping (Boyle et al., 1991) (see Figure 5-1B, spots a, b, and c).

0-8493-4573-1/94/$0.00+$.50
© 1994 by CRC Press, Inc.

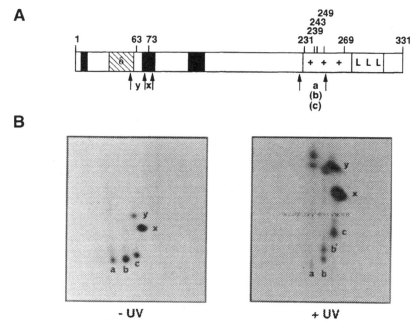

Figure 5-1 **(A)** Schematic diagram of the human c-Jun protein. Amino acids are numbered; black boxes indicate transactivation domains as identified by *in vivo* transactivation experiments (Angel et al., 1989); the dotted boxes represent the DNA binding domain, composed of the basic region (+) and the leucine zipper (L); the hatched box denotes the δ-domain, which is deleted in v-Jun (see also text); arrows indicate trypsin cleavage sites that lead to the appearance of phosphopeptides after *in vivo* labeling of cells with ^{32}P-orthophosphate; the phosphopeptides are indicated by lower case letters (a, b, c, x, and y) (see also text). **(B)** Autoradiogram of *in vivo* ^{32}P-labeled c-Jun protein, isolated by immunoprecipitation from UV-irradiated (+UV) and nontreated (–UV) T9 teratocarcinoma cells trantly transfected with a c-Jun expression plasmid. c-Jun was digested with trypsin and the peptides separated into two dimensions, according to Hunter and Sefton (1980).

In vitro studies using purified glycogen synthase kinase-3 (GSK-3) have suggested that Thr239, Ser243, and Ser249 are the main phosphoacceptor sites within this region (Boyle et al., 1991). However, recent *in vivo* labeling data point to Thr231 rather than Thr239, and together with Ser249 these two amino acids are *in vivo* substrates for nuclear casein kinase II (CKII) (Lin et al., 1992). The kinase that addresses Ser243 *in vivo* still must be identified.

Both reports, however, clearly define Ser243 as a major regulatory site within the DNA binding domain of c-Jun. Substitution of Ser243 to phenylalanine severely reduces phosphorylation of all adjacent sites and, subsequently, leads to higher DNA binding and transactivation activity of c-Jun (Boyle et al., 1991). Single mutations of the other phosphorylation sites (Thr231, Thr239, and Ser249) to alanines all lead to higher transactivation capability of Jun *in vivo;* but abolishing all phosphorylation events in this region, by mutating either all four sites or only Ser243 to alanine, results in the most potent transactivator (Hagmeyer et al., 1993).

In v-Jun, the retroviral counterpart of chicken c-Jun (Maki et al., 1987), the corresponding serine to human Ser243 is at position 216 and is mutated to phenylalanine.

Together with a second point mutation at position 242 (Cys to Ser), these substitutions constitute the only differences between chicken c-Jun and v-Jun in the C-terminal region. Cys242 has been identified as a target of redox regulation of DNA binding of Jun *in vitro*, leading to inhibition of DNA binding under oxidative conditions (Abate et al., 1990). It was, therefore, concluded that the lack of redox regulation of DNA binding leads to the higher transactivation potential of v-Jun, at least *in vitro*. However, mutating Cys252 in chicken c-Jun (which corresponds to Cys242 in v-Jun) alone, led to a much weaker transactivator when compared to v-Jun or c-Jun *in vivo* (Oehler et al., 1993). Upon introducing the additional Ser>Phe mutation at position 226 (corresponding to position 216 in v-Jun), full v-Jun-like transactivation characteristics were restored (Oehler et al., 1993), indicating that Ser226 of chicken c-Jun (Ser216 of v-Jun and Ser243 of human c-Jun) is involved in at least two different processes regulating DNA binding of Jun: phosphorylation and redox regulation.

The above-mentioned phosphorylation sites close to the DNA binding domain are targets for different signal transduction pathways, leading to changes in DNA binding and, hence, to altered transactivation ability of Jun. Hypophosphorylation of the DNA binding domain of c-Jun occurs upon stimulation of cells with agents that activate c-Jun's transactivation potential [e.g., phorbol ester tumor promoters (TPA) (Boyle et al., 1991), UV-C irradiation (Devary et al., 1992; Radler-Pohl et al., 1993), and transforming oncogenes (Smeal et al., 1991, 1992)], thereby increasing the affinity of Jun for its DNA recognition element. Tryptic phosphopeptide mapping of c-Jun from stimulated cells reveals a shift from tri- toward mono- and diphosphorylated isoforms of the peptide spanning amino acids 227–253 (see also Figure 5-1B). Whether this is because of the activation of a specific phosphatase or to inhibition of the corresponding kinase awaits further elucidation.

The N-terminal part of c-Jun is also subjected to complex phosphorylation events upon stimulation of resting cells with TPA, UV-C, growth factors, or oncoproteins. In resting cells, c-Jun is phosphorylated at low levels at Ser63 and Ser73 (see Figure 5-1A and B). All stimuli that lead to increased transactivating potential of Jun cause hyperphosphorylation at these sites (Smeal et al., 1991, 1992; Pulverer et al., 1991; Adler et al., 1992; Devary et al., 1992; Franklin et al., 1992; Radler-Pohl et al., 1993; Hagmeyer et al., 1993). Ser63 and Ser73 reside on separate tryptic peptides that are commonly referred to as peptides "x" (amino acids 71–78) and "y" (amino acids 57–70) (see Figure 5-1A). Depending on the stimulus and the cell type, additional phosphorylations can be detected (Pulverer et al., 1991; Hagmeyer et al., 1993; Radler-Pohl et al., 1993), whose locations have yet to be identified precisely.

Ser63 and Ser73 are directly involved in the transactivation process and are necessary for c-Jun to cooperate with Ha-Ras in the transformation of rat embryo fibroblasts (Smeal et al., 1991). Mutation of Ser63/Ser73 to nonphosphorylatable amino acids abolishes inducible transcriptional activity of c-Jun (Smeal et al., 1991; Franklin et al., 1992; Devary et al., 1992; Radler-Pohl et al., 1993). Activity is judged by transactivation of reporter genes by full-length c-Jun or chimeric proteins, where the N-terminal part of c-Jun is fused to a heterologous DNA binding portion.

Differences between two distinct Ser63/Ser73 double mutants concerning their influence on basal transactivation of c-Jun have been reported (Pulverer et al., 1991; Smeal et al., 1991; Franklin et al., 1992; Radler-Pohl et al., 1993). These are probably due to different amino acids used to substitute the serines (leucines vs. alanines). The three-dimensional conformation of the N-terminus, in addition to the increase of net negative charge by phosphorylation of Ser63/Ser73, may be important for enhanced transcriptional activity of c-Jun. The finding that exchanging Ser63/Ser73 to acidic amino acids does not lead to a stronger transactivator supports this hypothesis (unpublished results, cited in Karin and Smeal, 1992) (see Figure 5-2).

64

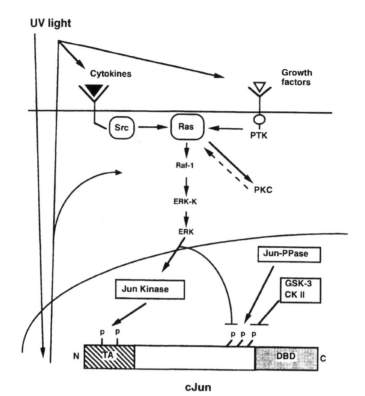

Figure 5-2 Schematic representation of the signal pathways leading to phosphorylation and dephosphorylation of Jun. See text for details. (Adapted from Karin and Smeal, 1992.)

Both serines reside within the *N*-terminal transactivation domains that have been delimited *in vivo* and *in vitro* (Angel et al., 1989; Bohmann and Tjian, 1989). Upon stimulation of the cells (e.g., with TPA), phosphorylation at these sites may catalyze other regulatory events occurring at the *N*-terminus of c-Jun, e.g., interactions with other proteins or with part of the c-Jun protein itself.

Deletion of the *N*-terminal "δ-domain" of c-Jun (amino acids 34–60) leads to hypophosphorylation of Ser63/Ser73 *in vivo* (Adler et al., 1992). This deletion is also found in v-Jun. Both proteins with δ deletions show increased transcriptional activity *in vitro* (Bohmann and Tjian, 1989), questioning the importance of phosphorylation of Ser63/Ser73 in the transactivation process. However, the inhibitory effect of the δ-domain seems to be cell type specific, suggesting that certain cell lines (e.g., HeLa) contain an inhibitor that interacts with the transactivation domains of Jun in a δ-dependent manner (Baichwal and Tjian, 1990). It is possible that this inhibitor is released upon phosphorylation of Ser63/Ser73.

When tested in other cells (e.g., the F9 mouse teratocarcinoma cell line), c-Jun and v-Jun have comparable transactivation characteristics (Angel et al., 1989; Baichwal and Tjian, 1990), and two-dimensional phosphopeptide mapping of *in vivo* labeled v-Jun reveals efficient phosphorylation of Ser73 (B. Hagmeyer, personal communication). Ser63 is also present in v-Jun. However, due to the deletion of the *N*-terminal trypsin cleavage site of peptide "y", which is located in the δ-domain (see Figure 5-1A), this phosphopeptide might escape detection during two-dimensional phosphopeptide mapping.

Phosphorylation of Ser63/Ser73 might also influence the interaction of the transactivation domains of c-Jun with components of the transcriptional machinery. One such example, the intermediary factor (p52/54), which specifically interacts with the transactivation domains of Jun and components of the RNA Pol II initiation complex, has been described (Oehler and Angel, 1992).

In conclusion, regulation of Jun's transactivation potential by phosphorylation takes place at two different levels: increased phosphorylation in the transactivation domain causes enhanced transcriptional activity, and enhanced binding to DNA is due to dephosphorylations in the DNA binding region.

c-JUN PROTEIN KINASES

As mentioned above, in uninduced cells, the c-Jun protein is phosphorylated by GSK-3, CKII, or another unidentified protein kinase at sites in the DNA binding region (residues Thr231, Thr239, Ser243, and Ser249). Treatment of cells with phorbol esters or expression of transforming oncogenes results in the dephosphorylation of at least one of these sites (Binétruy et al., 1991; Boyle et al., 1991), by stimulating a c-Jun phosphatase and/ or inhibiting the specific kinase (Goode et al., 1992). Negative regulation of c-Jun by GSK-3 can be observed *in vivo* using transfection experiments (de Groot et al., 1993; Nikolakaki et al., 1993). The activity and function of the GSK-3 was shown to be positively regulated by tyrosine phosphorylation *in vivo*. Resting cells show a constitutively active GSK-3 protein kinase, suggesting a mechanism of how the DNA binding activity of Jun is suppressed (Hughes et al., 1993). Moreover, the product of the *Drosophila melanogaster* gene, *shaggy (sgg)*, which shares homology to the mammalian GSK-3 protein, has a similar negative affect on c-Jun activity in mammalian cells (de Groot et al., 1993). Mutational analysis of *sgg* has revealed its crucial importance for the establishment of cell fate during various developmental events in *Drosophila* embryos, suggesting that GSK-3-dependent regulation of c-Jun function might be important for these processes (Bourois et al., 1989).

Considerable interest in CKII, which is an ubiquitous cyclic nucleotide-independent serine/threonine protein kinase, was generated by its ability to constitutively phosphorylate nuclear proteins that have been implicated in growth control; e.g., c-ErbA (Glineur et al., 1989), c-Myb (Luscher et al., 1990), the cAMP response element binding factor (CREB) (Lee et al., 1990), and the serum response element binding factor (SRF) (Manak et al., 1990).

The possible role of CKII for c-Jun activity *in vivo* was demonstrated by microinjection of an excess of synthetic peptides representing CKII substrates into target cells, which resulted in an activation of TRE-based transcription, most likely by competitive inhibition of c-Jun phosphorylation (Lin et al., 1992). Analysis of *in vitro* CKII phosphorylated c-Jun protein by manual Edman degradation showed that the major phosphorylation sites are Thr231 and Ser249. Since these sites are also found *in vivo* in ^{32}P-labeled c-Jun protein, the CKII kinase is another good candidate for the physiologic kinase regulating DNA binding of the c-Jun protein in resting cells.

CKII has been reported to be activated in response to mitogens including serum (Carroll and Marshak, 1989), phorbol ester (Carroll et al., 1988), insulin (Klarlund and Czech, 1988), and epidermal growth factor (Ackermann and Osheroff, 1989). These findings might appear to be contradictory, since the phosphorylations of the *C*-terminal sites of c-Jun negatively regulate DNA binding of Jun. However, CKII might not be the decisive player in regulating c-Jun's DNA binding activity in response to extracellular signals. Most importantly, as described above, Ser243 is not phosphorylated by CKII, but the phosphorylation of Thr231 and Ser249 by CKII is dependent on the phosphorylation status of Ser243. Therefore, the activity of CKII on the c-Jun protein depends on another protein kinase (possibly GSK-3, since it can phosphorylate Ser243, or another, unidentified protein kinase). This unknown protein kinase may be the actual target for regulation by extracellular signals, which, in cooperation with CKII, may be responsible for the phosphorylation-dependent regulation of DNA binding of Jun. Alternatively, a specific phosphatase might be activated by CKII, leading to dephosphorylation of Ser243, which in turn results in a complete loss of phosphorylation at the neighboring sites. However, the existence of such phosphatase(s) has yet to be confirmed.

In contrast to the negative effect of hyperphosphorylation on DNA binding, enhanced phosphorylation at the N-terminus is required for the activation of the transactivation function of Jun. Serines 63 and 73 are located close to a stretch of amino acids described as the "proline-rich region", which is required for transactivation and can serve as *in vitro* substrates for mitogen-activated protein (MAP) kinases (Pulverer et al., 1991). The finding that in extracts of TPA-treated cells a Jun kinase co-purifies with the pp42/44 and pp54 MAP kinases suggests that Jun may be phosphorylated by these protein kinases *in vivo* as well (Pulverer et al., 1991). In addition, a MAP-related protein kinase has been identified that is able to phosphorylate Jun at Ser243 *in vitro* (Alvarez et al., 1991).

MAP kinases are proline-directed serine/threonine kinases and are known by a variety of designations, named after the substrates that were used to detect their activity, such as microtubule-associated protein kinase 2 (MAP2-kinase), myelin basic protein kinase (MBPK), and ribosomal protein S6 kinase (p90rsk). Other names, such as extracellular signal-regulated kinase, indicate the various stimuli that lead to the rapid activation of these kinases, including nerve growth factor, epidermal growth factor, platelet-derived growth factor, fibroblast growth factor, cytokines, and insulin (Ray and Sturgill, 1988; Hoshi et al., 1988; Rossomando et al., 1989; Ahn et al., 1990; Gómez et al., 1990; Miyasaka et al., 1990; Boulton et al., 1991), as well as exposure of cells to DNA-damaging agents such as UV irradiaton and phorbol ester tumor promoters such as TPA (Radler-Pohl et al., 1993).

The MAP kinase family is composed of 40 to 54-kDa isoforms, requiring both tyrosine and threonine phosphorylation for maximal activation (Anderson et al., 1990). The regulatory phosphorylation sites in pp42mapk are Thr183 and Tyr185. Both residues are also present in pp44erk1. MAP kinases are phosphorylated by MAP kinase kinase, which acts here as a "dual specificity" kinase (Gómez and Cohen, 1991). The location of the phosphorylation residues close to subdomain VIII (a nucleotide binding region that is conserved in many protein kinases), suggests that phosphorylation within this region might render the enzyme more accessible to ATP. Regardless of the exact mechanism, enhanced phosphorylation may be responsible for an enhanced translocation of MAP kinase into the nucleus (Chen et al., 1992).

The role of MAP kinases in cellular signal transduction is supported by recent evidence for the involvement of Ras, Raf, and Src in the regulation of MAP kinases. These oncoproteins augment c-Jun activity by stimulating phosphorylation of Ser63 and Ser73 (Binétruy et al., 1991, Smeal et al., 1992). Several studies have shown the role of Ras in Raf activation (Wood et al., 1992; Zhang et al., 1993) as well as in the activation of MAP kinases by Raf (Kyriakis et al., 1992). In addition, there is evidence that Raf-1 kinase is directly phosphorylated by protein kinase Cα (PKCα) (Kölch et al., 1993).

Additional kinases that phosphorylate serines 63 and 73 were detected by their ability to associate with the transactivation domain of Jun. A 67-kDa protein was purified by affinity chromatography that phosphorylates a GST–c-Jun fusion protein as well as the complete c-Jun protein specifically at Ser63 and Ser73. Treatment of this putative c-Jun protein kinase with phosphatase 2A inhibited the c-Jun phosphorylation; incubation with PKC restored the protein kinase activity. These results suggest that p67 kinase might be regulated by these two components (Adler et al., 1992). However, the nature and structure of p67 kinase is still unknown.

CONCLUSION

Many different stimuli seem to funnel into common signaling pathways. It is only now that the signal transmission mechanisms of some of the key components are elucidated. Surely this is only a part of the story. However, recent insights into the pathways leading to changes in phosphorylation of the transcription factor c-Jun, have served as important paradigms for normal and oncogenic signaling in cells.

NOTE ADDED DURING PROOF

Recently serveral kinases have been identified upon their ability to specifically bind to and phosphotylate c-Jun at Ser63/Ser73. c-Jun N-terminal kinase 1 (JNK-1) (Dérijard et al., Cell, 76, 1025, 1994) and JNK-2 (Hibi et al., Genes Dev., 7, 2135, 1993) are activated in Ha-ras transformed or UV-irradiated cells. The group of p54 stress-activated protein kinases (SAPK's) (Kynakis et al., Nature, 369, 156, 1994) are addressed by different protein synthesis inhibitors, cyrokines and UV. Both, JNK's and SAPK's, share homologies to MAP-kinases and are similarly regulated by Thr/Tyr-specific phosphonylation.

REFERENCES

Abate, C., Patel, L., Rauscher, F., and Curran, T., Redox regulation of Fos and Jun DNA-binding activity in vitro, *Science*, 249, 1157, 1990.

Ackermann, P. and Osheroff, N., Regulation of casein kinase II activity by epidermal growth factor in human A-431 carcinoma cells, *J. Biol. Chem.*, 264, 11958, 1989.

Adler, V., Polotskaya, A., Wagner, F., and Kraft, A. S., Affinity-purified c-Jun amino-terminal protein kinase requires serine/threonine phosphorylation for activity, *J. Biol. Chem.*, 267, 17001, 1992.

Ahn, N. G., Weiel, J. E., Chan, C. P., and Krebs, E. G., Identification of multiple epidermal growth factor-stimulated protein serine/threonine kinases from Swiss 3T3 cells, *J. Biol. Chem.*, 265, 11487, 1990.

Alvarez, E., Northwood, I. C., Gonzalez, F. A., Latour, D. A., Seth, A., Abate, C., Curran, T., and Davies, R. J., Pro-Leu-Ser/Thr-Pro is a consensus primary sequence for substrate protein phosphorylation, *J. Biol. Chem.*, 266, 15277, 1991.

Anderson, N. G., Maller, J. L., Tonks, N. K., and Sturgill, T. W., Requirement for integration of signals from two distinct phosphorylation pathways for activation of MAP kinase, *Nature*, 343, 651, 1990.

Angel, P. and Karin, M., The role of Jun, Fos and the AP-1 complex in cell-proliferation and transformation, *Biochim. Biophys. Acta*, 1072, 129, 1991.

Angel, P., Allegretto, E. A., Okino, S. T., Hattori, K., Boyle, W. J., Hunter, T., and Karin, M., Oncogene jun encodes a sequence-specific trans-activator similar to AP-1, *Nature*, 332, 166, 1988.

Angel, P., Smeal, T., Meek, J., and Karin, M., Jun and v-Jun contain multiple regions that participate in transcriptional activation in an interdependent manner, *New Biol.*, 1, 35, 1989.

Baichwal. V. R. and Tjian., R., Control of cJun activity by interaction of a cell-specific inhibitor with regulatory domain δ: Differences between v- and c-Jun. *Cell*, 63. 815. 1990.

Binétruy. B.. Smeal. T.. and Karin. M.. Ha-Ras augments c-Jun activity and stimulates phosphorylation of its activation domain. *Nature*, 351. 122. 1991.

Bohmann. D. and Tjian. R.. Biochemical analysis of transcriptional activation by Jun: Differential activity of c- and v-Jun. *Cell*, 59, 709, 1989.

Boulton. T. G.. Nye. S. H.. Robbins. D. J.. Ip. N. Y.. Radziejewska. E.. Morgenbesser. S. D.. DePinho. R. A.. Panayotatos. N.. Cobb. M. H.. and Yancopoulos. G. D.. ERKs: A family of protein-serine/threonine kinases that are activated and tyrosine phosphorylated in response to insulin and NGF. *Cell*, 65. 663. 1991.

Bourois. M.. Heitzler. P.. El Messal. M.. and Simpson. P.. Mutant Drosophila embryos in which all cells adopt a neural fate. *Nature*, 341. 442. 1989.

Boyle. W. J.. Smeal. T.. Defize. L. H. K.. Angel. P.. Woodgett. J. R.. Karin. M.. and Hunter. T.. Activation of protein kinase C decreases phosphorylation of c-Jun at sites that negatively regulate its DNA-binding activity. *Cell*, 64. 573. 1991.

Carroll. D. and Marshak. D. R.. Serum-stimulated cell growth causes oscillations in casein kinase II activity, *J. Biol. Chem.*, 264, 7345. 1989.

Carroll. D., Santoro. N.. and Marshak. D. R.. Regulating cell growth: Casein kinase II-dependent phosphorylation of nuclear oncoproteins, *Cold Spring Harbor Symp. Quant. Biol.*, 53. 91. 1988.

Chen. R. H.. Sarneki. C.. and Blenis. J.. Nuclear localization and regulation of erk- and rsk-encoded protein kinases. *Mol. Cell. Biol.*, 12. 915. 1992.

Cooper. J. A.. Sefton. B. M.. and Hunter. T.. Diverse mitogenic agents induce the phosphorylation of two related 42.000-Dalton proteins on tyrosine in quiescent chick cells. *Mol. Cell. Biol.*, 4. 30. 1984.

de Groot. R.. Auwerx. J.. Bourouis. M.. and Sassone-Corsi. P.. Negative regulation of Jun/AP-1: Conserved function of glycogen synthase kinase 3 and the Drosophila kinase shaggy. *Oncogene*, 7. 841. 1993.

Devary. Y.. Gottlieb. R. A.. Smeal. T., and Karin. M.. The mammalian ultraviolet response is triggered by activation of Src tyrosine kinases. *Cell*, 71. 1081. 1992.

Franklin. C. C.. Sanchez. V., Wagner. F.. Woodgett. J. R.. and Kraft. A. S.. Phorbol ester-induced amino-terminal phosphorylation of human JUN but not JUNB regulates transcriptional activation. *Proc. Natl. Acad. Sci. U.S.A.*, 89. 7247. 1992.

Glineur. C.. Bailly. M.. and Ghysdael. J.. The c-Erb-A a-encoded thyroid hormone receptor is phosphorylated in its amino terminal domain by casein kinase II. *Oncogene*, 4. 1247. 1989.

Gómez. N. and Cohen. P.. Dissection of the protein kinase cascade by which nerve growth factor activates MAP kinase. *Nature*, 353. 170. 1991.

Gómez. N.. Tonks. N. K.. Morrison. C.. Harmar. T.. and Cohen. P.. Evidence for communication between nerve growth factor and protein tyrosine phosphorylation. *FEBS Lett.*, 271. 119. 1990.

Goode. N.. Hughes. K.. Woodgett. J. R.. and Parker. P. J.. Differential regulation of glycogen synthase kinase-3β by protein kinase C isotypes. *J. Biol. Chem.*, 267. 16878. 1992.

Hagmeyer. B.. König. H.. Herr. I.. Offringa. R.. Zantema. A.. van der Eb. A. J.. Herrlich. P.. and Angel. P.. Adenovirus E1A negatively and positively modulates transcription of AP-1 dependent genes by dimer-specific regulation of the DNA binding and transactivation activities of Jun. *EMBO J.*, 12. 3559. 1993.

Hoshi. M.. Nishida. E.. and Sakai. H.. Activation of a Ca2+-inhibitable protein kinase that phosphorylates microtubule-associated protein 2 *in vitro* by growth factors, phorbol esters, and serum in quiescent cultured human fibroblasts. *J. Biol. Chem.*, 263. 5396. 1988.

Hughes, K., Nikolakaki, E., Plyte, S. E., Totty, N. F., and Woodgett, J. R., Modulation of the glycogen synthase kinase-3 family by tyrosine phosphorylation, *EMBO J.*, 12, 803, 1993.

Hunter, T. and Sefton, B. M., Transforming gene product of Rous sarcoma virus phosphorylates tyrosine, *Proc. Natl. Acad. Sci. U.S.A.*, 77, 1311, 1980.

Karin, M. and Smeal, T., Control of transcription factors by signal transduction pathways: The beginning of the end, *TIBS*, 17, 418, 1992.

Klarlund, J. K. and Czech, M. P., Insulin-like growth factor I and insulin rapidly increase casein kinase II activity in BALB/c 3T3 fibroblasts, *J. Biol. Chem.*, 263, 15872, 1988.

Kölch, W., Heidecker, G., Kochs, G., Hummel, R., Vahidi, H., Mischak, H., Finkenzeller, G., Marmè, D., and Rapp, U. R., Protein kinase Ca activates RAF-1 by direct phosphorylation, *Nature*, 364, 249, 1993.

Kyriakis, J. M., App, H., Zhang, X., Banerjee, P., Brautigan, D. L., Rapp, U. R., and Avruch, J., Raf-1 activates MAP kinase-kinase, *Nature*, 358, 417, 1992.

Lee, C. G., Yun, Y., Hoeffler, J. P., and Habener, J. F., Cyclic-AMP-responsive transcription activation of CREB-237 involved independent phosphorylated subdomains, *EMBO J.*, 9, 4455, 1990.

Leevers, S. J. and Marshall, C. J., Activation of extracellular signal-regulated kinase, ERK2, by p21ras oncoprotein, *EMBO J.*, 11, 569, 1992.

Lin, A., Frost, J., Deng, T., Smeal, T., Al-Alawi, N., Kikkawa, U., Hunter, T., Brenner, D., and Karin, M., Casein kinase II is a negative regulator of c-Jun DNA binding and AP-1 activity, *Cell*, 70, 777, 1992.

Luscher, B., Christenson, E., Litchfield, D. W., Krebs, E. G., and Eisenman, R. N., Myb DNA binding inhibited by phosphorylation at a site deleted during oncogenic activation, *Nature*, 344, 517, 1990.

Maki, Y., Bos, T. J., Davis, C., Starbuck, M., and Vogt, P. K., Avian sarcoma virus 17 carries the jun oncogene, *Proc. Natl. Acad. Sci. U.S.A.*, 84, 2848, 1987.

Manak, J. R., de Bisschop, N., Kris, R. M., and Prywes, R., Casein kinase II enhances the DNA binding activity of serum response factor, *Genes Dev.*, 4, 955, 1990.

Miyasaka, T., Chao, M. V., Sherline, P., and Saltiel, A. R., Nerve growth factor stimulates a protein kinase in PC-12 cells that phosphorylates microtubule-associated protein-2, *J. Biol. Chem.*, 265, 4730, 1990.

Nikolakaki, E., Coffer, P. J., Hemelsoet, R., Woodgett, J. R., and Defize, L. H. K., Glycogen synthase kinase 3 phosphorylates Jun family members *in vitro* and negatively regulates their transactivating potential in intact cells, *Oncogene*, 8, 833, 1993.

Oehler, T. and Angel, P., A common intermediary factor (p52/54) recognizing "acidic-blob"-type domains is required for transcriptional activation by the Jun proteins, *Mol. Cell. Biol.*, 12, 5508, 1992.

Oehler, T., Pintzas, A., Stumm, S., Darling, A., Gillespie, D., and Angel, P., Mutation of a phosphorylation site in the DNA binding domain is required for redox-independent transactivation of AP1-dependent genes by v-Jun, *Oncogene*, 8, 1141, 1993.

Payne, D. M., Rossomando, A. J., Martino, P., Erickson, A. K., Her, J.-H., Shabanowitz, J., Hunt, D. F., Weber, M. J., and Sturgill, T. W., Identification of the regulatory phosphorylation sites in pp42/mitogen-activated protein kinase (MAP kinase), *EMBO J.*, 10, 885, 1991.

Pulverer, B. J., Kyriakis, J. M., Avruch, J., Nikolakaki, E., and Woodgett, J. R., Phosphorylation of c-jun mediated by MAP kinases, *Nature*, 353, 670, 1991.

Radler-Pohl, A., Sachsenmaier, C., Gebel, S., Auer, H.-P., Bruder, J. T., Rapp, U., Angel, P., Rahmsdorf, H. J., and Herrlich, P., UV-induced activation of AP-1 involves obligatory extranuclear steps including Raf-1 kinase, *EMBO J.*, 12, 1005, 1993.

Ray, L. B. and Sturgill, T. W., Characterization of insulin-stimulated microtubule-associated protein kinase. Rapid isolation and stabilization of a novel serine/threonine kinase from 3T3-L1 cells, *J. Biol. Chem.*, 263, 12721, 1988.

Rossomando, A. J., Payne, D. M., Weber, M. J., and Sturgill, T. W., Evidence that pp42, a major tyrosine kinase target protein, is a mitogen-activated serine/threonine protein kinase, *Proc. Natl. Acad. Sci. U.S.A.*, 86, 6940, 1989.

Smeal, T., Binétruy, B., Mercola, D. A., Birrer, M., and Karin, M., Oncogenic and transcriptional cooperation with Ha-ras requires phosphorylation of c-Jun on serines 63 and 73, *Nature*, 354, 494, 1991.

Smeal, T., Binétruy, B., Mercola, D., Grover-Bardwick, A., Heidecker, G., Rapp, U. R., and Karin, M., Oncoprotein mediated signalling cascade stimulates cJun activity by phosphorylation of serines 63 and 73, *Mol. Cell. Biol.*, 12, 3507, 1992.

Wood, K. W., Sarnecki, C., Roberts, T. M., and Blenis, J., Ras mediates nerve growth factor receptor modulation of three signal-transducing protein kinases: MAP-kinase, Raf-1, and RSK, *Cell*, 68, 1041, 1992.

Zhang, X., Settleman, J., Kyriakis, J. M., Takeuchi-Suzuki, E., Elledge, S. J., Marshall, M. S., Bruder, J. T., Rapp, U. R., and Avruch, J., Normal and oncogenic p27^tas proteins bind to the amino-terminal regulatory domain of c-Raf-1, Nature, 364, 308, 1993.

Chapter 6

Involvement of Phosphorylation and Dephosphorylation in the Control of AP-1 Activity

Peter E. Shaw and Hendrik Gille

CONTENTS

INTRODUCTION

The proliferative signals mimicked by a number of transforming oncogenes converge on and stimulate the activity of a nuclear transcription factor, first referred to as AP-1. Although correct, this concise introductory statement represents an untoward oversimplification of the complexity of intracellular signal transduction pathways and transcriptional regulation, not only in absolute terms, but also in terms of our present understanding of this fascinating subject. Not surprisingly, perhaps, it turns out that activation of AP-1 can be achieved by numerous seemingly independent mechanisms, which can clearly be usurped by transforming oncogenes. This chapter discusses what is known to date of the means by which AP-1 is constitutively activated by specific oncogenes and, by extrapolation, how AP-1 activity is potentiated in response to growth factors and mitogenic stimuli in untransformed cells.

AP-1 was originally portrayed as an essentially homogeneous protein factor with a relative molecular mass of 47 kDa and the ability to activate transcription *in vitro* when bound to specific sequence elements in the human metallothionein IIA promoter (Lee et al., 1987). It is now accepted that AP-1, as purified by affinity to its cognate recognition sequence, includes many related components. Nevertheless, the established consensus holds that either a homodimer of the c-jun gene product, Jun, or a heterodimer of Jun and the c-fos gene product, Fos, constitute AP-1 (Bohmann et al., 1987; Angel et al., 1988; Chiu et al., 1988; Curran and Franza, 1988; Halazonetis et al., 1988; Kouzarides and Ziff, 1988; Rauscher et al., 1988; Sassone-Corsi et al., 1988a).

In addition to the classical form of AP-1, both Fos and Jun dimerize with other family members, thus giving rise to a group of complexes with different biochemical properties (Abate et al., 1991; Hai and Curran, 1991; Kerppola and Curran, 1991). The actual state of Fos or Jun, therefore, depends on the availability of dimerization partners in a cell at any given time.

The function of AP-1 is controlled in part through the posttranslational modification, principally phosphorylation, of its two components, Jun and Fos. It has been known for some time that both proteins are multiply phosphorylated, but the functional relevance of this observation has only become apparent since the nature of the probable kinases responsible was established. Another major means of influencing the function of AP-1 is at the level of the two promoters. Both c-fos and c-jun are representative of the class of so-called immediate early genes, which are induced rapidly in response to mitogenic stimulation of cells. The regulation of the two promoters is discussed elsewhere in this book, but one aspect of promoter regulation, the posttranslational modification of transcription factors implicated in c-fos promoter function, will be addressed here in detail. Two recurrent themes in this chapter are, therefore, the phosphorylation of transcription factors by protein kinases and, to an extent reduced almost to inference, their dephosphorylation by protein phosphatases.

JUN AS A PHOSPHOPROTEIN

The complexities of Jun as a phosphoprotein warrant (and have) a chapter to themselves — Chapter 5. At this point, Jun can be summarized as being a target for multiple phosphorylation events and at the behest of several kinases. Depending on the sites in question, phosphorylation can modulate the function of the protein in different, indeed contradictory, ways. The mechanisms by which changes in function are induced remain to be elucidated.

FOS AS A PHOSPHOPROTEIN

Ample evidence has accrued in the literature to demonstrate that Fos is a phosphoprotein (Curan and Teich, 1982; Curran et al., 1984; Barber and Verma, 1987). In fact, the large difference between the predicted molecular weight of 42 kDa and its apparent size of 55 to 62 kDa can be attributed largely to phosphorylation. Phosphoserine is reported to account for 90% of the phosphate incorporation, and C-terminal truncations drastically reduce phosphorylation of Fos. v-Fos, which differs substantially from Fos at its C-terminal end beyond residue 322, is also less efficiently phosphorylated (Curran et al., 1984; Barber and Verma, 1987). As several C-terminal serine residues are present in consensus sites for cAMP-dependent protein kinases, this region of Fos has been proposed to be a target for phosphorylation by cAMP-dependent protein kinase A (PKA) (Abate et al., 1991; Tratner et al., 1992).

The C-terminus of Fos is involved in transrepression of the c-fos gene (Sassone-Corsi et al., 1988b; Wilson and Treisman, 1988; Lucibello et al., 1989; Gius et al., 1990). It contains three groups of serine residues (362–364, 368–369, and 371–374), and when any of these groups of serines is mutated to alanines, the resulting Fos protein is unable to transrepress following stimulation by either serum or TPA, while its transactivation potential remains unimpaired (Ofir et al., 1990). In this respect, the lesion can be reverted by introducing negatively charged residues in place of the critical serines (Ofir et al., 1990).

Two serines that are phosphorylated in vivo were subsequently mapped (362 and 373/374), and serine 362 was shown to be a substrate for PKA in vitro (Tratner et al., 1992). Although the ability of the mutant Fos protein lacking serines 362–364, to transrepress the c-fos promoter was impaired, its transforming potential increased (Tratner et al., 1992).

Sites other than those in the C-terminal portion of Fos are also phosphorylated in vivo (Abate et al., 1991). These sites have not been mapped but are proposed to lie within putative transcriptional activation regions of the protein (Abate et al., 1990). In addition to their bZip domains, Fos and Jun share two discrete regions of limited homology, referred to as HOB1 and HOB2, which appear to act as modular activation domains (Sutherland et al., 1992). A conserved serine/threonine in HOB1 is one of two residues in the N-terminal transactivation domain of Jun that are phosphorylated in response to mitogens, TPA, and transformation by Ha-Ras (Smeal et al., 1991). Fos is presumably phosphorylated *in vivo* at the analogous site. Different kinases have been shown to phosphorylate the sites in Jun *in vitro* (Pulverer et al., 1991; Adler et al., 1992). This issue is discussed at length in Chapter 5, but, to extend the above analogy, the kinase for HOB1 in Fos is also obscure at present.

POTENTIAL FOS OR JUN PROTEIN KINASES

The identification of a specific protein kinase for a given nuclear substrate is a serious problem in the elucidation of signal transduction pathways. The crux lies in providing *in vivo* confirmation for what may be readily observed *in vitro*. For example, there is *in vitro* evidence for phosphorylation of Fos by a number of kinases, including cAMP-dependent PKA (Abate et al., 1991; Tratner et al., 1992), p34cdc2, protein kinase C (PKC) (Abate et al., 1991), mitogen-activated protein (MAP) kinases, and p90rsk (Chen et al., 1992). None of these observations has been confirmed *in vivo*. In the case of Jun, phosphorylation of the N-terminal serines 63 and 73 *in vivo* in response to mitogenic stimulation or in cells transformed with Ha-ras has been correlated with the stimulation of Jun's transactivation potential (Binétruy et al., 1991; Boulton et al., 1991; Pulverer et al., 1991; Baker et al., 1992; Devary et al., 1992; Smeal et al., 1992; Pulverer et al., 1993; Radler-Pohl et al., 1993).

The nature of the kinase(s) responsible is still an open matter, but the family of MAP kinases, or extracellular signal-regulated kinases (ERKs) (Kozma and Thomas, 1992), is likely to harbor candidates for which both Fos and Jun are targets. Indeed, Jun has been reported to be phosphorylated by MAP1 and MAP2 kinases (Pulverer et al., 1991, 1993). This would seem to make sense, as MAP kinases are known to be activated rapidly by many of the growth factors and mitogens that induce Fos expression and activate AP-1 responsive genes (Kozma and Thomas, 1992). They are also known to phosphorylate and activate p90rsk, the 90-kDa ribosomal S6 kinase II, *in vitro* (Sturgill et al., 1988). However, another subgroup of MAP kinases has recently been described that phosphorylate Jun at the N-terminal and are activated by UV light and stress (Kyriakis et al., 1994; Dérijard et al., 1994).

THE MAP KINASE SIGNAL TRANSDUCTION PATHWAY

Recent work supports the concept of an activation cascade consisting of several kinases (see Figure 6-1) that are activated serially in response to mitogenic stimulation. Because MAP kinases require phosphorylation on both threonine and tyrosine residues for activation, they were thought to play a key role as integrators of signals generated by both tyrosine kinases and serine/threonine kinases (Anderson et al., 1990b). However, the simultaneous discovery that the activator of MAP kinases is a "dual specificity" kinase and is itself activated by serine/threonine phosphorylation confirms that MAP kinases lie downstream of several other signaling components (Crews and Erikson, 1992; Kosako et al., 1992; Matsuda et al., 1992; Nakielny et al., 1992; Rossomando et al., 1992; Shirakabe et al., 1992).

The MAP kinase family is not restricted to vertebrates. The closely related enzymes FUS3, KSS1 (*Saccharomyces cerevisiae*) and spk1 (*Schizosaccharomyces pombe*) exist

74

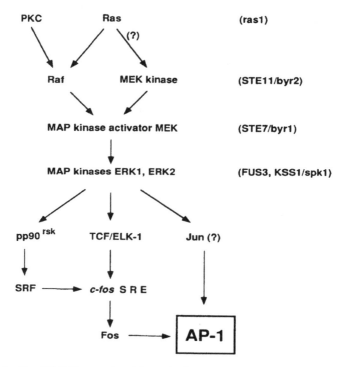

Figure 6-1 The MAP kinase signal transduction pathway as it is presently understood. The homologous yeast kinases (*S. pombe* in lower case) are shown in brackets on the right. See text for details.

in yeast. They are phosphorylated by STE7 and byr1, respectively (Gartner et al., 1992). Mutations in STE7 or byr1 result in loss of response to pheromones and sterility (Kurjan, 1992). On cloning of MAP kinase activators (MAP kinase kinase or MEK for MAP/ERK kinase) from various sources (Crews et al., 1992; Kosako et al., 1993; Wu et al., 1993), STE7 and byr1 were recognized as yeast counterparts (Kosako et al., 1993, Haystead et al., 1993). Thus the MAP kinase pathway has been conserved during evolution.

The mammalian signaling pathway may bifurcate upstream of the MAP kinase activator MEK. With the availability of MEK and the knowledge that it requires serine/threonine phosphorylation to be active, its suitability as a substrate for phosphorylation by Raf kinase could be demonstrated. Both MAP kinase and MEK are constitutively active in Raf-transformed COS-1 and NIH3T3 cells, and Raf kinase, purified by immunoprecipitation from activated cells or expressed as a fusion protein in bacteria, phosphorylates and activates MEK (Dent et al., 1992; Howe et al., 1992; Kyriakis et al., 1992). However, Raf bears no homology to the yeast STE7/byr1 kinases STE11 and byr2 (Howe et al., 1992). In light of this, there is evidence for the existence of a second MEK kinase that may be homologous to STE11 and byr2. In PC12 cells, Ras does not fully activate MAP kinase and Raf does not activate MAP kinase at all. Furthermore, in Rat 1a cells, neither Ras nor Raf stimulate MAP kinase (Gallego et al., 1992). These observations point to the existence of a second mode of activation for MEK and MAP kinases that is independent of Raf. In fact, a MEK kinase that is homologous to STE11/byr2 has recently been cloned and characterized (Lange-Carter et al., 1993).

The mode of regulation of Raf kinase is obscure. In v-raf, which is a constitutively active kinase, gag sequences replace the N-terminal half of the protein (Rapp et al., 1983). This region of Raf contains two domains, CR1 and CR2, that are implicated in the regulation of Raf. CR1 contains a cysteine finger motif and, when expressed independently, it shows a dominant negative phenotype, apparently by titrating a Raf activating factor (Kölch et al., 1991; Bruder et al., 1992). Furthermore, highly transforming deletion mutants of Raf lack an N-terminal segment, including CR2, that is multiply phosphorylated *in vivo* (McGrew et al., 1992). Phosphorylation of Raf is associated with increased kinase activity (Morrison et al., 1988; Blackshear et al., 1990; Kovacina et al., 1990), but as Raf undergoes autophosphorylation, the question of cause or effect is unclear. Nevertheless, several types of the TPA-inducible PKC have recently been shown to phosphorylate and activate Raf kinase in insect cells and *in vitro* (Sözeri et al., 1992).

Ras has been implicated in the activation of Raf kinase as well as MAP kinases and p90rsk by growth factors and phorbol esters in several cell lines (Leevers and Marshall, 1992; Szeberenyi et al., 1992; Thomas et al., 1992; Troppmair et al., 1992; Wood et al., 1992), and in *Drosophila*, the Raf homologue Draf has been shown to function downstream of Ras in the sevenless signal transduction pathway (Dickson et al., 1992). Moreover, recent experiments have shown that Ras and Raf undergo physical interactions, although the functional consequences of the association are not yet understood (Vojtek et al., 1993; Warne et al., 1993).

Activation of the MAP kinase pathway is associated in other ways with the augmentation of AP-1 activity. The c-fos gene is rapidly and transiently induced in response to serum growth factors and mitogens. This pattern of expression is dependent on the serum response element (SRE) (Treisman, 1985, 1986), a promoter element that is bound by a multicomponent transcription factor complex that includes the serum response factor (SRF) (Norman et al., 1988) and the ternary complex factor (TCF) (Shaw et al., 1989), which is homologous to Elk-1, an ets protein (Hipskind et al., 1991). It has been shown that overexpression of c-Ha-Ras and oncogenic Ha-Ras leads to transcription activation from the c-fos promoter (Schönthal et al., 1988). Furthermore, the introduction of activated Raf kinase into cells leads to elevated c-fos expression by a mechanism involving the SRE (Kaibuchi et al., 1989; Jamal and Ziff, 1990).

TCF/Elk-1 appears to be a nuclear target for MAP kinases as phosphorylation by MAP kinases *in vitro* stimulates SRF-dependent DNA binding (Gille et al., 1992). Specific stimulation of the MAP kinase pathway by EGF in swiss 3T3 cells induces c-fos expression and stimulates TCF as assayed in nuclear extracts (Malik et al., 1991; Gille et al., 1992). MAP kinases phosphorylate TCF/Elk-1 *in vitro* at several serine and threonine residues, all of which lie within the C-terminal cyanogen bromide fragment of the protein. These sites appear to correspond to those that are phosphorylated *in vivo* in serum-stimulated cells (Marais et al., 1992; Gille and Shaw, unpublished results). This suggests that TCF/Elk-1 is a critical target in the c-fos promoter for activation by the growth factor-dependent MAP kinase pathway.

Activation of the MAP kinase pathway may also lead to phosphorylation of SRF by other kinases (Misra et al., 1991), including p90rsk, a growth factor-dependent kinase that is activated by MAP kinases (Anderson et al., 1990; Ahn et al., 1991; Chen et al., 1991). Although the functional significance of this event is not immediately apparent, phosphorylation of the p90rsk site does occur *in vivo* (Smeal et al., 1992; Rivera et al., 1993) and may affect the rate at which SRF associates with the DNA, much like phosphorylation of SRF by casein kinase II at adjacent sites (Manak et al., 1990; Manak and Prywes, 1991; Marais et al., 1992; Janknecht et al., 1992). Thus, there appear to be a number of options by which activation of the MAP kinase pathway might potentiate AP-1 activity.

DOMINANT KINASE MUTANTS

One way to demonstrate the involvement of a kinase in a signal transduction pathway is to invoke the use of dominant mutants of the kinase, insofar as they are available, much as mutants of the small GTP-binding protein p21ras have been used to influence signal transmission (Feig and Cooper, 1988; Cai et al., 1990; Szeberenyi et al., 1990; Leevers and Marshall, 1992; Wood et al., 1992). Various kinase mutants that are either constitutively active or inhibitory are known and it is conceivable that more complex mutants could be generated, such as substrate specificity changes, with which, instead of simply switching a signal on or off, redirection might be possible.

One kinase for which this approach appears to have been effective is Raf, which exists in constitutively active forms as the oncogenic v-raf (Rapp et al., 1983) and as a truncated protein in which the N-terminal regulatory domain is lacking, the result of molecular manipulation (Kölch et al., 1991; Heidecker et al., 1990). Alternatively, both kinase-defective Raf mutants and regulatory domain fragments of Raf act as dominant inhibitory proteins (Kölch et al., 1991; Bruder et al., 1992; Devary et al., 1992; Radler-Pohl et al., 1993). In NIH3T3 cells transformed by v-raf, both MEK and MAP kinases are activated (Dent et al., 1992; Howe et al., 1992). In contrast, dominant negative mutants of Raf revert the phenotype of Raf-transformed fibroblasts and, to a lesser degree, Ras-transformed NIH3T3 cells (Kölch et al., 1991).

It is only a matter of time until analogous experiments with dominant forms of MAP kinase activator, MAP kinase, and p90rsk are described. Interestingly, no oncogenic forms of these kinases have been reported to date, which suggests that the generation of constitutively active forms of these kinases may not be trivial.

PHOSPHATASES

One characteristic the numerous protein kinases mentioned above have in common with their substrates is that their activity is controlled by phosphorylation. In this respect, the antithesis of a kinase is a phosphatase.

In general, although not without exception, it seems that phosphorylation is synonymous with activation and that phosphatases may provide a constant negative force against which kinases have to contend. In such a simplistic scenario, protein phosphatases would be constitutively active, maintaining phosphorylation-dependent kinases and their substrates in the dephosphorylated, inactive form or reverting them to the inactive form from the active state at a constant rate.

Overexpression of the wild-type receptor-like protein tyrosine phosphatase (PTPα) in embryonal fibroblasts results in chronic dephosphorylation of the negative regulatory Tyr527 of pp60src. This results in persistent activation of the kinase and concomitant cell transformation (Zheng et al., 1992). Thus, simply elevating the expression of the phosphatase results in a phenotypic change, which implies that there is no negative regulation of PTPα.

However, this is not irrefutably the case: there are indications that the activity of phosphatases in the cell is controlled. As an example, in quiescent cells, Jun is phosphorylated at several adjacent sites that become dephosphorylated rapidly on activation of PKC (Boyle et al., 1991). This induced dephosphorylation is coincident with an enhancement of AP-1 binding activity $in\ vitro$ (Boyle et al., 1991). Inhibition of phosphorylation by amino acid substitutions results in a Jun protein that exhibits strongly enhanced transactivation potential $in\ vivo$, compared to wild-type c-Jun (P. Angel, personal communication). Furthermore, dual specificity phosphatases that specifically dephosphorylate and inactivate MAP kinases have recently been described (Sun et al., 1993; Ward et al., 1994). Either way, phosphatases are essential for regulation of intracellular functions by phosphorylation.

Two protein phosphatases, PP-1 and PP-2A, are known to be involved in the regulation of growth factor and mitogen induced signal transduction pathways in eucaryotic cells (Cohen, 1989). The activity of PP-1 is regulated by a specific phosphatase inhibitor (Huang and Glinsmann, 1976; Hemmings et al., 1984; Strålfors et al., 1985), which in turn is subject to the antagonistic control of cAMP-dependent PKA and either of the two phosphatases PP-2A and PP-2B (Hubbard and Cohen, 1989).

The regulatory mechanisms governing PP-2A are complicated. The holoenzyme exists as a common core, consisting of a 36-kDa catalytic subunit and a 65-kDa regulatory subunit. A third, variable subunit can be associated with the dimer, and its presence influences substrate specificity and phosphatase activity of the core subunits (Shenolikar and Nairn, 1991). For example, it was shown recently that in *Drosophila*, the gene encoding the 55-kDa variable subunit of PP-2A (PR55) rescues mutations in *aar* (abnormal anaphase resolution), a locus required for mitosis (Mayer-Jaekel et al., 1993).

Okadaic acid is a potent tumor promoter that specifically inhibits protein phosphatases PP-1 and PP-2A (Suganuma et al., 1988; Bialojan and Takai, 1988). Treatment of NIH3T3 cells with okadaic acid leads to the activation of c-fos expression (Schönthal et al., 1991), which suggests that these phosphatases are important for maintaining the gene in its repressed state. However, the kinetics differ markedly from the rapid induction seen in response to growth factors and mitogens, and are seemingly independent thereof. Treatment of fibroblasts with okadaic acid also activates MAP kinases, indicating that they are maintained in the inactive state by an okadaic acid-sensitive phosphatase in quiescent cells (Gotoh et al., 1990). However, the activation of MAP kinases is transient even in the presence of okadaic acid.

In agreement with these observations, parallel treatment of cells with serum and okadaic acid does not prevent repression of the c-fos gene after serum stimulation. Instead, two waves of c-fos expression are observed that can be shifted relative to each other by administering serum and okadaic acid at different times (Shaw and Gille, unpublished observations). Therefore, although the okadaic acid-sensitive phosphatases PP-1 and PP-2A appear to be responsible for keeping the c-fos gene in its state of "poised" repression, they are not involved in the rapid, Fos-dependent transrepression of the gene following induction. The delayed expression of Fos in response to prolonged okadaic acid treatment may reflect the elevated Fos expression that has been observed in association with tissue regression and apoptotic cell death (Buttyan et al., 1988; Colombel et al., 1992).

EFFECT OF PHOSPHORYLATION ON SUBSTRATES

A question that remains completely open is how serine/threonine phosphorylation results in the potentiation of AP-1 activity, the TCF/Elk-1-mediated transcriptional activation of the c-fos promoter, or the activation of any of the numerous kinases involved in growth signal transmission.

One alternative is to increase the net negative charge of a protein domain by phosphorylation. Negatively charged domains, or "acid blobs", have been implicated in the activation function of numerous transcriptional activators from yeast to mammalian systems (Ptashne, 1988). The potential importance of negative charge is demonstrated in the case of transrepression by Fos (Ofir et al., 1990). However, the introduction of negatively charged residues into ERK2 in place of the threonine and tyrosine residues that are phosphorylated by MEK does not produce an enzymatically or biologically active kinase (Ebert and Cobb, unpublished; Kortenjann and Shaw, unpublished).

A second possibility is that phosphoserine and phosphothreonine residues in phosphorylation-regulated kinases and transcription factors may provide key components of recognition sites for as yet unknown factors, with which interaction is indispensable for function. This is known to be the case with receptor tyrosine kinases, in which phosphotyrosine residues have been shown to function as sites of interaction with Src homology (SH2) domains (Anderson et al., 1990; Matsuda et al., 1990; Mayer and Hanafusa, 1990; Fantl et al., 1992). These are present in proteins involved in the generation and propagation of intracellular signals, such as the GTPase activating protein rasGAP (Trahey and McCormick, 1987; Vogel et al., 1988), GRB2 (Lowenstein et al., 1992), and, of course, p60src (Anderson et al., 1990a; Cantley et al., 1991). A potential analogy is suggested by the recent report of a phosphorylation-dependent signal on Jun targeting Fos in *trans* for rapid degradation (Papavassiliou et al., 1992).

A third alternative is that of phosphorylation leading to conformational change and, hence, to altered function. Nonreceptor tyrosine kinases provide examples of this phenomenon insofar as phosphorylation of a C-terminal regulatory tyrosine promotes the intramolecular association of the phosphotyrosine with an SH2 domain, thereby effectively blocking productive, intermolecular interactions with other SH2 domains (Cantley et al., 1991). It appears that phosphorylation of the C-terminal regulatory domain of TCF/Elk-1 induces a conformational change in the protein to relieve inhibition. Removal of the domain, which is multiply phosphorylated *in vivo* and efficiently phosphorylated by MAP kinase *in vitro* (Gille et al., 1992), allows SRF-dependent DNA binding *in vitro* in the absence of phosphorylation (Gille and Shaw, unpublished results). However, C-terminally deleted forms of TCF/Elk-1 are not constitutively active (Kortenjann et al., 1994).

CONCLUSIONS

The activity of AP-1, the nuclear transcription factor that comprises the c-fos and c-jun gene products, is subject to modulation by phosphorylation at various levels. Both genes are induced in response to growth factors and mitogens that trigger several intracellular signaling cascades operative in different cell types. Furthermore, the Fos and Jun proteins are multiply phosphorylated (and dephosphorylated) by various kinases (and phosphatases) in response to the same stimuli. The MAP kinase pathway appears to be an important transduction pathway for the activation of AP-1.

ACKNOWLEDGMENTS

We wish to thank those who shared their results with us prior to publication, Monika Kortenjann for comments and suggestions, and Gabriele Prosch for assistance with the manuscript. The authors' research is supported by the Max Planck Gesellschaft and the Deutsche Forschungsgemeinschaft.

REFERENCES

Abate, C., Luk, D., Gagne, E., Roeder, R. G., and Curran, T., Fos and Jun cooperate in transcriptional regulation via heterologous activation domains. *Mol. Cell. Biol.*, 10, 5532, 1990.

Abate, C., Marshak, D. A., and Curran, T., Fos is phosphorylated by p34^{cdc2}, cAMP-dependent protein kinase and protein kinase C at multiple sites clustered within regulatory regions, *Oncogene*, 6, 2179, 1991.

Adler, V., Polotskaya, A., Wagner, F., and Kraft, A. S., Affinity-purified c-Jun amino-terminal protein kinase requires serine/threonine phosphorylation for activity, *J. Biol. Chem.*, 267, 17001, 1992.

Ahn, N. G., Seger, R., Bratlien, R. L., Diltz, C. D., Tonks, N. K., and Krebs, E. G, Multiple components in an epidermal growth factor-stimulated protein kinase cascade, *J. Biol. Chem.*, 266, 4220, 1991.

Anderson, D., Koch, C. A., Grey, L., Ellis, C., Moran, M. F., and Pawson, T., Binding of SH2 domains of PLC-γ1, GAP and src to activated growth factor receptors, *Science*, 250, 979, 1990a.

Anderson, N. G., Maller, J. L., Tonks, N. K., and Sturgill, T. W., Requirement for integration of signals from two distinct phosphorylation pathways for activation of MAP kinase, *Nature*, 343, 651, 1990b.

Angel, P., Allegretto, E. A., Okino, S. T., Hattori, K., Boyle, W. J., Hunter, T., and Karin, M., Oncogene jun encodes a sequence-specific trans-activator similar to AP-1, *Nature*, 332, 166, 1988.

Baker, S. J., Kerppola, T. K., Luk, D., Vandenberg, M. T., Marshak, D. R., Curran, T., and Abate, C., Jun is phosphorylated by several protein kinases at the same sites that are modified in serum stimulated fibroblasts. *Mol. Cell. Biol.*, 12, 4694, 1992.

Barber, J. R. and Verma, I. M., Modification of *fos* proteins: Phosphorylation of c-*fos*, but not v-*fos*, is stimulated by 12-tetradecanoyl-phorbol-13-acetate and serum, *Mol. Cell. Biol.*, 7, 2201, 1987.

Bialojan, C. and Takai, A., Inhibitory effect of a marine-sponge toxin, okadaic acid, on protein phosphatases. *Biochem. J.*, 256, 283, 1988.

Binétruy, B., Smeal, T., and Karin, M., Ha-Ras augments c-Jun activity and stimulates phosphorylation of its activation domain, *Nature*, 351, 122, 1991.

Blackshear, P. J., Haupt, D. M., App. H., and Rapp. U. R., Insulin activates the Raf-1 protein kinase. *J. Biol. Chem.*, 265, 12131, 1990.

Bohmann, D., Bos, T. J., Admon, A., Nishimura, T., Vogt, P. K., and Tjian, R., Human proto-oncogene c-jun encodes a DNA binding protein with structural and functional properties of transcription factor AP-1, *Science*, 238, 1386, 1987.

Boulton, T. G., Nye, S. H., Robbins, D. J., Ip, N. Y., Radziejewska, E., Morgenbesser, S. D., DePinho, R. A., Panayotatos, N., Cobb, M. H., and Yancopoulos, G. D., ERKs: A family of protein-serine/threonine kinases that are activated and tyrosine phosphorylated in response to insulin and NGF, *Cell*, 65, 663, 1991.

Boyle, W. J., Smeal, T., Defize, L. H. K., Angel, P., Woodgett, J. R., Karin, M., and Hunter, T., Activation of protein kinase C decreases phosphorylation of c-Jun at sites that negatively regulate its DNA-binding activity. *Cell*, 64, 573, 1991.

Bruder, J. T., Heidecker, G., and Rapp, U. R., Serum-, TPA-, and Ras-induced expression from Ap-1/Ets-driven promoters requires Raf-1 kinase. *Genes Dev.*, 6, 545, 1992.

Buttyan, R., Zakeri, Z., Lockshin, R., and Wolgemuth, D., Cascade induction of c-*fos*, c-*myc* and heat shock 70K transcripts during regression of the rat ventral prostate gland, *Mol. Endocrinol.*, 2, 650, 1988.

Cai, H., Szeberenyi, J., and Cooper, G. M., Effect of a dominant inhibitory Ha-*ras* mutation on mitogenic signal transduction in NIH 3T3 cells, *Mol. Cell. Biol.*, 10, 5314, 1990.

Cantley, L. C., Auger, K. R., Carpenter, C., Duckworth, B., Graziani, A., Kapeller, R., Soltoff, S., Oncogenes and signal transduction. *Cell*, 64, 281, 1991.

Chen, R.-H., Chung, J., and Blenis, J., Regulation of pp90^{rsk} phosphorylation and S6 phosphotransferase activity in Swiss 3T3 cells by growth factor-, phorbol ester-, and cyclic AMP-mediated signal transduction. *Mol. Cell. Biol.*, 11, 1861, 1991.

Chen, R.-H., Sarnecki, C., and Blenis, J., Nuclear localization and regulation of *erk-* and *rsk-* encoded protein kinases, *Mol. Cell. Biol.*, 12, 915, 1992.

Chiu, R., Boyle, W. J., Meek, J., Smeal, T., Hunter, T., and Karin, M., The c-*fos* protein interacts with c-Jun/AP-1 to stimulate transcription of AP-1 responsive genes, *Cell*, 54, 541, 1988.

Cohen, P., The structure and regulation of protein phosphatases, *Annu. Rev. Biochem.*, 58, 453, 1989.

Colombel, M., Olsson, C. A., Ng, P.-Y., and Buttyan, R., Hormone-regulated apoptosie results from reentry of differentiated prostate cells onto a defective cell cycle, *Cancer Res.*, 52, 4313, 1992.

Crews, C. M. and Erikson, R. L., Purification of a murine protein-tyrosine/threonine kinase that phosphorylates and activates the Erk-1 gene product, *Proc. Natl. Acad. Sci. U.S.A.*, 89, 8205, 1992.

Crews, C. M., Alessandrini, A., and Erikson, R. L., The primary structure of MEK, a protein kinase that phosphorylates and activates the Erk-1 gene product, *Science*, 258, 478, 1992.

Curran, T. and Franza, B. R., Fos and Jun: The AP-1 connection, *Cell*, 55, 395, 1988.

Curran, T. and Teich, N. M., Candidate product of the FBJ murine osteosarcoma virus oncogene: characterization of a 55,000 dalton phosphoprotein, *J. Virol.*, 42, 114, 1982.

Curran, T., Miller, A. D., Zokas, L., and Verma, I. M., Viral and cellular *fos* proteins: A comparative analysis, *Cell*, 36, 259, 1984.

Dent, P., Haser, W., Haystead, T. A. J., Vincent, L. A., Roberts, T. M., and Sturgill, T. W., Activation of mitogen-activated protein kinase kinase by v-Raf in NIH 3T3 cells and in vitro, *Science*, 257, 1404, 1992.

Dérijard, B., Hibi, M., Wu, I.-H., Barrett, T., Su, B., Deng, T., Karin, M., and Davis, R. J., JNK1: A protein kinase stimulated by UV light and Ha-Ras that binds and phosphorylates the c-Jun activation domain, *Cell*, 76, 1025, 1994.

Devary, Y., Gottlieb, R. A., Smeal, T., and Karin, M., The mammalian ultraviolet response is triggered by activation of Src tyrosine kinases, *Cell*, 71, 1081, 1992.

Dickson, B., Sprenger, F., Morrison, D., and Hafen, E., Raf functions downstream of Ras1 in the Sevenless signal transduction pathway, *Nature*, 360, 600, 1992.

Fantl, W. J., Escobedo, J. A., Martin, G. A., Turck, C. W., del Rosario, M., McCormick, F., and Williams, L. T., Distinct phosphotyrosines on a growth factor receptor bind to specific molecules that mediate different signalling pathways, *Cell*, 69, 413, 1992.

Feig, L. A. and Cooper, G. M., Inhibition of NIH 3T3 cell proliferation by a mutant *ras* protein with preferential affinity for GDP, *Mol. Cell. Biol.*, 8, 3235, 1988.

Gallego, C., Gupta, S. K., Heasley, L. E., Qian, N. X., and Johnson, G. L., Mitogen-activated protein kinase activation resulting from selective oncogene expression in NIH 3T3 and rat 1a cells, *Proc. Natl. Acad. Sci. U.S.A.*, 89, 7355, 1992.

Gartner, A., Nasmyth, K., and Ammerer, G., Signal transduction in *Saccharomyces cerevisiae* requires tyrosine and threonine phosphorylation of FUS3 and KSS1, *Genes Dev.*, 6, 1280, 1992.

Gille, H., Sharrocks, A. D., and Shaw, P. E., Phosphorylation of transcription factor p62[TCF] by MAP kinase stimulates ternary complex formation at c-fos promoter, *Nature*, 358, 414, 1992.

Gius, D., Cao, X., Rauscher, F. J., Cohen, D. R., Curran, T., and Sukhatme, V. P., Transcriptional activation and repression by fos are independent functions: The C terminus represses immediate-early gene expression via CArG elements, *Mol. Cell. Biol.*, 10, 4243, 1990.

Gotoh, Y., Nishida, E., and Sakai, H., Okadaic acid activates microtubule-associated protein kinase in quiescent fibroblastic cells, *Eur. J. Biochem.*, 193, 671, 1990.

Hai, T. and Curran, T., Cross-family dimerization of transcription factors Fos/Jun and ATF/CREB alters DNA binding specificity, *Proc. Natl. Acad. Sci. U.S.A.*, 88, 3720, 1991.

Halazonetis, T. D., Georgopoulos, K., Greenberg, M. E., and Leder, P., c-Jun dimerizes with itself and with c-*fos*, forming complexes of different DNA binding affinities, *Cell*, 55, 917, 1988.

Haystead, C. M. M., Wu, J., Gregory, P., Sturgill, T. W., and Haystead, T. A. J., Functional expression of a MAP kinase kinase in COS cells and recognition by an anti-STE7/byr1 antibody, *FEBS Lett.*, 317, 12, 1993.

Heidecker, G., Huleihel, M., Cleveland, J. L., Kolch, W., Beck, T. W., Lloyd, P., Pawson, T., and Rapp, U. R., Mutational activation of c-raf-1 and definition of the minimal transforming sequence. *Mol. Cell. Biol.*, 10, 2503, 1990.

Hemmings, H. C., Greengard, P., Lim Tung, H. Y., and Cohen, P., DARPP-32, a dopamine-regulated neuronal phosphoprotein, is a potent inhibitor of protein phosphatase-1. *Nature*, 310, 503, 1984.

Hipskind, R. A., Rao, V. N., Mueller, C. G. F., Reddy, E. S. P., and Nordheim, A., Ets-related protein Elk-1 is homologous to the c-*fos* regulatory factor p62[TCF], *Nature*, 354, 531, 1991.

Howe, L. R., Leevers, S. J., Gomez, N., Nakielny, S., Cohen, P., and Marshall, C. J., Activation of the MAP kinase pathway by the protein kinase raf, *Cell*, 71, 335, 1992.

Huang, F. L. and Glinsmann, W. H., Separation and characterization of two phosphorylase phosphatase inhibitors from rabbit skeletal muscle, *Eur. J. Biochem.*, 70, 419, 1976.

Hubbard, M. J. and Cohen, P., Regulation of protein phosphatase-1$_G$ from rabbit skeletal muscle. 1. Phosphorylation by cAMP-dependent protein kinase at site 2 releases catalytic subunit from the glycogen-bound holoenzyme. *Eur. J. Biochem.*, 186, 701, 1989.

Jamal, S. and Ziff, E., Transactivation of c-fos and beta-actin genes by raf as a step in early response to transmembrane signals, *Nature*, 344, 463, 1990.

Janknecht, R., Hipskind, R. A., Houthaeve, T., Nordheim, A., and Stunnenberg, H. G., Identification of multiple SRF N-terminal phosphorylation sites affecting DNA binding properties, *EMBO J.*, 11, 1045, 1992.

Kaibuchi, K., Fukumoto, Y., Oku, N., Hori, Y., Yamamoto, T., Toyoshima, K., and Takai, Y., Activation of the serum response element and 12-*O*-tetradecanoylphorbol-13-acetate response element by the activated c-raf-1 protein in a manner independent of protein kinase C. *J. Biol. Chem.*, 264, 20855, 1989.

Kerppola, T. K. and Curran, T., Fos-Jun heterodimers and Jun homodimers bend DNA in opposite orientations: Implications for transcription factor cooperativity, *Cell*, 66, 317, 1991.

Kölch, W., Heidecker, G., Lloyd, P., and Rapp, U. R., Raf-1 protein kinase is required for growth of induced NIH/3T3 cells, *Nature*, 349, 426, 1991.

Kortenjann, M., Thomae, O., and Shaw, P.E., Inhibition of v-Raf dependent c-*fos* expression and transformation by a kinase-defective mutant of the MAP Kinase ERK2. *Mol. Cell. Biol.*, in press.

Kosako, H., Gotoh, Y., Matsuda, S., Ishikawa, M., and Nishida, E., *Xenopus* MAP kinase activator is a serine/threonine/tyrosine kinase activated by threonine phosphorylation, *EMBO J.*, 11, 2903, 1992.

Kosako, H., Nishida, E., and Gotoh, Y., cDNA cloning of MAP kinase kinase reveals cascade pathways in yeasts to vertebrates. *EMBO J.*, 12, 787, 1993.

Kouzarides, T. and Ziff, E., The role of the leucine zipper in the fos-jun interaction, *Nature*, 336, 646, 1988.

Kovacina, K. S., Yonezawa, K., Bräutigan, D. L., Tonks, N. K., Rapp, U. R., and Roth, R. A., Insulin activates the kinase activity of the Raf-1 proto-oncogene by increasing its serine phosphorylation, *J. Biol. Chem.*, 265, 12115, 1990.

Kozma, S. C. and Thomas, G., Serine/threonine kinases in the propagation of the early mitogen response, *Rev. Physiol., Biochem. Pharmacol.*, 119, 124, 1992.

Kurjan J., Pheromone response in yeast, *Annu. Rev. Biochem.*, 61, 1097, 1992.

Kyriakis, J. M., App, H., Zhang, X., Banerjee, P., Brautigan, D. L., Rapp, U. R., and Avruch, J., Raf-1 activates MAP kinase kinase, *Nature*, 358, 417, 1992.

Kyriakis, J. M., Banerjee, P., Nikolakaki, E., Dai, T., Rubie, E. A., Ahmad, M. F., Avruch, J., and Woodgett, J. R., The stress-activated protein kinase subfamily of c-Jun kinases. *Nature*, 369, 156, 1994.

Lange-Carter, C. A., Pleiman, C. M., Gardner, A. M., Blumer, K. J., and Johnson, G. L., A divergence in the MAP kinase regulatory network defined by MEK kinase and Raf. *Science*, 260, 315, 1993.

Lee, W., Mitchell, P., and Tjian, T. Purified transcription factor AP-1 interacts with TPA-inducible enhancer elements, *Cell*, 49, 741, 1987.

Leevers, S. J. and Marshall, C. J., Activation of extracellular signal-regulated kinase, ERK2, by p21ras oncoprotein, *EMBO J.*, 11, 569, 1992.

Lowenstein, E. J., Daly, R. J., Batzer, A. G., Li, W., Margolis, B., Lammers, R., Ullrich, A., Skolnik, E. Y., Bar-Sagi, D., and Schlessinger, J., The SH2 and SH3 domain-containing protein GRB2 links receptor tyrosine kinases to ras signaling, *Cell*, 70, 431, 1992.

Lucibello, F. C., Lowag, C., Neuberg, M., and Müller, R., *Trans*-repression of the mouse c-*fos* promoter: A novel mechanism of Fos-mediated *trans*-regulation, *Cell*, 59, 999, 1989.

Malik, R. K., Roe, M. W., and Blackshear, P. J., Epidermal growth factor and other mitogens induce binding of a protein complex to the *c-fos* serum response element in human Astrocytoma and other cells, *J. Biol. Chem.*, 266, 8576, 1991.

Manak, J. R. and Prywes, R., Mutation of serum response factor phosphorylation sites and the mechanism by which its DNA-binding activity is increased by casein kinase II, *Mol. Cell. Biol.*, 11, 3652, 1991.

Manak, J. R., de Bisschop, N., Kris, R. M., and Prywes, R., Casein kinase II enhances the DNA binding activity of serum response factor, *Genes Dev.*, 4, 955, 1990.

Marais, R. M., Hsuan, J. J., McGuigan, C., Wynne, J., and Treisman, R., Casein kinase II phosphorylation increases the rate of serum response factor-binding site exchange, *EMBO J.*, 11, 97, 1992.

Marais, R., Wynne, J., and Treisman, R., The SRF accessory protein Elk-1 contains a growth factor-regulated transcriptional activation domain, *Cell*, 73, 381, 1993.

Matsuda, M., Mayer, B. J., Fukui, Y., and Hanafusa, H., Binding of the oncoprotein P47gag-crk to a broad range of phosphotyrosine-containing proteins, *Science*, 248, 1537, 1990.

Matsuda, S., Kosako, H., Takenaka, K., Moriyama, K., Sakai, H., Akiyama, T., Gotoh, Y., and Nishida, E., *Xenopus* MAP kinase activator: Identification and function as a key intermediate in the phosphorylation cascade, *EMBO J.*, 11, 973, 1992.

Mayer, B. J. and Hanafusa, H., Association of the v-crk oncogene product with phosphotyrosine-containing proteins and protein kinase activity, *Proc. Natl. Acad. Sci. U.S.A.*, 87, 2638, 1990.

Mayer-Jaekel, R. E., Ohkura, H., Gomes, R., Sunkel, C. E., Baumgartner, S., Hemmings, B. A., and Glover, D. M., The 55-kDa regulatory subunit of drosophila protein phosphatase 2A is required for anaphase, *Cell*, 72, 621, 1993.

McGrew, B. R., Nichols, D. W., Stanton, V. P., Jr., Cai, H., Whorf, R. C., Patel, V., Cooper, G. M., and Laudano, A. P., Phosphorylation occurs in the amino terminus of the Raf-1 protein, *Oncogene*, 7, 33, 1992.

Misra, R. P., Rivera, V. M., Wang, J. M., Fan, P.-D., and Greenberg, M. E. The serum response factor is extensively modified by phosphorylation following its synthesis in serum-stimulated fibroblasts, *Mol. Cell. Biol.*, 11, 4545, 1991.

Morrison, D. K., Kaplan, D. R., Rapp, U. R., and Roberts, T. M., Signal transduction from membrane to cytoplasm: Growth factors and membrane-bound oncogenes products increase Raf-1 phosphorylation and associated protein kinase activity, *Proc. Natl. Acad. Sci. U.S.A.*, 85, 8855, 1988.

Nakielny, S., Cohen. P., Wu, J., and Sturgill, T., MAP kinase activator from insulin-stimulated skeletal muscle is a protein threonine/tyrosine kinase, *EMBO J.*, 11, 2123, 1992.

Norman, C., Runswick, M., Pollock, R., and Treisman, R., Isolation and properties of cDNA clones encoding SRF, a transcription factor that binds to the c-*fos* serum response element, *Cell*, 55, 989, 1988.

Ofir, R., Dwarki, V. J., Rashid, D., and Verma, I. M., Phosphorylation of the C terminus of fos protein is required for transcriptional transrepression of the c-*fos* promoter, *Nature*, 348, 80, 1990.

Papavassiliou, A. G., Treier, M., Chavrier, C., and Bohmann, D., Targeted degradation of c-fos, but not v-fos, by a phosphorylation-dependent signal on c-Jun. *Science*, 258, 1941, 1992.

Ptashne, M., How eukaryotic transcriptional activators work, *Nature*, 335, 683, 1988.

Pulverer, B. J., Kyriakis, J. M., Avruch, J., Nikolakaki, E., and Woodgett, J. R., Phosphorylation of c-jun mediated by MAP kinases, *Nature*, 353, 670, 1991.

Pulverer, B. J., Hughes, K., Franklin, C. C., Kraft, A. S., Leevers, S. J., and Woodgett, J. R., Co-purification of mitogen-activated protein kinases with phorbol ester-induced c-Jun kinase activity in U937 cells, *Oncogene*, 8, 407, 1993.

Radler-Pohl, A., Sachsenmaier, C., Gebel, S., Auer, H.-P., Bruder, J. T., Rapp, U., Angel, P., Rahmsdorf, H. J., and Herrlich, P., UV-induced activation of AP-1 involves obligatory extranuclear steps including Raf-1 kinase, *EMBO J.*, 12, 1005, 1993.

Rapp, U. R., Reynolds, F. H., and Stephenson, J. R., New mammalian transforming virus: Demonstration of a polypeptide gene product, *J. Virol.*, 45, 914, 1983.

Rauscher, F. J., Cohen, D. R., Curran, T., Bos, T. J., Vogt, P. K., Bohmann, D., Tijan, R., and Franza, B. R., Fos-associated protein p39 is the product of the *jun* proto-oncogene, *Science*, 240, 1010, 1988.

Rivera, V. M., Miranti, C. K., Misra, R. P., Ginty, D. D., Chen, R.-H., Blenis, J., and Greenberg, M. E., A growth factor-induced kinase phosphorylates the serum response factor at a site that regulates its DNA-binding activity, *Mol. Cell. Biol.*, 13, 6260, 1993.

Rossomando, A., Wu, J., Weber, M. J., and Sturgill, T. W., The phorbol ester-dependent activator of the mitogen-activated protein kinase p42^{mapk} is a kinase with specificity for the threonine and tyrosine regulatory sites, *Proc. Natl. Acad. Sci. U.S.A.*, 89, 5221, 1992.

Ryseck, R. P. and Bravo, R., cJUN, JUN B and JUN D differ in their binding affinities to AP-1 and CRE consensus sequences: Effect of FOS proteins, *Oncogene*, 6, 533, 1991.

Sassone-Corsi, P., Lamph, W. W., Kamps, M., and Verma, I. M., *fos*-associated cellular p39 is related to nuclear transcription factor AP-1, *Cell*, 54, 553, 1988a.

Sassone-Corsi, P., Sisson, C., and Verma, I. M., Transcriptional autoregulation of the proto-oncogene *fos*, *Nature*, 334, 314, 1988b

Schönthal, A., Herrlich, P., Rahmsdorf, H. J., and Ponta, H., Requirement for *fos* gene expression in the transcriptional activation of collagenase by other oncogenes and phorbol esters, *Cell*, 54, 325, 1988.

Schönthal, A., Tsukitani, Y., and Feramisco, J. R., Transcriptional and post-transcriptional regulation c-*fos* expression by the tumor promoter okadaic acid, *Oncogene*, 6, 423, 1991.

Shaw, P. E., Schröter, H., and Nordheim, A., The ability of a ternary complex to form over the serum response element correlates with serum inducibility of the human c-fos promoter, *Cell*, 56, 563, 1989.

Shenolikar, S. and Nairn, A. C., Protein phosphatases: Recent progress, in *Advances in Second Messengers and Phosphoprotein Research*, vol. 25, Greengard, P. and Robinson, G. A., Eds., Raven Press, New York, 1991.

Shirakabe, K., Gotoh, Y., and Nishida, E., A mitogen-activated protein (MAP) kinase activating factor in mammalian mitogen-stimulated cells is homologous to *Xenopus* M phase MAP kinase activator, *J. Biol. Chem.*, 267, 16685, 1992.

Smeal, T., Binétruy, B., Mercola, D. A., Birrer, M., and Karin, M., Oncogenic and transcriptional cooperation with HA-Ras requires phosphorylation of c-Jun on serines 63 and 73. *Nature*, 354, 494, 1991.

Smeal, T., Binetruy, B., Mercola, D., Grover-Bardwick, A., Heidecker, G., Rapp, U. R., Karin, M., Oncoprotein-mediated signalling cascade stimulates c-Jun activity by phosphorylation of serines 63 and 73. *Mol. Cell. Biol.*, 12, 3507, 1992.

Sözeri, O., Vollmer, K., Liyanage, M., Frith, D., Kour, G., Mark, G. E., III, and Stabel, S., Activation of the c-raf protein kinase by protein kinase C phosphorylation, *Oncogene*, 7, 2259, 1992.

Strålfors, P., Hiraga, A., and Cohen, P., The protein phosphatases of the glycogen-bound form of protein phosphatase-1 from rabbit skeletal muscle. *Eur. J. Biochem.*, 149, 295, 1985.

Sturgill, T. W., Ray, L. B., Erikson, E., and Maller, J. L., Insulin-stimulated MAP-2 kinase phosphorylates and activates ribosomal protein S6 kinase II. *Nature*, 334, 715, 1988.

Suganuma, M., Fujiki, H., Suguri, H., Yoshizawa, S., Hirota, M., Nakayasu, M., Ojika, M., Wakamatsu, K., Yamada, K., and Sugimura, T., Okadaic acid: An additional non-phorbol-12-tetradecanoate-13-acetate-type tumor promoter. *Proc. Natl. Acad. Sci. U.S.A.*, 85, 1768, 1988.

Sun, H., Charles, C. H., Lau, L. F., and Tonks, N. K., MKP-1 (3CH134), an immediate early gene product, is a dual specificity phosphatase that dephosphorylates MAP kinase *in vivo*, *Cell*, 75, 487, 1993.

Sutherland, J. A., Cook, A., Bannister, A. J., and Kouzarides, T., Conserved motifs in Fos and Jun define a new class of activation domain. *Genes Dev.*, 6, 1810, 1992.

Szeberenyi, J., Cai, H., and Cooper, G. M., Effect of a dominant inhibitory Ha-*ras* mutation on neuronal differentiation of PC12 cells. *Mol. Cell. Biol.*, 10, 5324, 1990.

Szeberenyi, J., Erhardt, P., Cai, H., and Cooper, G. M., Role of Ras in signal transduction from the nerve growth factor receptor: Relationship to protein kinase C, calcium and cyclic AMP. *Oncogene*, 7, 2105, 1992.

Thomas, S. M., DeMarco, M., D'Arcangelo, G., Halegoua, S., and Brugge, J. S., Ras is essential for nerve growth factor- and phorbol ester-induced tyrosine phosphorylation of MAP kinases. *Cell*, 68, 1031, 1992.

Trahey, M. and McCormick, F., A cytoplasmic protein stimulates normal N-ras p21 GTPase, but does not affect oncogenic mutants. *Science*, 242, 1697, 1987.

Tratner, I., Ofir, R., and Verma, I. M., Alteration of a cyclic AMP-dependent protein kinase phosphorylation site in the c-fos protein augments its transforming potential. *Mol. Cell. Biol.*, 12, 998, 1992.

Treisman, R., Transient accumulation of c-fos RNA following serum stimulation requires a conserved 5' element and c-fos 3' sequences. *Cell*, 42, 889, 1985.

Treisman, R., Identification of a protein-binding site that mediates transcriptional response of the c-fos gene to serum factors. *Cell*, 46, 567, 1986.

Troppmair, J., Bruder, J. T., App, H., Cai, H., Liptak, L., Szeberenyi, J., Cooper, G. M., and Rapp, U. R., Ras controls coupling of growth factor receptors and protein kinase C in the membrane to Raf-1 and B-Raf protein serine kinases in the cytosol. *Oncogene*, 7, 1867, 1992.

Vogel, U. S., Dixon, R. A. F., Schaber, M. D., Diehl, R. E., Marshall, M. S., Scolnick, E. M., Sigal, I. S., and Gibbs, J. B., Cloning of bovine GAP and its interaction with oncogenic ras p21, *Nature*, 335, 90, 1988.

Vojtek, A. B., Hollenberg, S. M., and Cooper, J. A., Mammalian Ras interacts directly with the serine/threonine kinase Raf. *Cell*, 74, 205, 1993.

Ward, Y., Gupta, S., Jensen, P., Wartmann, M., Davis, R. J., and Kelly, K., Control of MAP kinase activation by the mitogen-induced threonine/tyrosine phosphatase PAC1, *Nature*, 367, 651, 1994.

Warne, P. H., Viciana, P. R., and Downward, J., Direct interaction of Ras and the amino-terminal region of Raf-1 *in vitro, Nature,* 364, 352, 1993.

Wilson, T. and Treisman, R., Fos C-terminal mutations block down-regulation of c-*fos* transcription following serum stimulation. *EMBO J.,* 7, 4193, 1988.

Wood, K. W., Sarnecki, C., Roberts, T. M., and Blenis, J., *Ras* mediates nerve growth factor receptor modulation of three signal-transducing protein kinases: MAP kinase, raf-1, and RSK. *Cell,* 68, 1041, 1992.

Wu, J., Harrison, J. K., Vincent, L. A., Haystead, C., Haystead, T. A. J., Michel, H., Hunt, D. F., Lynch, K. R., Sturgill, T. W., Molecular structure of a protein-tyrosine/threonine kinase activating p42 mitogen-activated protein (MAP) kinase: MAP kinase kinase. *Proc. Natl. Acad. Sci. U.S.A.,* 90, 173, 1993.

Zheng, X. M., Wang, Y., and Pallen, C. J., Cell transformation and activation of pp60c-src by overexpression of a protein tyrosine phosphatase, *Nature,* 359:336, 1992.

c-Fos Protein Transport into the Nucleus

Pierre Roux, Serge Carillo, Jean-Marie Blanchard, Philippe Jeanteur, and Marc Piechaczyk

CONTENTS

INTRODUCTION

Evidence has accumulated indicating that the transport of numerous karyophilic proteins into the nucleus is a highly selective process that depends on the presence of active internal nuclear localization signals (NLS). According to our current knowledge of the process, NLS are recognized by cytoplasmic NLS-binding proteins that, in a second step, deliver nuclear proteins to nuclear pores. The pore opens and the complex of NLS-binding and -bearing proteins passes through. While delivery to nuclear pores is presumably energy independent, ATP is required for movement into the nucleus. There, the complex dissociates and the NLS-binding proteins can recycle to the cytoplasm. Nuclear transport is not constitutive for all proteins. An increasing number of reports have indicated that the activity of some proteins is regulated at the level of their transport from the cytoplasm to the nucleus at certain stages of the cell cycle or in response to developmental cues, hormones, or growth factors (Silver, 1991; Garcia-Bustos et al., 1991). Some reports point to the existence of NLS within the cFos protein, the action of which may be regulated via extracellular signals by mechanisms that remain to be characterized. Interestingly, the mutated Fos proteins that have been transduced by the two murine osteosarcomagenic retroviruses, FBR and FBJ (Curran, 1988), evade nuclear transport control, thus suggesting that nuclear transport processes contribute to their tumorigenic potential.

c-FOS PROTEIN IS A NLS-BEARING PROTEIN

Tratner and Verma (1991) demonstrated the existence of at least two NLS in the cFos protein. Only one of them has been characterized molecularly. It lies in the central part of the molecule between amino acids 139 and 161 (out of 380) within the basic domain that has been shown to interact with the AP-1 DNA motif (Ransone and Verma, 1990). This observation is in agreement with a previous observation by Jenuwein and Müller (1987) that deletion mutants retaining only the central part of the protein can enter the nucleus. The demonstration that this region contains an actual NLS comes from the observation that, when inserted into the coding sequence of the chicken pyruvate kinase

0-8493-4573-1/94/$0.00+$.50

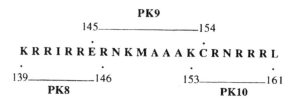

EFFECT ON PYRUVATE KINASE NUCLEAR ACCUMULATION

PK10 > PK8, PK9

Figure 7-1 Composite NLS of c-Fos protein. The c-Fos protein NLS characterized by Tratner and Verma (1991) lies in the basic domain involved in binding to DNA. Grafting of all the indicated peptides on the coding sequence of the pyruvate kinase entails nuclear localization of the enzyme. However, the effect of pK10 is stronger than that of PK8 and PK9.

(PK), it drags the chimeric protein into the nucleus of simian Cos7 cells in a transfection assay. However, its deletion does not entail the retention of the mutated cFos protein into the cytoplasm, thus pointing to the existence of at least one other active NLS. In this respect, c-Fos protein resembles other oncogenic or anti-oncogenic proteins such as c-Myc (Dang and Lee, 1988) or p53 (Addison et al., 1990; Shaulsky et al., 1990), where two and three NLS, respectively, have been characterized. Nevertheless, the situation of c-Fos departs from that of its most studied transcription partner, c-Jun, the transport of which appears to depend on a single NLS residing within the DNA binding domain (Chida and Vogt, 1992). Interestingly, when the c-Fos protein basic domain is split into three regions, each fragment confers karyophilic properties to the chicken pyruvate kinase, although with different efficiencies (see Figure 7-1). This suggests that the effect of the entire NLS is mediated through the additive effects of smaller units. If true, this NLS is different from, on the one hand, simple basic NLS (e.g., SV40 virus large T antigen), which are short motifs of basic amino acids, and, on the other hand, bipartite NLS (e.g., nucleoplasmin) (Garcia-Bustos et al., 1991; Robbins et al., 1991), which are composed of two basic regions separated by a short linker but whose mutations in either basic region inhibit activity.

Attempts to identify other c-Fos protein NLS have been made but have failed. Outside the DNA binding domain, two regions encompassing amino acids 201–205 and 124–130 display similarity with SV40 large T antigen NLS. However, none of them is able to drive pyruvate kinase into the nucleus (Dang and Lee, 1989; Tratner and Verma, 1991). Since no other sequence of c-Fos protein has NLS characteristics, the additional NLS may be of a novel type, perhaps representing an interaction of noncontiguous regions, the action of which could not be reconstituted in the various chimeric pyruvate kinase proteins that have been tested (Tratner and Verma, 1991). Alternatively, since c-Fos protein transport into the nucleus is regulated (see below), it is also possible that both NLS and motifs for cytoplasmic retention are colocalized within the protein, the latter being dominant over the former when fused to the pyruvate kinase sequence.

c-FOS AND c-JUN PROTEINS TRANSPORT IS REGULATED

The idea that c-Fos protein transport into the nucleus undergoes regulation by extracellular signals comes from the observation that c-Fos immunostaining concomitantly disappears from the nucleus and increases in the cytoplasm of mouse fibroblasts either transfected or microinjected with a constitutive *c-fos* gene, when cultured in a medium totally devoid of serum (Figure 7-2) (Roux et al., 1990; Vriz et al., 1992; Kölch et al., 1992). Unpublished experiments by our group have also shown that serum factors are not the only agents able to trigger c-Fos protein translocation into the nucleus *in vitro*. Cell–cell contacts and/or intercellular signals are critical since retention of the protein in the cytoplasm is much more efficient at low cell density. Interactions with the extracellular matrix are also likely to be important. Cytoplasmic localization of cFos protein is more easily obtained when cells are grown on polylysine-coated dishes than when grown on gelatin, fibronectin, or collagen. This observation is in agreement with the finding that laminin, or a 19-mer peptide from the long arm of its A chain, which promotes cell adhesion, is able to fully induce *c-fos* gene expression, including the nuclear localization of the protein (Kubota et al., 1992). Stresses, such as small variations in temperature and pH, are nuclear translocation inducers. It is worth noting that heat shock induces *c-fos* gene expression in addition to nuclear localization of the protein (Andrews et al., 1987). Finally, pricking of cells can also stimulate c-Fos protein nuclear transport. When serum-starved cells are microinjected with expression vectors, cFos protein accumulates transiently within the nucleus for 1 to 1.5 hours before being quantitatively found in the cytoplasm.

A number of questions arise concerning this phenomenon. Are the other components of the AP-1 complex involved? Is it an *in vitro* artefact, or is it a real new posttranslational regulatory step of biological relevance? Is c-Fos protein transported back from the nucleus to the cytoplasm or is it degraded within the nucleus, whereas newly synthesized proteins are retained within the cytoplasm? What is the fate of the protein in the cytoplasm?

Serum deprivation experiments conducted with mouse fibroblasts transfected with a constitutively transcribed *c-jun* gene indicate that cJun protein is subjected to the same regulation as c-Fos (Roux et al., unpublished observation) (see Figure 7-2C and D). This observation apparently differs from that of Chida and Vogt (1992), with the chicken cJun protein in chicken embryo fibroblasts. However, it must be emphasized that starvation experiments by these authors were conducted in the presence of 0.4% chicken and 1% calf serum and that trace amounts of serum permit cFos protein nuclear transport in the equivalent mouse system. The situation has not been investigated for the other members of the *fos* (Fos-B, Fra-1, and Fra-2) and *jun* (Jun-B and Jun-D) families. However, exclusive cytoplasmic localization of c-Myc protein was found both in serum-starved fibroblasts (Vriz et al., 1992) and in Xenopus oocytes before fertilization (Gusse et al., 1989), suggesting that the control of nuclear import may concern a number of cell division-involved transcription factors.

A number of observations suggest that the cytoplasmic retention of c-Fos protein is not limited to transfected mouse fibroblasts. Stachowiak et al. (1990) observed that basal levels of c-Fos protein are exclusively cytoplasmic in mouse adrenal medullary cells in primary culture. Interestingly, when *c-fos* gene expression is induced by angiotensin in these cells, c-Fos protein first accumulates to high levels in the cytoplasm before being transported into the nucleus. This suggests that the inhibition of transport is relieved only as a second step of the induction process. Furthermore, occasional accumulation of cFos protein in the cytoplasm has been detected in primary cultures of mouse kidney glomerular cells (Guyon, unpublished results), in the basal epithelial cells of large airways (Demoly, unpublished results), and occasionally in keratinocytes of human skin biopsies (Basset-Seguin, unpublished results). The latter two observations support the *in vivo* relevance of nuclear transport control.

Questions relative to the fate of *c-fos* protein in serum-deprived cells concern its status both within the cytoplasm and within the nucleus. Whether the nuclear disappearance of

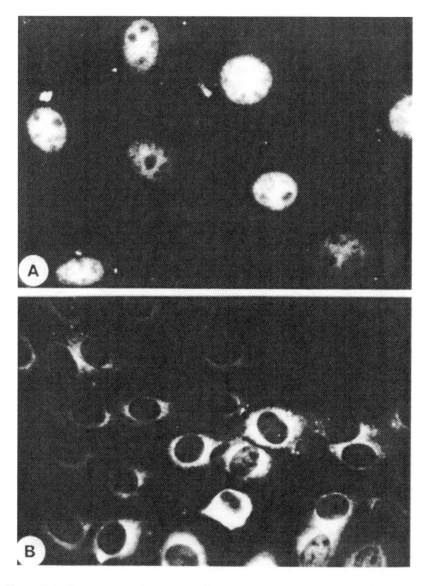

Figure 7-2 Serum deprivation entails c-Fos and c-Jun protein accumulation into the cytoplasm of mouse fibroblasts. L tk⁻ cells were transfected with either the PM43.1 human *c-fos*-expressing vector (Roux et al., 1990) or the pAH119-pJ3 mouse *c-jun*-expressing vector (kind gift from Dr. R. Bravo) to obtain constitutive expression of the corresponding proteins. Starvation experiments were conducted in the conditions described in Roux et al. (1990): after three extensive washings cells were incubated in a medium absolutely devoid of serum. Immunostaining of c-Fos and c-Jun proteins was performed with specific rabbit antisera (kind gifts from Drs. B. Verrier, A. Sergeant, and M. Karin). (A) control c-Fos-expressing L tk⁻ cells; (B) serum-starved c-Fos-expressing L tk⁻ cells; (C) control c-Jun-expressing L tk⁻ cells; (D) serum-starved c-Jun-expressing L tk⁻ cells.

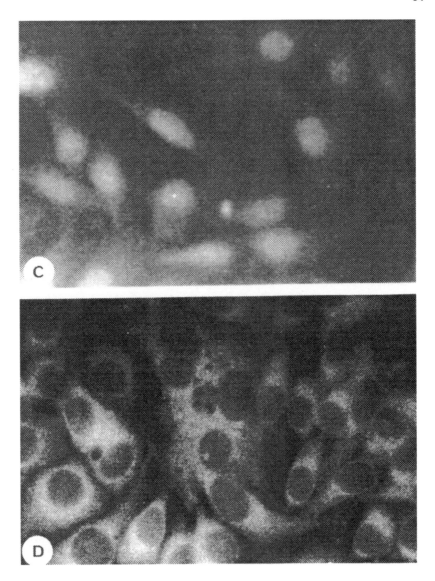

Figure 7-2   (continued)

c-Fos protein in serum-deprived cells reflects its return into the cytoplasm in a modified form or its degradation within the nucleus has not been established. Although the second possibility offers the advantage of conceptual simplicity, it has not yet been formally demonstrated whether c-Fos protein can actually be degraded within the nucleus, regardless of the physiological conditions considered. This issue is, thus, left open. Cell fractionation and assay of c-Fos protein by Western blotting show that there is as much intact protein in the nucleus of cells grown in the presence of serum as in the cytoplasm of serum-deprived ones (Roux et al., 1990). This indicates, first, that cytoplasmic immunostaining does not correspond to partially degraded protein, which would perhaps be unable to migrate to the nucleus, and, second, that the turnover of c-Fos protein within the nucleus of cells cultured in the presence of serum and within the cytoplasm of serum-deprived cells is comparable. Inhibition of protein synthesis using cycloheximide has conclusively demonstrated that c-Fos protein is actively degraded in the cytoplasm of serum-deprived cells (Roux et al., 1990). We have recently shown that the process is initiated by calcium-dependent proteases (Carillo et al., 1994). Altogether, these observations indicate that the neosynthesized c-Fos protein is not stored in a stable cytoplasmic, functionally inactivated form, as is the case for c-Myc protein in unfertilized *Xenopus* oocytes (Gusse et al., 1989) and the glucocorticoid receptor in the absence of hormone (Picard and Yamamoto, 1987). Finally, it is not yet known whether cytoplasmic c-Fos protein is retained in the cytoplasm of serum-starved cells because it is anchored to cytoplasmic structures or because its NLS are masked by cognate proteins. The identification of all c-Fos NLS and cytoplasmic c-Fos-interacting proteins will be necessary to resolve this question.

ACTIVATORS AND INHIBITORS OF c-FOS PROTEIN TRANSPORT INTO THE NUCLEUS

Two protein kinases are clearly involved in the activation of c-Fos protein transport into the nucleus. When serum-deprived cells are treated with agents that either mimic c-AMP, such as 8-Br-c-AMP, or elevate intracellular concentrations of c-AMP, such as cholera toxin or isoproterenol, a rapid and quantitative nuclear localization of c-Fos protein is observed, thus suggesting a role for protein kinase A (Roux et al., 1990). More recently, Kölch et al. (1992), using *raf* revertant cells deficient in the induction of early response genes by TPA and serum, have shown that the serine/threonine kinase encoded by the *c-raf* proto-oncogene, in addition to being necessary for AP-1 transcriptional activity, is also required for activating c-Fos protein nuclear translocation. Using cycloheximide, which is a reversible inhibitor of protein synthesis, we have postulated the existence of a labile protein acting as an inhibitor of transport (Roux et al., 1990). However, a more recent report by Edward's and Mahadevan (1992) has raised another interesting possibility. These authors have shown that protein synthesis inhibitors, and particularly cycloheximide, can act on their own, i.e., independently of blocking protein neosynthesis, as agonists in the phosphorylation of a stable pp33/pp15 protein whose role is still unknown but might be linked to the induction of genes like *c-fos* and *c-jun*, at least when assayed at the level of mRNA accumulation. It would be interesting to test the hypothesis of multiple roles for pp33/pp15 in the regulation of *c-fos* gene expression. Auwerx and Sassone-Corsi (1991) characterized a dominant inhibitor of Fos/Jun complex transcriptional activity, called IP-1. This protein, which is both nuclear and cytoplasmic, exhibits rapid turnover and is inactivated in the presence of various kinases, including protein kinase A, protein kinase C, and calcium/calmodulin-dependent kinase (Auwerx and Sassone-Corsi, 1991, 1992). It is, thus, tempting to speculate that this molecule corresponds to the putative labile inhibitor of transport and, consequently, possesses a dual function like c-Raf. However, that IP-1 is not inhibited in the presence of a functional *v-raf* oncogene (Auwerx and Sassone-Corsi, 1992) suggests, if the speculation is true, that alternative pathways lead to the activation of nuclear transport. In support of this idea, it is worth noting that epidermal growth factor has been reported to induce the *c-fos* gene, including the nuclear localization of the protein

(Bravo et al., 1985), without a notable increase in the intracellular concentration of c-AMP (Dumont et al., 1989).

v-FOS^{FBR} AND v-FOS^{FBJ} PROTEINS EVADE CONTROL OF NUCLEAR TRANSPORT

The *c-fos* gene originally was found in mutated oncogenic forms in two murine retroviruses, FBJ and FBR (Curran, 1988). Five amino acid changes, plus a frame shift changing the last 48 amino acids, have accumulated in v-Fos^{FBJ} protein. The situation is more complex for v-Fos^{FBR}. This protein is truncated at both ends: 310 amino acids from the *gag* protein of the parental retrovirus substitute for the first 24 c-Fos amino acids, and 8 amino acids from a genomic locus, called *fox*, substitute for the last 98 c-Fos amino acids. In addition, there are five point mutations and two in-frame deletions of 9 and 13 amino acids (Curran, 1988). Both proteins were found to constitutively enter the nucleus of serum-starved cells (Roux et al., 1990). Moreover, whereas c-Fos protein nuclear translocation is impeded in *raf* revertant cells, v-Fos^{FBR} protein is transported efficiently (Kölch et al., 1992). Both observations show that structural alterations are responsible for the loss of nuclear transport control and, therefore, raise the possibility that, although not constituting the primary basis for oncogenicity, the latter might strengthen the tumorigenic effects of transduced viral Fos proteins. Chida and Vogt (1992) have reported a very different situation in the case of the Jun protein carried by the chicken ASV-17 retrovirus. The viral protein is a *gag* fusion protein, having acquired mutations that display a cell cycle-dependent accumulation within the nucleus, whereas c-Jun protein is nuclear irrespective of the cell cycle phase, under the conditions that were tested. The transport is increased in the G2 phase and decreased in G1 and S because of a point mutation within the single Jun NLS located within the DNA binding domain. As in the case of viral Fos proteins that constitutively enter the nucleus, the actual contribution of this phenomenon to the tumorigenic potential of ASV-17 *jun* gene remains to be demonstrated experimentally.

CONCLUSION

The transport of c-Fos protein into the nucleus is regulated by various extracellular signals. This control operates in addition to several others affecting transcription initiation and elongation, transcripts and protein degradation, and posttranslational modifications of the protein, in order to rapidly and efficiently shutoff gene activity. The multiplicity of regulation levels likely explains why the constitutive activation of the *c-fos* gene has not yet been found to be involved in the generation of naturally occurring tumors. Although protein kinase A and the *c-raf* proto-oncogene product are likely to be involved in the stimulation of transport, one cannot exclude that alternative pathways of activation also exist. Interestingly, mutations in FBJ and FBR retroviruses allow viral proteins to evade the nuclear transport control and are valuable for identifying the c-Fos protein domains responsible for cytoplasmic retention.

ACKNOWLEDGMENTS

This work was supported by grants from the CNRS, ARC, ANRS, AFM, Fondation pour la Recherche Médicale and Ligue contre le Cancer. We are grateful to our colleagues from the URA CNRS 1191 for critical reading of the manuscript.

REFERENCES

Addison. C.. Jenkins, J. R.. and Stürzbecher. H.-W.. The p53 nuclear localization signal is structurally linked to a p34^{cdc2} kinase motif. *Oncogene*, 5, 423, 1990.

Andrews, C.. Harding, M. A.. Calvet, J. P.. and Adamson, E. D.. The heat shock response in HeLa cells is accompanied by elevated expression of the *c-fos* proto-oncogene. *Mol. Cell. Biol.*, 7, 3452, 1987.

Auwerx. J. and Sassone-Corsi. P.. IP-1: A dominant inhibitor of Fos/Jun whose activity is modulated by phosphorylation. *Cell*, 64, 983, 1991.

Auwerx. J. and Sassone-Corsi. P.. AP-1 (Fos/Jun) regulation by IP-1: Effect of signal transduction pathways and cell growth. *Oncogene*, 7, 2271, 1992.

Bravo. R.. Burckhardt. J.. Curran, T., and Müller. R... Stimulation and inhibition of growth by EGF in different A431 cell clones is accompanied by the rapid induction of *c-fos* and *c-myc* proto-oncogenes, *EMBO J.*, 4, 1193, 1985.

Carillo, S., Pariat. M.. Steff. A.-M.. Roux. P.. Etieane-Julan, M.. Lorca, T., and Piechaczyk. M.. Differential sensitivity of FOS and JUN family members to calpains. *Oncogene*, 9, 1679, 1994.

Chida. K. and Vogt. K. P.. Nuclear translocation of viral Jun but not of cellular Jun is cell cycle dependent. *Proc. Natl. Acad. Sci. U.S.A.*, 89, 4290, 1992.

Curran. T.. The fos oncogene. in *Oncogene Handbook*, Reddy. E. P.. Skalka. A. M.. and Curran. T.. Eds.. Elsevier. New York, pp. 307–325, 1988.

Dang, C. V. and Lee. W. M. F.. Identification of the human c-Myc protein nuclear translocation signals. *Mol. Cell Biol.*, 8, 4048, 1988.

Dang. C. V. and Lee. W. M. F.. Nuclear and nucleolar targeting sequences of c-erbA. c-myb. N-myc. p53, Hsp 70 and HIV Tat protein. *J. Biol. Chem.*, 264, 18019, 1989.

Dumont. J. E.. Jauniaux, J. C.. and Roger. P. P.. The cyclic AMP-mediated stimulation of cell proliferation. *Trends Biochem. Sci.*, 14, 67, 1989.

Edwards, D. R. and Mahadevan, L.. Protein synthesis inhibitors differentially superinduce *c-fos* and *c-jun* by three distinct mechanisms: Lack of evidence for a labile repressor, *EMBO J.*, 11, 2415, 1992.

Garcia-Bustos. J.. Heitman. J.. and Hall, M. N.. Nuclear protein localization. *Biochim. Biophys. Acta*, 1071, 83, 1991.

Gusse. M.. Ghysdael. J.. Evan. G.. Soussi. T.. and Méchali. M.. Translocation of a store of maternal c-Myc protein into nuclei during early development. *Mol. Cell Biol.*, 9, 5395, 1989.

Jenuwein. T. and Müller. R.. Structure-function analysis of Fos protein: A single aminoacid change activates the immortalizing potential of *c-fos*, *Cell*, 48, 647, 1987.

Kölch. , W.. Heidecker. G.. Troppmair. J.. Yanagihara. K.. Bassin. R. H.. and Rapp. U. R.. Raf revertant cells resist transformation by non-nuclear oncogenes and are deficient in the induction of early response genes by TPA and serum. *Oncogene*, 8, 361, 1993.

Kubota. S., Tashiro. K., and Yamada. Y.. Signaling site of laminin with mitogenic activity, *J. Biol. Chem.*, 267, 4285, 1992.

Picard. D. and Yamamoto. K. R.. Two signals mediate hormone-dependent nuclear localization of the glucocorticoid receptor. *EMBO J.*, 6, 3330, 1987.

Ransone. L. J. and Verma. I. M.. Nuclear proto-oncogenes Fos and Jun, *Annu. Rev. Cell Biol.*, 6, 305, 1990.

Robbins. J.. Dilworth, S. M.. Laskey. R. A.. and Dingwall, C., Two interdependent domains in nucleoplasmin nuclear targeting sequence: Identification of a class of bi-partite nuclear targeting sequence. *Cell*, 64, 615, 1991.

Roux. P.. Blanchard. J.-M.. Fernandez. A.. Lamb. N.. Jeanteur. P.. and Piechaczyk. M.. Nuclear localization of cFos. but not vFos proteins. is controlled by extracellular signals. *Cell*, 63, 341, 1990.

Shaulsky, G., Goldfinger, N., Ben-Ze'ev, A., and Roter, V., Nuclear accumulation of p53 protein is mediated by several nuclear localization signals and plays a role in tumorigenesis. *Mol. Cell. Biol.,* 10, 6565, 1990.

Silver, P. A., How proteins enter the nucleus, *Cell,* 64, 489, 1991.

Stachowiak, M. K., Sar, M., Tuominen, H. K., Jiang, S. A., Iaradola, M. J., Popisner, A. M., and Hong, J. S., Stimulation of adrenal medullary cells *in vivo* and *in vitro* induces expression of *c-fos* proto-oncogene, *Oncogene,* 5, 69, 1990.

Tratner, I. and Verma, I. M., Identification of a nuclear targeting sequence in the cFos protein, *Oncogene,* 6, 2049, 1991.

Vriz, S., Lemaitre, J.-M., Leibovici, M., Thierry, N., and Méchali, M., Comparative analysis of the intracellular localization of c-Myc, c-Fos, and replicative proteins during cell cycle progression, *Mol. Cell Biol.,* 12, 3548, 1992.

Chapter 8

Transcriptional Regulation of the Human c-*fos* Proto-Oncogene

Alfred Nordheim, Ralf Janknecht, and Robert A. Hipskind

CONTENTS

THE c-*fos* PROTO-ONCOGENE: A MODEL IMMEDIATE EARLY GENE

The c-*fos* proto-oncogene is rapidly induced to maximal transcriptional activity when quiescent cells are stimulated to grow by the addition of serum or individual growth factors (Greenberg and Ziff, 1984; Kruijer et al., 1984; Müller et al., 1984). This response is transient, as c-*fos* promoter activity is again diminished within 30 to 40 min after stimulation. Whereas induction does not require de novo protein synthesis, the promoter shut-off does, since it does not occur in the presence of the protein synthesis inhibitors cycloheximide, emetine, or anisomycin (Greenberg et al., 1986). Treatment with such inhibitors in the absence of serum can also lead to a slow induction of the gene. These characteristics of transcriptional regulation, together with the instability of *fos* mRNA,

0-8493-4573-1/94/$0.00+$.50
© 1994 by CRC Press, Inc.

establish c-*fos* as the prototype for immediate early genes (IEGs), which are defined by their stimulation upon mitogenic signaling (Lau and Nathans, 1987; Almendral et al., 1988; Bravo, 1990).

The induction of this class of genes, comprising approximately 100 members (Bravo, 1990), is not only a necessary trigger for cells to exit G_0 and enter into the cell cycle, but also is implicated in some differentiation pathways, neurotransmitter signaling, and excitation of neurons. A component in these signaling pathways appears to be an increase in the cytoplasmic concentrations of second messengers, such as cAMP, Ca^{2+}, or diacylglycerol (Sheng and Greenberg, 1990). The structure and function of the c-*fos* gene has been reviewed (Cohen and Curran, 1989), as has been the wide variety of stimuli leading to the above-mentioned transcriptional response (Sheng and Greenberg, 1990; Rivera and Greenberg, 1990; Angel and Karin, 1991).

CELL TYPE- AND ORGAN-SPECIFIC c-*fos* EXPRESSION

As reviewed in Chapter 12 and elsewhere (Cohen and Curran, 1989) c-*fos* can be induced transcriptionally in a variety of cell types, including fibroblasts, pheochromocytoma (PC12) cells, embryonal carcinoma (EC) cells, amnion cells, leukocytes, hematopoietic cells, neurons, and osteogenic progenitor cells. Studies using *in vitro* differentiation systems, organ culture, *in situ* hybridization, and transgenic mice have provided insight into the specific expression of c-*fos* in developmental pathways, particularly those leading to bone formation and hematopoiesis. A role for c-*fos* induction has been postulated in mediating both cyclical and long-term responses of neuronal cells as a consequence of external stimuli (Morgan et al., 1987; Sheng and Greenberg, 1990). Although experiments using cells in culture implicated c-*fos* induction as essential for cell proliferation and progression through the cell cycle, the *in situ* results, particularly with *fos*-negative mice (Johnson et al., 1992; Wang et al., 1992), belie this. Also seemingly contradictory is the role played by c-*fos* in programmed cell death (Colotta et al., 1992). Nevertheless, c-*fos* appears to function in the special situation of growth-arrested cells reentering the active cell cycle, i.e., undergoing the G_0 to G_1 transition.

c-*fos* ACTIVATION DURING THE G_0 TO G_1 TRANSITION

c-*fos* can be induced by a plethora of stimuli, such as serum, individual growth factors (e.g., NGF, EGF, PDGF), TNFα, TGFβ, interleukins, tumor-promoting phorbol esters, neurotransmitters, electrical excitation, Ca^{2+} influx, increased cAMP levels, cell wounding, and UV radiation (Angel and Karin, 1991).

The rapid transcriptional induction of c-*fos* upon mitogen stimulation of growth factor-deprived G_0 cells can be measured directly at the transcriptional level and is seen as soon as 6 min after stimulation of NIH3T3 cells or A431 epithelial carcinoma cells (Greenberg and Ziff, 1984; Stewart et al., 1990). After this burst of activity the c-*fos* promoter is silenced again within 30 min. Similar kinetics are found in many different cell types. Northern analysis shows rapid accumulation of c-*fos* mRNA, followed by an equally rapid decline in the steady state levels of this mRNA approx. 40 min post-induction (Figure 8-1). Likewise, the Fos protein has a short half-life that is influenced by its modification and that of its partner c-Jun (Papavassiliou et al., 1992).

The stimuli listed above induce c-*fos* via various signal transduction pathways and, consequently, may target the c-*fos* promoter at different sites within the upstream regulatory region. The identification and functional characterization of these individual *cis*-elements has led to important insights into the mechanisms by which signal transduction governs transcriptional regulation.

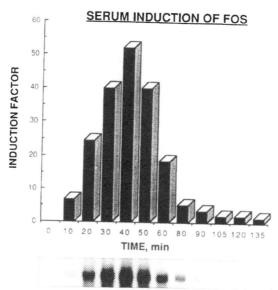

SERUM INDUCTION OF FOS

Figure 8-1 Typical induction/repression kinetics of c-*fos*. mRNA levels are determined by Northern blotting at the indicated times after treatment of starved A431 cells with serum.

cis-ELEMENTS IN THE c-*fos* PROMOTER

The collective work of many laboratories has identified multiple *cis*-elements in the c-*fos* promoter that contribute to both basal and induced transcriptional activity of *fos* (Cohen and Curran, 1989; Rivera and Greenberg, 1990). The arrangement of these elements and their cognate binding factors is shown in Figure 8-2.

BASAL ELEMENTS

Basal promoter activity is determined by the TATA element, the –60 region, the direct repeats at –95, the –170 region, and the serum response element (SRE), centered around coordinate –310, together with its flanking sequences (Gilman et al., 1986; Fisch et al., 1987; Runkel et al., 1991; Lucibello et al., 1991). Three additional basal elements apparently lie upstream of the SRE (Lucibello et al., 1991).

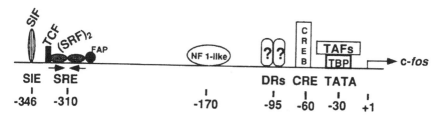

Figure 8-2 Representation of the human c-*fos* upstream control region and its regulatory *cis*-elements. The cognate transcription factors are indicated. Not shown are the many different DNA binding proteins described to interact with the SRE. See text for further details.

ELEMENTS MEDIATING cAMP AND Ca²⁺ INDUCTION

The –60 element contains a sequence resembling a cAMP response element (CRE), and can be bound by the cAMP response element binding factor (CREB), at least *in vitro* (Sassone-Corsi et al., 1988; Berkowitz et al., 1989). This element is essential but not fully sufficient for c-*fos* stimulation by both cAMP and increased Ca²⁺ (Büscher et al., 1988; Sassone-Corsi et al., 1988; Sheng et al., 1988). The convergence of these two distinct pathways at the CREB site of c-*fos* is apparently achieved at the level of CREB phosphorylation at residue Ser133, resulting in its ability to activate transcription (Sheng et al., 1990, 1991). This function of CREB may be enhanced by promoter sequences located around and upstream of the SRE, as well as intragenic sequences (Berkowitz et al., 1989; Metz and Ziff. 1991; Härtig et al., 1991).

ELEMENTS MEDIATING PDGF INDUCTION

A conditioned medium from v-*sis* transformed cells, which contains an analog of the PDGF β chain. and PDGF itself induce the c-*fos* promoter. This occurs via the v-*sis* conditioned medium induction element (SIE) located upstream of the SRE at –346 (Hayes et al., 1987; Wagner et al., 1990). On stimulation of cells with a conditioned medium, EGF, or PDGF, the transcription factor SIF (v-*sis*-inducible factor) is activated to bind to this element (Hayes et al., 1987; Wagner et al., 1990). Interestingly, genomic footprinting in A431 cells induced with EGF also showed a change around the SIE (Herrera et al., 1989), although this induced binding has yet to be shown *in vitro*. The ability of the SIE to confer sis/PDGF inducibility onto the c-*fos* promoter requires promoter proximal elements between –100 and –57 (Wagner et al., 1990). PDGF induction can also be mediated by the SRE (Büscher et al., 1988). In the wild-type c-*fos* promoter the two elements probably function together in response to the PDGF-triggered signal. The inducible binding of SIF to the SIE is in striking contrast to the constitutive presence of factors bound to the SRE (Herrera et al, 1989; see below), indicating two potentially different mechanisms for transcriptional stimulation via these two elements.

SERUM GROWTH FACTOR RESPONSIVE ELEMENT

Activation of the c-*fos* promoter with serum requires the presence of the SRE, positioned at –300 in the promoter (Treisman, 1985, 1986, 1990. 1992; Gilman et al., 1986; Greenberg et al., 1987).This *cis*-element was originally identified as the binding site of serum response factor (SRF), whose interaction with the SRE appears to be intimately involved in serum stimulation (Treisman, 1985, 1987; Greenberg et al., 1986; Gilman. 1988). The SRE mediates promoter stimulation not only by serum but also by a variety of other growth factors (Table 8-1), and can confer inducibility to heterologous promoters (Treisman, 1985). Since the SRE represents the major regulatory element in the c-*fos* promoter, it is discussed in greater detail below.

ELEMENTS FOR OTHER INDUCTION PATHWAYS

Stimulation of c-*fos* by phorbol esters (e.g., TPA), ionizing radiation, and UV irradiation is also mediated by the SRE (Table 8-1) (Büscher et al., 1988; Siegfried and Ziff, 1989; Datta et al., 1992). Similarly, this element is responsible for generating the response to treatment with cycloheximide (Greenberg et al., 1986; Subramaniam et al., 1989; König et al., 1989). On the other hand. membrane depolarization, as induced in neuronally differentiated PC12 cells by K⁺ influx, results in c-*fos* stimulation via the cAMP response elements, mainly the –60 element (Greenberg et al., 1986). This is also true for the cholinergic agonist nicotine (Sheng et al., 1988). These effects can be explained by increased intracellular Ca²⁺ upon membrane depolarization. In undifferentiated PC12 cells the same elements mediate transcriptional stimulation with elevated K⁺, treatment

TABLE 8-1   *c-fos* Regulators Acting Through the SRE

Inducers	Ref.
Serum	Büscher et al., 1988; Treisman, 1986; Siegfried and Ziff, 1989
PDGF	Büscher et al., 1988; Siegfried and Ziff, 1989
TPA	Büscher et al., 1988; Siegfried and Ziff, 1989; Fisch et al., 1987
cAMP	Berkowitz et al., 1989; Fisch et al., 1987 (however: Gilman, 1988)
Ca^{2+}	Fisch et al., 1987
UV	Büscher et al., 1988
Insulin	Stumpo et al., 1988
NGF	Visvader et al., 1988
EGF	Malik et al., 1991; Fisch et al., 1987
Cycloheximide	Subramaniam et al., 1989
v-*Raf*	Kaibuchi et al., 1989; Jamal and Ziff, 1990
v-*abl*	Hori et al., 1990
PKC	Gilman, 1988; Graham and Gilman, 1991
MAPK	Gille et al., 1992
Tax 1 (HTLV-1)	Fujii et al., 1992
Oxidant (H_2O_2)	Nose et al., 1991
Antioxidant (PDTC)	Meyer et al., 1993
IL-2	Hatakeyama et al., 1992
IL-3	Hatakeyama et al., 1992
EPO	Hatakeyama et al., 1992
Inhibitors	
Fos	Subramaniam et al., 1989; König et al., 1989; Sassone-Corsi et al., 1988; Wilson and Treisman, 1988; Schönthal et al., 1988; Schönthal et al., 1989; Gius et al., 1990; Ofir et al., 1990; Lucibello et al., 1989; Leung and Miyamoto, 1989; Shaw et al., 1989; Rivera et al., 1990
FRA	Gius et al., 1990
Retinoic acid	Busam et al., 1992

with the Ca^{2+} channel agonist BAY K8644, or Ba^{2+} (Morgan and Curran, 1986; Curran and Morgan, 1986; Sheng et al., 1986).

THE c-*fos* SERUM RESPONSE ELEMENT

SRE

The core of the SRE is generated by the palindromic DNA sequence $CC(A/T)_6GG$ (or CArG box). The central importance of the SRE is evidenced by the broad spectrum of signaling agents (Table 8-1), including serum, that exert their effect via this element. This indicates that either several signaling pathways converge at the same functional point on the SRE or the SRE is a multifunctional structure able to receive different signals. Consistent with the latter possibility, mutations across the SRE showed different responses to PKC-dependent and -independent activation (Gilman. 1988; Graham and Gilman, 1991). Many different proteins have been identified to specifically interact with this sequence (Treisman, 1992), which offers the possibility for regulation of the SRE in various cell types. SREs are important in the regulatory regions of many other IEGs, notably EGR-1 (Sukhatme et al., 1988), and the CArG core element is an important promoter element of muscle-specific genes (Treisman, 1990). In addition to activation, the SRE also mediates autoregulatory shutoff by c-Fos protein (see below).

SRF

The SRE is specifically recognized and stably bound by a homodimer of SRF, a ubiquitous nuclear phosphoprotein (Treisman, 1986, 1987; Prywes and Roeder, 1987; Schröter et al., 1987; Norman et al., 1988; Prywes et al., 1988; Misra et al., 1991). The crucial role fulfilled by SRF for SRE-dependent *c-fos* induction has been amply demonstrated by mutational analysis (Treisman, 1986; Greenberg et al., 1987; Gilman, 1988) and by microinjection of anti-SRF antibodies (Gauthier-Rouvière et al., 1991). The SRF gene belongs to a multigene family of related transcription factors (Norman et al., 1988; Pollock and Treisman, 1991). An internal SRF domain, the MADS box (Schwarz-Sommer et al., 1990), shares strong homology with transcription factors found in *Saccharomyces cerevisiae* (MCM1, ARG80) (Dolan and Fields, 1991) and plants (*Deficiens, Agamous, Globosa*) (Sommer et al., 1990; Yanofsky et al., 1990; Tröbner et al., 1992). The internal region spanning SRF residues 132–222, which includes the MADS homology, encodes the three functions of dimerization, specific DNA binding, and interaction with ternary complex factors (Norman et al., 1988; Schröter et al., 1990; Mueller and Nordheim, 1991). Phosphorylation of SRF at up to five serine residues located aminoterminal to this internal DNA binding domain enhances the exchange rate of SRF with the SRE but has little affect on the overall DNA binding affinity (Marais et al., 1992; Janknecht et al., 1992). SRF is also modified posttranslationally by glycosylation (Schröter et al., 1990; Reason et al., 1992), but no function for this alteration has yet been found. Further details about SRF can be found in recent reviews (Treisman and Ammerer, 1992; Treisman, 1993).

SRE-DEPENDENT TERNARY COMPLEX

The binary SRE–SRF complex interacts with another binding activity termed ternary complex factor (TCF), thereby generating a stable ternary complex (Shaw et al., 1989b; Schröter et al., 1990; see below). The binding pattern seen in genomic footprinting studies is consistent with the ternary complex forming *in vivo* (Herrera et al., 1989; König, 1991), and it has been postulated to represent a target within the c-*fos* promoter for the signal transduction cascade (Shaw et al., 1989b).

TERNARY COMPLEX FORMATION AT THE c-*fos* SRE: INTERACTIONS BETWEEN SRF AND ETS PROTEINS

IDENTIFICATION AND CHARACTERIZATION OF TCFs

The identification of p62^{TCF} (Shaw et al., 1989b; Schröter et al., 1990) and the demonstration of its ability to form a ternary complex with the SRE–SRF binary complex provided new insight into both the structural and functional complexity of the SRE. The functional importance of ternary complex formation *in vivo* was indicated by mutational analysis of the TCF contact site at the 5' end of the SRE (Shaw et al., 1989b; Graham and Gilman, 1991) and by genomic footprinting (Herrera et al., 1989; König, 1991). TCFs by themselves cannot bind to the SRE with high affinity, but require interaction with DNA-bound SRF. Thus, they make both protein–protein and protein–DNA contacts (Shaw et al., 1989b; Shaw, 1992). The cloning by Dalton and Treisman (1992) of SRF-interacting factors led to the identification of TCFs as members of the Ets family of transcription factors. To date the family of TCFs comprises cDNA clones encoding Elk-1 (Hipskind et al., 1991), SAP-1a, SAP-1b, and SAP-2 (Dalton and Treisman, 1992), and ERP-1 (T. Libermann, personal communication). Comparing Elk-1, SAP-1a, and SAP-1b, the greatest resemblance to $p62^{TCF}$ is displayed by Elk-1 (Pingoud et al., submitted). All TCFs share three regions of homology (Dalton and Treisman, 1992), which include the amino-terminal Ets domain (region A), the SRF-interaction domain (region B) (Janknecht and Nordheim, 1992; Treisman et al., 1992; Rao and Reddy, 1992), and a putative transactivation domain (region C). Homology region B also acts to inhibit autonomous DNA binding by the Ets domain (Treisman et al., 1992; Rao and Reddy, 1992).

FUNCTIONAL MODEL FOR TERNARY COMPLEXES AT THE SRE

Genomic footprinting of the c-*fos* SRE in a variety of cells (Herrera et al., 1989; König, 1991) provided evidence for ternary complex formation *in vivo* and also revealed continuous occupancy of the SRE during the entire course of the c-*fos* transcriptional activation/repression cycle (Herrera et al., 1989). In the yeast *S. cerevisiae*, ternary complexes assembled by the SRF homologue Mcm1 are critical in the regulation of pheromone-induced gene activation and repression (Dolan and Fields, 1991). Many members of the signal transduction pathway have been conserved between yeast and mammals (Dolan and Fields, 1991; Kosako et al., 1993). Similarly, an important role may be postulated for SRF–TCF interactions at the c-*fos* SRE during signal-dependent c-*fos* regulation. Both c-*fos* induction and repression (see below) appear to require the SRE. Phosphorylation and dephosphorylation of SRE-interacting proteins may be induced upon signal-triggered stimulation of kinases and phosphatases, thereby leading to c-*fos* activation and repression (Figure 8-3). In fact, we have been able to show a transient phosphorylation of TCF that closely parallels c-*fos* promoter activity (Zinck et al., 1993). This modification did not alter the ability of the SRF-dependent ternary complex to form, and suggests, together with the genomic footprinting results, that this transient phosphorylation may occur *in situ* on the SRE-bound transcription factors. Alternative models of an inducible formation of ternary complexes have also been proposed (Malik et al., 1991; Gille et al., 1992). Regardless of which model is correct, such modifications are likely to affect the protein conformations of the participating factors, thereby leading to stimulation or inhibition of the general transcriptional machinery.

**MITOGENIC SIGNAL CASCADES
TARGETING THE c-*fos* PROMOTER**

The identification and characterization of separate regulatory *cis*-elements in the c-*fos* promoter has aided the definition of signal transduction pathways that target c-*fos*, namely the pathways triggered by PDGF, cAMP, Ca^{2+}, phorbol esters, and serum growth factors. Except for the SIE, the target promoter elements appear to be constitutively occupied *in vivo*. This suggests that various pathways may target DNA-bound proteins *in situ*. Although protein kinases participating in these signaling cascades are being identified, in no instance has a complete pathway been delineated, nor has the actual activation mechanism for transcriptional stimulation of RNA polymerase II been elucidated.

PDGF-TRIGGERED PATHWAY

Signaling by PDGF probably targets the SRE, as do other serum growth factors (see below), and the factors binding to the SIE sequence at -346 (Hayes et al., 1987). The latter mechanism is the only one identified so far that leads to an induced binding of a

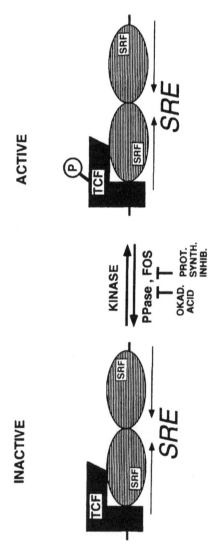

Figure 8-3 Ternary complexes involving the *c-fos* SRE at the inactive and active transcriptional states of the promoter. The model shown indicates occupancy of the SRE by SRF and TCF before, during, and after transcriptional stimulation (Herrera et al., 1989) and emphasizes different degrees of TCF phosphorylation during the *c-fos* induction cycle (Zinck et al., 1993). The action of kinases and phosphatases is thereby implied. The role of Fos in transrepression is indicated, as are the effects of inhibitors of phosphatases (okadaic acid) and protein synthesis. Further description is provided in the text.

regulatory protein to the c-*fos* promoter *in vitro* or *in vivo* (Hayes et al., 1987; Herrera et al., 1989). Nuclear translocation of the SIE factor involved may be a direct consequence of its phosphorylation by a cytoplasmic protein tyrosine kinase (M. Gilman, personal communication).

cAMP-STIMULATED PATHWAY
Increasing the cytoplasmic levels of cAMP leads to the activation and translocation of the catalytic subunit of protein kinase A (PKA) to the nucleus. This mediates the activation of the transcription factor CREB upon phosphorylation at Ser133. Although this scenario has been characterized most fully in the model system of somatostatin gene induction (Gonzalez et al., 1991), it bears striking similarity to the c-*fos* promoter, where the major CRE is located at promoter coordinate –60. This simple model is considerably complicated by the wide spectrum of CRE-interacting proteins (Ziff, 1990).

Ca^{2+}-STIMULATED PATHWAY
Membrane depolarization of PC12 cells by K$^+$, leading to elevated intracellular concentrations of Ca^{2+}, stimulates Ca^{2+}-calmodulin-dependent protein kinases (CaM kinase). CaM kinases I and II were shown to phosphorylate CREB at the same Ser residue modified by cAMP-activated protein kinase A (Sheng et al., 1991). This finding provides a rationale for the observation that cAMP- and Ca^{2+}-activated signaling pathways converge at the CRE of the c-*fos* promoter, thereby implicating proteins of the CREB family as integrating Ca^{2+} and cAMP signals.

PATHWAYS INVOLVING ACTIVATED MEMBRANE-BOUND RECEPTORS
The major target within the c-*fos* promoter for growth factor signaling is the SRE. Growth factor signals are transmitted through interaction with the corresponding receptors, leading to the activation of both PKC-dependent and -independent pathways. Signaling pathways that lead to activation of MAP kinase (MAPK) (Leevers and Marshall, 1992) have been postulated to stimulate c-*fos* transcription via the SRE (Gille et al., 1992). v-*raf*-stimulated signal transduction also belongs to this category (Kaibuchi et al., 1989; Jamal and Ziff, 1990), as might v-*abl*-mediated (Hori et al., 1990) and v-*fps*-mediated (Alexandropoulos et al., 1992) signaling. The proteins bound to the SRE, namely SRF and TCF, represent the likely targets of the incoming, SRE-directed signals.

Mutations interfering with the ability of SRF to interact with the SRE block PKC-dependent and -independent stimulation (Gilman, 1988; Graham and Gilman, 1991). Although an increase in SRF binding was reported at induction (Prywes and Roeder, 1986), this has not proven to be reproducible. However, increased phosphorylation of SRF was observed upon EGF stimulation of virus-transformed mouse embryo fibroblasts (Schalasta and Doppler, 1990). This effect could be blocked by interfering with diacylglycerol formation and thereby indirectly inhibiting protein kinase C (PKC). Induced phosphorylation within SRF has yet to be mapped and it is, therefore, unclear whether it involves the serines identified previously (Marais et al., 1992; Janknecht et al., 1992).

TCFs have been shown to contribute to the efficiency of the serum response (Shaw et al., 1989b). It also has been argued that PKC-dependent pathways require TCF interaction at the SRE, whereas PKC-independent pathways may not (Graham and Gilman, 1991). Our own observation that induced phosphorylation of TCF occurs in tight temporal correlation with the activity of the c-*fos* promoter, provides strong evidence for the involvement of TCFs in the mechanism of promoter induction (Zinck et al., 1993). This finding, together with the observation that TCFs are efficient substrates for MAPK (Gille et al., 1992), suggests that the TCF component of the SRE ternary complex represents a major mitogen signal entry point within the c-*fos* promoter.

DOWNREGULATION OF c-*fos* TRANSCRIPTION

The transient nature of c-*fos* transcriptional activity is a characteristic feature shared by most immediate early genes, which ensures their expression at a narrow time interval at the G_0 to G_1 transition. Rapid transcriptional silencing of the c-*fos* gene to the low prestimulation levels is seen within 30 min after promoter induction. This down regulation must function at two levels: first, maintenance of basal repression, and second, reversal of c-*fos* induction ("de-induction") after mitogen stimulation. Inhibition of protein synthesis relieves basal repression (Subramaniam et al., 1989) and superinduces c-*fos* after serum stimulation (Greenberg et al., 1986; Subramaniam et al., 1989). The latter reflects interference with postinductional repression, since it leads to prolonged promoter activity. This shut-off involves the autoregulatory effect of c-Fos, as determined in transient transfection studies (Sassone-Corsi et al., 1988; Wilson and Treisman, 1988; Schönthal et al., 1988, 1989). The carboxy terminal protein domain plays a crucial role in this transrepressing activity of Fos (Sassone-Corsi et al., 1988; Wilson and Treisman, 1988; Gius et al., 1990) and requires at least the last 27 amino acids (Wilson and Treisman, 1988; Gius et al., 1990). These C-terminal sequences and their functions are conserved in Fos-related proteins, particularly Fra-1, and are sufficient for mediating downregulation, since fusing the Fos C-terminus to an unrelated protein endows it with autoregulatory potential (Gius et al., 1990).

The C-terminal domain of Fos is phosphorylated on several sites, and this modification appears to be essential for autoregulatory shut-off (Wilson and Treisman, 1988; Ofir et al., 1990). In addition, C-terminal Fos mutants that remove phosphorylation sites and display increased protein stability are defective in autorepression. This type of alteration is also found in transforming v-*fos*. Interestingly, Fos mutants that prevent downregulation are dominant (Wilson and Treisman, 1988), and this effect requires replacement of the C-terminus with other protein sequences rather than its removal (Wilson and Treisman, 1988). Fos mutants impaired for heterodimerization (Gius et al., 1990) or DNA binding (Lucibello et al., 1889; Gius et al., 1990) can still transrepress, although a role has been postulated for Fos heterodimerization in autorepression (Schönthal et al., 1989; Lucibello et al., 1989).

Several reports indicate that Fos transrepression is effected through the SRE (Sassone-Corsi et al., 1988; Subramaniam et al., 1989; König et al., 1989; Lucibello et al., 1989; Leung and Miyamoto, 1989; Shaw et al., 1989a; Gius et al., 1990; Rivera et al., 1990). It appears unlikely that a direct interaction of Fos with SRE DNA sequences is responsible for this effect, since no binding of Fos to the SRE or its associated proteins has been observed, and Fos mutants exist that are defective in DNA binding while fully competent for autoregulation (Lucibello et al., 1989; Rivera et al., 1990). Although the mechanism is still unclear, some evidence implicates SRF binding as crucial for Fos autoregulation (Leung and Miyamoto, 1989; Shaw et al., 1989a; Rivera et al., 1990).

Recent results indicate TCF activity is affected conformationally by phosphorylation events that correlate temporally with c-*fos* promoter activity (Zinck et al., 1993). This finding would explain the requirement for SRE–SRF interactions in Fos-mediated downregulation and provides new clues to understanding this process. The affects of okadaic acid in blocking c-*fos* downregulation (Schönthal et al., 1991) can now be understood in this context by implicating deregulated TCF phosphorylation. This suggests the direct involvement of protein phosphatase activity in c-*fos* downregulation (Figure 8-3).

In addition to the SRE, other *cis*-elements in the fos promoter have been postulated to mediate autoregulatory transrepression, including AP-1-like binding sites at –295 and –60 (Wilson and Treisman, 1988; Schönthal et al., 1989; König et al., 1989). Furthermore, c-*fos* transcription may be repressed by nonautoregulatory pathways, possibly involving p105 Rb (Robbins et al., 1990), CREM (Foulkes et al., 1991), or a block to transcriptional elongation (Lamb et al., 1990). The physiological roles played by these latter mechanisms still must be determined.

OUTLOOK

We have seen considerable progress in the understanding of c-*fos* transcriptional regulation in the last several years. Recent advances in the identification and cDNA cloning of new c-*fos* regulatory proteins, such as SRF and TCF (Elk-1, SAP-1), will permit the elucidation of SRE-dependent and -independent molecular mechanisms by which c-*fos* transcription can be regulated upon mitogenic signaling. Evidence now suggests that the SRE ternary complex represents a final target of activated signal transduction cascades. Consistent with this, the protein components of the ternary complex, namely TCFs and possibly SRF itself, appear to be direct substrates for signaling-dependent phosphorylation and dephosphorylation. Once the regulatory role played by these modification events has been analyzed, this should unravel how such transient changes stimulate RNA polymerase II transcription from distant regulatory elements. The development of *in vitro* transcription systems (Norman et al., 1988; Prywes et al., 1988; Hipskind and Nordheim, 1991a, 1991b) will aid in the dissection and reconstruction of functional aspects of these regulatory mechanisms. A molecular understanding of eukaryotic signal-regulated gene activity can thus be derived.

ACKNOWLEDGMENT

F. Stewart provided the original northern blot shown in Figure 8-1. We thank A. Borchert for secretarial help in the preparation of this manuscript and M. Cahill for comments. The work in the authors' laboratory is funded by the DFG (No. 120/7-2) and the Fonds der Chemischen Industrie.

REFERENCES

Alexandropoulos, K., Qureshi, S. A., Rim, M., Sukhatme, V. P., and Forster, D. A., v-Fps-responsiveness in the Egr-1 promoter is mediated by serum response elements, *Nucleic Acids Res.*, 20, 2355, 1992.

Almendral, J. M., Sommer, D., Macdonald-Bravo, H., Burckhardt, J. H., Perera, J., and Bravo, R., Complexity of the early genetic response to growth factors in mouse fibroblasts, *Mol. Cell. Biol.*, 8, 2140, 1988.

Angel, P. and Karin, M., The role of Jun, Fos and the AP-1 complex in cell-proliferation and transformation, *Biochim. Biophys. Acta*, 1072, 129, 1991.

Berkowitz, L. A., Riabowol, K. T., and Gilman, M. Z., Multiple sequence elements of a single functional class are required for cyclic AMP responsiveness of the mouse c-*fos* promoter, *Mol. Cell. Biol.*, 9, 4272, 1989.

Bravo, R., Genes induced during the G_0/G_1 transition in mouse fibroblasts, *Semin. Cancer Biol.*, 1, 37, 1990.

Busam, K. J., Roberts, A. B., and Sporn, M. B., Inhibition of mitogen-induced c-*fos* expression in melanoma cells by retinoic acid involves the serum response element, *J. Biol. Chem.*, 267, 19971, 1992.

Büscher, M., Rahmsdorf, H. J., Litfin, M., Karin, M., and Herrlich, P., Activation of the c-*fos* gene by UV and phorbol ester: Different signal transduction pathways converge to the same enhancer element, *Oncogene*, 3, 301, 1988.

Cohen, D. R. and Curran, T., The structure and function of the *fos* proto-oncogene, *CRC Criti. Rev. Oncogenesis*, 1, 65, 1989.

Colotta, F., Polentarutti, N., Sironi, M., and Montovani, A., Expression and involvement of c-fos and c-jun protooncogenes in programmed cell death induced by growth factor deprivation in lymphoid cell lines, *J. Biol. Chem.*, 267, 18278, 1992.

Curran, T. and Morgan, J. I., Barium modulates c-*fos* expression and post-translational modification, *Proc. Natl. Acad. Sci. U.S.A.*, 83, 8521, 1986.

Dalton, S. and Treisman, R., Characterization of SAP-1, a protein recruited by serum response factor to the c-*fos* serum response element, *Cell*, 68, 597, 1992.

Datta, R., Rubin, E., Sukhatme, V., Qureshi, S., Hallahan, D., Weichselbaum, R. R., and Kufe, D. W., Ionizing radiation activates transcription of the EGR1 gene via CArG elements, *Proc. Natl. Acad. Sci. U.S.A.*, 89, 10149, 1992.

Dolan, J. W. and Fields, S., Cell-type-specific transcription in yeast, *Biochim. Biophys. Acta*, 1088, 155, 1991.

Fisch, T. M., Prywes, R., and Roeder, R. G., c-*fos* sequences required for basal expression and induction by EGF, TPA and Ca^{2+} ionophore, *Mol. Cell. Biol.*, 7, 3490, 1987.

Fisch, T. M., Prywes, R., and Roeder, R. G., Multiple sequence elements in the c-*fos* 5′ flanking region mediate induction by cAMP. *Genes Dev.*, 2, 391, 1988.

Foulkes, N. S., Laoide, B. M., Schlotter, F., and Sassone-Corsi, P., Transcriptional antagonist cAMP-responsive element modulator (CREM) down-regulates c-fos cAMP-induced expression, *Proc. Natl. Acad. Sci. U.S.A.*, 88, 5448, 1991.

Fujii, M., Tsuchiya, H., Chuhjo, T., Akizawa, T., and Seiki, M., Interaction of HTLV-1 Tax 1 with p67[SRF] causes the aberrant induction of cellular immediate early genes through CArG boxes, *Genes Dev.*, 6, 2066, 1992.

Gauthier-Rouvière, C., Basset, M., Blanchard, J.-M., Cavadore, J.-C., Fernandez, A., and Lamb, N. J. C., Casein kinase II induces c-fos expression via the serum response element pathway and p67[SRF] phosphorylation in living fibroblasts, *EMBO J.*, 10, 2921, 1991.

Gille, H., Sharrocks, A. D., and Shaw, P. E., Phosphorylation of transcription factor p62[TCF] by MAP kinase stimulates ternary complex formation at c-*fos* promoter, *Nature*, 358, 414, 1992.

Gilman, M. Z., The c-*fos* serum response element responds to protein kinase C-dependent and -independent signals but not to cyclic AMP, *Genes Dev.*, 2, 394, 1988.

Gilman, M. Z., Wilson, R. N., and Weinberg, R. A., Multiple protein-binding sites in the 5′-flanking region regulate c-*fos* expression, *Mol. Cell. Biol.*, 6, 4305, 1986.

Gius, D., Cao, X., Rauscher, F. J., III, Cohen, D. R., Curran, T., and Sukhatme, V. P., Transcriptional activation and repression by Fos are independent functions: The C terminus represses immediate-early gene expression via CArG elements, *Mol. Cell. Biol.*, 10, 4243, 1990.

Gonzalez, G. A., Menzel, P., Leonard, J., Fischer, W. H., and Montminy, M. R., Characterization of motifs which are critical for activity of the cyclic AMP-responsive transcription factor CREB, *Mol. Cell. Biol.*, 11, 1306, 1991.

Graham, R. and Gilman, M., Distinct protein targets for signals acting at the c-*fos* serum response element, *Science*, 251, 189, 1991.

Greenberg, M. E. and Ziff, E. B., Stimulation of 3T3 cells induces transcription of the c-*fos* proto-oncogene, *Nature*, 311, 433, 1984.

Greenberg, M. E., Hermanowski, A. L., and Ziff, E. B., Effect of protein synthesis inhibitors on growth factor activation of c-*fos*, c-*myc*, and actin gene transcription, *Mol. Cell. Biol.*, 6, 1050, 1986a.

Greenberg, M. E., Ziff, E. B., and Green, L. A., Stimulation of neuronal acetylcholine receptors induces rapid gene transcription, *Science*, 234, 80, 1986b.

Greenberg, M. E., Siegfried, Z., and Ziff, E. B., Mutation of the c-*fos* gene dyad symmetry element inhibits serum inducibility of transcription *in vivo* and the nuclear regulatory factor binding *in vitro*, *Mol. Cell. Biol.*, 7, 1217, 1987.

Härtig, E., Loncarevic, I. F., Büscher, M., Herrlich, P., and Rahmsdorf, H. J., A new cAMP response element in the transcribed region of the human c-*fos* gene, *Nucleic Acids Res.*, 19, 4153, 1991.

Hatakeyama, M., Kawahara, A., Mori, H., Shibuya, H., and Taniguchi, T., c-*fos* gene induction by interleukin 2: Identification of the critical cytoplasmic regions within the interleukin 2 receptor β chain, *Proc. Natl. Acad. Sci. U.S.A.*, 89, 2022, 1992.

Hayes, T. E., Kitchen, A. M., and Cochran, B. H., Inducible binding of a factor to the c-*fos* regulatory region, *Proc. Natl. Acad. Sci. U.S.A.*, 84, 1272, 1987.

Herrera, R. E., Shaw, P. E., and Nordheim, A., Occupation of the c-*fos* serum response element *in vivo* by a multi-protein complex is unaltered by growth factor induction, *Nature*, 340, 68, 1989.

Hipskind, R. A. and Nordheim, A., Functional dissection *in vitro* of the human c-*fos* promoter, *J. Biol. Chem.*, 266, 19583, 1991a.

Hipskind, R. A. and Nordheim, A., *In vitro* transcriptional analysis of the human c-*fos* proto-oncogene, *J. Biol. Chem.*, 266, 19572, 1991b.

Hipskind, R. A., Rao, V. N., Mueller, C. G. F., Reddy, E. S. P., and Nordheim, A., The Ets-related protein Elk-1 is homologous to the c-*fos* regulatory factor p62^TCF, *Nature*, 354, 531, 1991.

Hori, Y., Kaibuchi, K., Fukumoto, Y., Oku, N., and Takai, Y., Activation of the serum-response and TPA-response elements by expression of the v-*abl* protein: Comparison of the mode of action of the v-*abl* protein with those of protein kinase C, cyclic AMP-dependent protein kinase, and the activated c-*raf* protein, *Oncogene*, 5, 1201, 1990.

Jamal, S. and Ziff, E., Transactivation of c-*fos* and β-actin genes by *raf* as a step in early response to transmembrane signals, *Nature*, 344, 463, 1990.

Janknecht, R. and Nordheim, A., Elk-1 protein domains required for direct and SRF-assisted DNA-binding, *Nucleic Acids Res.*, 20, 3317, 1992.

Janknecht, R., Hipskind, R. A., Houthaeve, T., Nordheim, A., and Stunnenberg, H. G., Identification of multiple SRF N-terminal phosphorylation sites affecting DNA binding properties, *EMBO J.*, 11, 1045, 1992.

Johnson, R. S., Spiegelman, B. M., and Papaloannou, V., Pleiotrophic effects of a null mutation in the c-*fos* proto-oncogene, *Cell*, 71, 577, 1992.

Kaibuchi, K., Fukumoto, Y., Oku, N., Hori, Y., Yamamoto, T., Toyoshima, K., and Takai, Y., Activation of the serum response element and 12-*O*-tetradecanoylphorbol-13-acetate response element by the activated c-*raf*-1 protein in a manner independent of protein kinase C, *J. Biol. Chem.*, 264, 20855, 1989.

König, H., Cell-type specific multiprotein complex formation over the c-*fos* serum response element *in vivo*: Ternary complex formation is not required for the induction of c-*fos*, *Nucleic Acids Res.*, 19, 3607, 1991.

König, H., Ponta, H., Rahmsdorf, U., Büscher, M., Schönthal, A., Rahmsdorf, H. J., and Herrlich, P., Autoregulation of *fos*: The dyad symmetry element as the major target of repression, *EMBO J.*, 8, 2559, 1989.

Kosako, H., Nishida, E., and Gotoh, Y., cDNA cloning of MAP kinase kinase reveals kinase cascade pathways in yeast to vertebrates, *EMBO J.*, 12, 787, 1993.

Kruijer, W., Cooper, J. A., Hunter, T., and Verma, I. M., Platelet-derived growth factor induces rapid but transient expression of the c-*fos* gene and protein, *Nature*, 312, 711, 1984.

Lamb, N. J. C., Fernandez, A., Tourkine, N., Jeanteur, P., and Blanchard, J.-M., Demonstration in living cells of an intragenic negative regulatory element within the rodent c-*fos* gene, *Cell*, 61, 485, 1990.

Lau. L. F. and Nathans. D.. Expression of a set of growth-related immediate-early genes in BALB/c 3T3 cells: Coordinate regulation with c-*fos* or c-*myc*, *Proc. Natl. Acad. Sci. U.S.A.*, 84. 1182. 1987.

Leevers. S. J. and Marshall. C. J.. MAP kinase regulation — the oncogene connection. *Trends Cell Biol.*, 2. 283. 1992.

Leung. S. and Miyamoto. N. G.. Point mutational analysis of the human c-*fos* serum response factor binding site. *Nucleic Acids. Res.*, 17. 1177. 1989.

Lucibello. F. C.. Lowag. C.. Neuberg. M., and Müller. R.. Trans-repression of the mouse c-*fos* promoter: A novel mechanism of Fos-mediated trans-regulation. *Cell*, 59. 999. 1989.

Lucibello. F. C.. Ehlert. F., and Müller. R.. Multiple interdependent regulatory sites in the mouse c-*fos* promoter determine basal level transcription:Cell type-specific effects. *Nucleic Acids Res.*. 19. 3583. 1991.

Malik. R. K.. Roe. M. W., and Blackshear. P. J.. Epidermal growth factor and other mitogens induce binding of a protein complex to the c-*fos* serum response element in human astrocytoma and other cells. *J. Biol. Chem.*, 266. 8576. 1991.

Marais. R. M.. Hsuan. J. J.. McGuigan. C.. Wynne. J.. and Treisman. R.. Casein kinase II phosphorylation increases the rate of serum response factor-binding site exchange. *EMBO J.*, 11. 97. 1992.

Metz. R. and Ziff. E.. cAMP stimulates the C/EBP-related transcription factor rNFIL-6 to translocate to the nucleus and induce c-*fos* transcription. *Genes Dev.*, 5. 1754. 1991.

Meyer. M.. Schreck. R.. and Bauerle. P. A.. H_2O_2 and antioxidants have opposite effects on activation of NF-κB and AP-1 in intact cells: AP-1 as secondary antioxidant-responsive factor. *EMBO J.*, 12. 2005. 1993.

Misra. R. P.. Rivera. V. M.. Wang. J. M.. Fan. P.-D.. and Greenberg. M. E.. The serum response factor is extensively modified by phosphorylation following its synthesis in serum-stimulated fibroblasts. *Mol. Cell. Biol.*, 11. 4545. 1991.

Morgan. J. I. and Curran. T.. Role of ion flux in the control of gene expression. *Nature*, 322. 552. 1986.

Morgan. J. I.. Cohen. D. R.. Hempstead. J. L.. and Curran. T.. Mapping patterns of c-fos expression in the central nervous system after seizure. *Science*, 237. 192. 1987.

Mueller. C. G. F. and Nordheim. A.. A protein domain conserved between yeast MCM1 and human SRF directs ternary complex formation. *EMBO J.*, 10. 4219. 1991.

Müller. R.. Bravo. R.. Burckhardt. J.. and Curran. T.. Induction of c-*fos* gene and protein by growth factors precedes activation of c-*myc*, *Nature*, 312. 716. 1984.

Norman. C.. Runswick. M.. Pollock. R.. and Treisman. R.. Isolation and properties of cDNA clones encoding SRF. a transcription factor that binds to the c-*fos* serum response element. *Cell*, 55. 989. 1988.

Nose. K.. Shibanuma. M.. Kikuchi. K.. Kageyama. H.. Sakiyama. S.. and Kuroki. T.. Transcriptional activation of early-response genes by hydrogen peroxide in a mouse osteoblastic cell line. *Eur. J. Biochem.*, 201. 99. 1991.

Ofir. R.. Dwarki. V. J.. Rashid. D.. and Verma. I. M.. Phosphorylation of the C terminus of Fos protein is required for transcriptional transrepression of the c-*fos* promoter. *Nature*, 348. 80. 1990.

Papavassiliou. A. G.. Treier. M.. Chavrier. C.. and Bohmann. D.. Targeted degradation of c-Fos. but not v-Fos. by a phosphorylation-dependent signal on c-Jun. *Science*, 258. 1941. 1992.

Pingaud. V.. Zinck. R.. Hipskind. R. A.. Janknecht. R.. and Nordheim. A.. Heterogeneity of ternary complex factors in hela cell nuclear extracts. submitted.

Pollock. R. and Treisman. R.. Human SRF-related proteins: DNA-binding properties and potential regulatory targets. *Genes Dev.*, 5. 2327. 1991.

Prywes, R. and Roeder. R. G.. Inducible binding of a factor to the c-*fos* enhancer. *Cell*, 47, 777, 1986.

Prywes, R. and Roeder. R. G.. Purification of the c-*fos* enhancer-binding protein. *Mol. Cell. Biol.*, 7, 3482, 1987.

Prywes, R., Dutta, A., Cromlish. J. A.. and Roeder. R. G.. Phosphorylation of serum response factor, a factor that binds to the serum response element of the c-fos enhancer. *Proc. Natl. Acad. Sci. U.S.A.*, 85, 7206. 1988a.

Prywes, R., Fisch. R. M.. and Roeder. R. G.. Transcriptional regulation of c-*fos*, *Cold Spring Harbor Symp. Quant. Biol.*, LIII. 739. 1988b.

Rao, V. N. and Reddy. E. S. P.. *elk*-1 domains responsible for autonomous DNA binding, SRE:SRF interaction and negative regulation of DNA binding. *Oncogene*, 7, 2335. 1992.

Reason. A. J., Morris, H. R., Panico. M., Marais, R., Treisman. R. H., Haltiwanger. R. S., Hart, G. W., Kelly, W. G., and Dell. A.. Localisation of O-GlcNAc modification on the serum response transcription factor, *J. Biol. Chem.*, 267, 16911, 1992.

Rivera. V. M. and Greenberg, M. E.. The ups and downs of fos regulation. *New Biol.*, 2, 751, 1990.

Rivera, V. M., Sheng. M.. and Greenberg. M. E.. The inner core of the serum response element mediates both the rapid induction and subsequent repression of c-*fos* transcription following serum stimulation. *Genes Dev.*, 4, 255, 1990.

Robbins, P. D., Horowitz, J. M., and Mulligan. R. C.. Negative regulation of human c-*fos* expression by the retinoblastoma gene product. *Nature*, 346, 668, 1990.

Runkel, L., Shaw, P. E., Herrera. R. E.. Hipskind. R. A.. and Nordheim. A.. Multiple basal promoter elements determine the level of human c-*fos* transcription. *Mol. Cell. Biol.*, 11, 1270, 1991.

Sassone-Corsi. P., Sisson. J. C.. and Verma. I. M., Transcriptional autoregulation of the proto-oncogene *fos*, *Nature*, 334, 314, 1988a.

Sassone-Corsi, P., Visvader. J., Ferland. L., Mellon. P. L.. and Verma. I. M.. Induction of proto-oncogene fos transcription through the adenylate cyclase pathway: Characterization of a cAMP-responsive element. *Genes Dev.*, 2, 1529. 1988b.

Schalasta. G. and Doppler. C.. Inhibition of c-*fos* transcription and phosphorylation of the serum response factor by an inhibitor of phospholipase C-type reactions. *Mol. Cell. Biol.*, 10, 5558, 1990.

Schönthal, A., Herrlich, P., Rahmsdorf, H. J., and Ponta, H.. Requirement for *fos* gene expression in the transcriptional activation of collagenase by other oncogenes and phorbol esters. *Cell*, 54, 325, 1988.

Schönthal, A., Büscher, M., Angel, P., Rahmsdorf, H. J., Ponta, H., Hattori, K., Chiu, R., Karin, M., and Herrlich, P., The Fos and Jun/AP-1 proteins are involved in the downregulation of Fos transcription. *Oncogene*, 4, 629, 1989.

Schönthal, A., Tsukitani, Y.. and Feramisco, J. R.. Transcriptional and post-transcriptional regulation of c-*fos* expression by the tumor promoter okadaic acid. *Oncogene*, 6, 423, 1991.

Schröter, H., Shaw, P. E., and Nordheim, A.. Purification of intercalator-released p67, a polypeptide that interacts specifically with the c-*fos* serum response element, *Nucleic Acids Res.*, 15, 10145, 1987.

Schröter, H., Mueller, C. G. F., Meese, K., and Nordheim, A.. Synergism in ternary complex formation between the dimeric glycoprotein p67^{SRF}, polypeptide p62^{TCF} and the c-*fos* serum response element. *EMBO J.*, 9, 1123, 1990.

Schwarz-Sommer, Z., Huijser, P., Nacken, W., Saedler, H., and Sommer, H., Genetic control of flower development by homeoric genes in *Antirrhinum majus*, *Science*, 250, 931, 1990.

Shaw, P. E., Ternary complex formation over the c-*fos* serum response element p62^{TCF} exhibits dual component specificity with contacts to DNA and an extended structure in the DNA-binding domain of p67^{SRF}, *EMBO J.*, 11, 3011, 1992.

Shaw, P. E., Frasch, S. and Nordheim, A., Repression of c-*fos* transcription is mediated through p67^{SRF} bound to the SRE. *EMBO J.*, 8. 2567, 1989a.

Shaw, P. E., Schröter, H., and Nordheim, A., The ability of a ternary complex to form over the serum response element correlates with serum inducibility of the human c-*fos* promoter, *Cell*, 56, 563, 1989b.

Sheng, M. and Greenberg, M. E., The regulation and function of c-*fos* and other immediate early genes in the nervous system, *Neuron*, 4, 477, 1990.

Sheng, M., Dougan, S. T., McFadden, G., and Greenberg, M. E., Calcium and growth factor pathways of c-*fos* transcriptional activation require distinct upstream regulatory sequences, *Mol. Cell. Biol.*, 8, 2787, 1988.

Sheng, M., McFadden, G., and Greenberg, M. E., Membrane depolarization and calcium induce c-*fos* transcription via phosphorylation of transcription factor CREB, *Neuron*, 4, 571, 1990.

Sheng, M., Thompson, M. A., and Greenberg, M. E., CREB: A Ca^{2+}-regulated transcription factor phosphorylated by calmodulin-dependent kinases, *Science,* 252, 1427, 1991.

Siegfried, Z. and Ziff, E. B., Transcription activation by serum, PDGF, and TPA through the c-*fos* DSE: Cell type specific requirements for induction, *Oncogene*, 4, 3, 1989.

Sommer, H., Beltran, J.-P., Huijser, O., Pape, H., Lönnig, W.-E., Saedler, H., and Schwarz-Sommer, Z., *Deficiens*, a homeotic gene involved in the control of flower morphogenesis in *Antirrhinum majus:* The protein shows homology to transcription factors, *EMBO J.*, 9, 605, 1990.

Stewart, A. F., Herrera, R. E., and Nordheim, A., Rapid induction of c-*fos* transcription reveals quantitative linkage of RNA polymerase II and DNA topoisomerase I enzyme activities, *Cell*, 60, 141, 1990.

Stumpo, D., Stewart, T. N., Gilman, M. Z., and Blackshear, P. J., Identification of c-*fos* sequences involved in induction by insulin and phorbol esters, *J. Biol. Chem.*, 263, 1611, 1988.

Subramaniam, M., Schmidt, L. J., Crutchfield, C. E., III, and Getz, M. J., Negative regulation of serum-responsive enhancer elements, *Nature*, 340, 64, 1989.

Sukhatme, V. P., Cao, X., Chang, L. C., Tsai-Morris, C.-H., Stamenkovich, D., Ferreira, P. C. P., Cohen, D. R., Edwards, S. A., Shows, T. B., Curran, T., Le Beau, M. M., and Adamson E. D., A zinc finger-encoding gene coregulated with c-*fos* during growth and differentiation, and after cellular depolarization, *Cell*, 53, 37, 1988.

Treisman, R., Transient accumulation of c-*fos* RNA following serum stimulation requires a conserved 5′element and c-*fos* 3′sequences, *Cell*, 42, 889, 1985.

Treisman, R., Identification of a protein-binding site that mediates transcriptional response of the c-*fos* gene to serum factors, *Cell*, 46, 567, 1986.

Treisman, R., Identification and purification of a polypeptide that binds to the c-*fos* serum response element, *EMBO J.*, 6, 2711, 1987.

Treisman, R., The SRE: A growth factor responsive transcriptional regulator, *Semin. Cancer Biol.*, 1, 47, 1990.

Treisman, R., The serum response element, *TIBS*, 17, 423, 1992.

Treisman, R., Structure and function of serum response factor, in *Transcriptional Regulation*, Cold Spring Harbor Laboratory Press, 881, 1992.

Treisman, R. and Ammerer, G., The SRF and MCM1 transcription factors, *Curr. Opinion Gen. Dev.*, 2, 221, 1992.

Treisman, R., Marais, R., and Wynne, J., Spatial flexibility in ternary complexes between SRF and its accessory proteins, *EMBO J.*, 11, 4631, 1992.

Tröbner, W., Ramirez, L., Motte P., Hue, I., Huijser, P., Lönnig, W.-E., Saedler, H., Sommer, H., and Schwarz-Sommer, S., GLOBOSA: A homeotic gene which interacts with DEFICIENS in the control of *Antirrhinum* floral organogenesis, *EMBO J.*, 11, 4693, 1992.

Visvader, J., Sassonne-Corsi, P., and Verma, I. M., Two adjacent promoter elements mediate nerve growth factor activation of the c-*fos* gene and bind different nuclear factors, *Proc. Natl. Acad. Sci. U.S.A.*, 85, 9474, 1988.

Wagner, B. J., Hayes, T. E., Hoban, C. J., and Cochran, B. H., The SIF binding element confers *sis*/PDGF inducibility onto the c-*fos* promoter, *EMBO J.*, 9, 4477, 1990.

Wang, Z.-Q., Ovitt, C., Grigoriadis, A. E., Möhle-Steinlein, U., Rüther, U., and Wagner, E. F., Bone and haematopoietic defects in mice lacking c-fos, *Nature*, 360, 741, 1992.

Wilson, T. and Treisman, R., Fos C-terminal mutations block down-regulation of c-*fos* transcription following serum stimulation, *EMBO J.*, 7, 4193, 1988.

Yanofsky, M. F., Ma, H., Bowman, J. L., Drews, G. N., Feldmann, K. A., and Meyerowitz, E. M., The protein encoded by the *Arabidopsis* homeotic gene *agamous* resembles transcription factors, *Nature*, 346, 35, 1990.

Ziff, E. B., Transcription factors: A new family gathers at the cAMP response site, *TIG*, 6, 69, 1990.

Zinck, R., Hipskind, R. A., Pingoud, V., and Nordheim, A., c-*fos* transcriptional activation and repression correlate temporally with the phosphorylation status of TCF, *EMBO J.*, 12, 2377, 1993.

Chapter 9

Structure and Regulation of the *c-jun* Promoter

F. Mechta and M. Yaniv

CONTENTS

INTRODUCTION

Cell proliferation is regulated by interaction of extracellular growth factors with specific receptors present on the cell surface. In the absence of mitogens, cells cease to proliferate and enter a quiescent state, called G_0. Reexposure to growth factors reverses this arrest, causes cells to reenter the cycle, and induces DNA synthesis 8 to 16 hours later. The mitogenic stimulus provided by extracellular growth factors binding to their receptors is transduced through a cascade of second messengers, protein kinases, and phosphatases from the cell surface to the nucleus. As ultimate targets, nuclear proto-oncogenes play a central role in the mitogenic signal transduction pathway. Their implication in cell proliferation was first postulated during a study of the *c-myc* proto-oncogene, whose increased transcription is observed rapidly in response of cells to mitogens (Stiles, 1985). This rapid and transient increased expression after growth stimulation of cells defines a new class of genes, the immediate early genes (IEGs) or competence genes (Lau and Nathans, 1987; Almendral et al., 1988). The *c-jun* proto-oncogene, the cellular homologue of the avian *v-jun* gene, was first identified in murine fibroblasts as a member of this family of genes (Ryseck et al., 1988; Quantin and Breathnach, 1988). Many IEGs, including *c-jun*, encode transcription factors that trigger the expression of genes further downstream whose products are involved more directly with cell proliferation, such as remodeling the extracellular matrix and preparing the cell for DNA synthesis. This second class of genes, including proteases like transin or collagenase and heavy metal binding proteins such as metallothionein, is stimulated more slowly but with prolonged duration after mitogenic stimuli (Angel and Karin, 1991; Rahmsdorf and Herrlich, 1991). The kinetics of induction of these genes are delayed but follow the kinetics of IEG induction, assuring the long-term transcriptional response of cells to growth factors. Proto-oncogenes such as *c-jun* or *c-fos* are also key players in cell differentiation. Finally, they can be induced in transformed cells by oncogenes exerting their affect at the cell membrane or in the cytoplasm (Angel and Karin, 1991).

To better understand the involvement of the *jun* gene family in cellular proliferation, differentiation, or neoplastic transformation, many laboratories have studied the regulation of *c-jun* expression in response to distinct growth factors, differentiation inducers, or transforming agents.

0-8493-4573-1/94/$0.00+$.50
© 1994 by CRC Press. Inc.

REGULATION OF *c-jun* GENE EXPRESSION

A nonexhaustive summary of agents and events that are linked to *c-jun* transcriptional activation is given in Table 9-1.

As mentioned above, *c-jun* activation is frequently associated with mitogenic stimulation. For example, addition of epidermal growth factor (EGF) to rat1 fibroblasts (Quantin and Breathnach, 1988), serum or TPA to mouse 3T3Balb/c or NIH3T3 fibroblasts (Ryseck et al., 1988) or human HeLa or HepG2 cells (Angel et al., 1988), and platelet derived growth factor (PDGF) to C3H10T1/2 fibroblasts (Zwiller et al., 1991), all result in increased abundance of Jun. Exposure of adult rat hepatocytes to transforming growth factor α (TGFα), (Brenner et al., 1989a), or of dog thyrocytes to thyrotropin (TSH) (Reuse et al., 1991), also leading to activation of *c-jun* transcription. Furthermore, stimulation of *c-jun* gene expression is associated with the mitogenic action of tumor necrosis factor α (TNFα) in human AF2 fibroblasts. This lymphokine is secreted by macrophages in response to inflammation, infection, or neoplasia and induces collagenase production. Increased collagenase expression is mediated by the AP-1 transcription factor and results from prolonged *c-jun* transcription (Brenner et al., 1989a, 1989b). Moreover, following partial hepatectomy, normally quiescent hepatic cells reenter the cell cycle; in response to this mitogenic signal, *c-jun* expression is highly induced (Sobczack et al., 1989; Morello et al., 1990; Alcorn et al., 1990).

Enhanced *c-jun* gene expression is not always directly correlated with induction of cell proliferation. For instance, the multifunctional polypeptide transforming growth factor β (TGFβ) enhances *c-jun* production in different cell types, which either stimulates growth (mouse embryo fibroblasts, AKR-2B), inhibits growth (human lung adenocarcinoma, A549) or does not affect it (human erythroleukemia, K562) (Pertovaara et al., 1989). Moreover, cytokines like interleukin 6 (IL6), interferons α and β, and TGFβ1 inhibit M1 myeloblastic cell growth but differentially regulate *c-jun* expression. Furthermore, our work with derivatives of hamster CCL39 fibroblasts show that certain effectors can stimulate an immediate early response, including *c-jun* induction, although this is not sufficient to bring the cells to S phase. This point is well illustrated in a recent study comparing muscarinic acetylcholine (mAch) and thrombin receptors in human 1321N1 astrocytoma cells. Thrombin is mitogenic in these cells, whereas the mAchR agonist carbachol (CCH) is not. With short exposure, both thrombin and CCH strongly stimulate *c-jun* expression. However, only thrombin induces a secondary sustained increase in *c-jun* synthesis. This biphasic induction of *c-jun* expression would be essential for efficient induction of AP-1 activity and probably for mitogenesis of 1321N1 cells (Trejo et al., 1992). These observations can explain at least some examples where early *c-jun* activation is observed in the absence of growth stimulation. However, such studies cannot explain the role of *c-jun* stimulation by agents that inhibit cell growth. In such cases, *c-jun* stimulation is often linked to the induction of a differentiation pathway.

Increased transcription of *c-jun* proto-oncogene was also observed in transformed cells. This was documented for fibroblasts transformed with v-mos, the oncoprotein of Moloney murine sarcoma virus (Schönthal and Feramisco, 1990), with polyomavirus middle-sized tumor (PmT) antigen (Schönthal et al., 1992), with the oncoprotein Ha-ras (Sistonen et al., 1989), and with the early region 1 transcription unit (E1A) of human adenovirus (van Dam et al., 1990).

The rat pheochromocytoma (PC12) cell line responds to nerve growth factor (NGF) or Ras oncoproteins by stopping cell division and acquiring the characteristics of sympathic neurons (Wu et al., 1989). Phorbol esters or granulocyte/macrophage colony stimulating factor (G/CSF) induce the differentiation of human leukemic cells U937 toward macrophages. In both cases, induction of differentiation pathways is accompanied by transient

Table 9-1 **Induction of *c-jun* Expression by Distinct Growth Factors, Transforming Oncogenes, and Differentiating Agents**

Stimulus	Agents	Cell type	Ref.
Mitosis	EGF	rat1 fibroblasts	Quantin and Breathnach, 1988
	TNFα	Human AF2 fibroblasts	Brenner et al., 1989b
	Serum, TPA	Mouse 3T3Balb/c, NIH3T3 fibroblasts	Ryseck et al., 1988
	PDGF	Mouse C3H10T1/2 fibroblasts	Zweiller et al., 1991
	TGFα	Rat hepatocytes	Brenner et al., 1989a
	TSH	Dog thyrocytes	Reuse et al., 1991
	TGFβ	Mouse embryo fibroblasts, AKR-2B	Pertovaara et al., 1989
	Hepatectomy	Hepatocytes	Sobczack et al., 1989; Morello et al., 1990; Alcorn et al., 1990
Transformation	v-mos	Mouse NIH3T3 fibroblasts	Schönthal et al., 1992
	PymT		Sistonen et al., 1989
	Ha-ras		van Dam et al., 1990
	E1A		Wu et al., 1989
Differentiation	IL6	M1 myeloblastic cells → macrophages	Trejo et al., 1992
	NGF	Rat pheochromocytoma (PC12) → neurons	Unlap et al., 1992
	Phorbol esters, G/CSF	Human leukemia U397 → macrophages	Hass et al., 1991
	RA	Embryonal stem cells P19 → mesoderm and endoderm	Yang-Yen et al., 1990
	E1A	Embryonal stem cells F9 → endoderm	Yamaguchi-Iwai et al., 1990

Note: Many agents that induce mitosis, transformation, or differentiation of various cell types lead to the transcriptional activation and increased expression of *c-jun*, as a primary genetic response to the activated signal transduction pathway. Abbreviations used are as follows: EGF, epidermal growth factor; TNFα, tumor necrosis factor α; TPA, 12-*O*-tetradecanoyl phorbol-13-acetate; PDGF, platelet-derived growth factor; TGF, transforming growth factor; TSH, thyrotropin; v-mos, the transforming oncoprotein of Moloney murine sarcoma virus; PymT, the polyomavirus middle-sized tumor antigen; Ha-ras, the oncoprotein Harvey-ras; E1A, the early region 1 of the human adenovirus transcription unit; NGF, nerve growth factor; G/CSF, granulocyte/macrophage colony stimulating factor; and RA, retinoic acid.

activation of *c-jun* expression (Unlap et al., 1992). The glucocorticoid dexamethasone inhibits the TPA-mediated increase of c-Jun-dependent collagenase transcription (Offringa et al., 1990) and of *c-jun* transcription and monocytic differentiation of U937 leukemia cells (Hass et al., 1991). Transcription of *c-jun* is also increased after differentiation of P19 malignant teratocarcinoma emoryonal stem cells (EC) by retinoic acid (RA) (de Groot et al., 1990). Moreover, the constitutive expression of *c-jun* alone causes P19 cells to acquire a differentiated phenotype, which suggests that *c-jun* has an important role in induction of this differentiation process and perhaps in early mammalian development. A similar induction of *c-jun* is observed in F9 teratocarcinoma cells upon exposure to RA (Yang-Yen et al., 1990; Yamaguchi-Iwai et al., 1990) or by expression of adenovirus E1A (Kitabayashi et al., 1991).

Exposure of mammalian cells to DNA damaging agents such as ultraviolet (UV) irradiation, H_2O_2, high temperatures, or heavy metal ions leads to activation of a genetic response, which is thought to protect the exposed cells against permanent damage. It has been shown that the UV response includes rapid and preferential stimulation of *c-jun* transcription (Devary et al., 1991; Stein et al., 1992). Another case involves the antimitotic DNA damaging agent 1-β-D arabinofuranosylcytosine (Ara-C), often used in cancer therapy. This nucleotide analogue prevents DNA replication of myeloid leukemia cells but also stimulates the transcription of *c-jun* (Brach et al., 1992b).

While the kinetics and amplitude of *c-jun* induction vary among the different stimuli, increased *c-jun* expression is a recurrent phenomenon in many distinct signal transduction pathways. However, only in a limited number of cases have attempts been made to attribute a direct functional role to the stimulation of *c-jun* synthesis. Kovary and Bravo (1991) have shown that injection of anti-c-Jun antibodies blocks exit from G_0 and transition through the S phase of Swiss 3T3 fibroblasts. These results together with the finding that overexpression of native c-Jun transforms chicken embryo fibroblasts (Castellazzi et al., 1991), supports the notion that c-Jun has a direct role in the control of cell growth. This is further supported by the observation that a truncated c-Jun, which lacks its transactivation domain, slows normal cell growth, probably by acting as a negative dominant transcription factor (Castellazzi et al., 1991). All these experiments strongly suggest that increased c-Jun activity is essential at least for fibroblast growth.

STRUCTURE OF *c-jun* PROMOTERS ISOLATED FROM DIFFERENT MAMMALIAN SPECIES

Hattori et al. (1988), in an attempt to define the precise targets of the various stimuli discussed above, isolated the human *c-jun* gene and its 5′ flanking region (31) (Figure 9-1). Curiously, *c-jun*, *jun-B*, and *jun-D* contain no introns in their respective coding sequences.

DNase I footprinting experiments using HeLa whole-cell extracts led to the identification of several protected sites located within the 5′ transcription control region of the human *c-jun* gene (Angel et al., 1988). Comparison of these sites with the consensus sequences specifically recognized by different DNA binding proteins revealed two variant nonconventional TATA box sequences ($5'_{-34}AGATAAG_{-28}3'$ and $5'_{-58}TATTTTTA_{-52}3'$). They are 29 and 52 bp, respectively, upstream of the major transcriptional start site (determined by RNase protection experiments and denoted by +1 in Figure 1) surrounded by a cluster of transcription initiation sites. It is plausible that the first TATA-like sequence (−29 bp) directs transcription from the major start site (+1 bp), whereas the second TATA-like sequence (−52) promotes transcription from two minor start sites, at positions −17 and −21 bp.

The upstream TATA-like sequence (−52) is overlapped by an RSRF binding site: $5'_{-59}CTATTTTTAG_{-50}3'$, first identified in the murine *c-jun* promoter and perfectly conserved in the human sequence (Han et al., 1992).

A short sequence located upstream of the distal TATA sequences (between positions −72 and −63 bp) deviates from the consensus AP-1 binding site (TGACTCA) by the insertion of a single nucleotide ($5'_{-72}GTGAC\underline{A}TCAT_{-63}3'$). *In vitro* DNA binding experiments confirmed that this sequence is in fact recognized by the AP-1 transcription factor

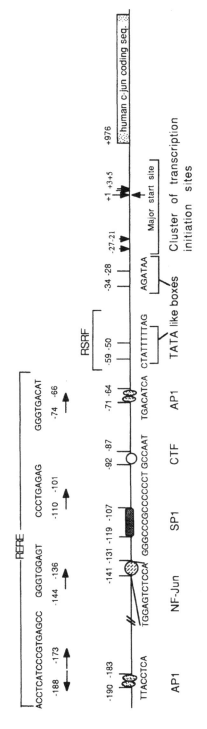

Figure 9-1 Structure of the human *c-jun* promoter. The first to be cloned and studied, the human *c-jun* promoter presents a number of different specific protein recognition sequences: two variant TATA-like boxes, two potential AP-1 binding sites, and consensus CAAT and GC motifs known, respectively, as CTF and SP-1 binding sites. A related serum responsive factor (RSRF) binding site overlaps the upper TATA-like box. This sequence, identified in the murine *c-jun* promoter, is perfectly conserved in the human promoter sequence. The study of the *c-jun* promoter leads to the identification of two new recognition sequences, NF-Jun and RERE motifs. Nuclear factor-Jun (NF-Jun) is a leukemia-specific factor that induces *c-jun* transcription in response to hematopoietic differentiating agents. The RERE defines the RA- and E1A-mediated *c-jun* induction in embryonal stem cells and consists of 5 variants of the same motif (represented by the short arrows).

and functions as a high-affinity AP-1 binding site (Angel et al., 1988). Another potential AP-1-like sequence was identified further upstream ($5'_{-190}$TTACCTCA$_{-183}3'$) (van Dam et al., 1990; Stein et al., 1992). Upstream of the first AP-1 recognition sequence, the human *c-jun* promoter contains consensus CAAT ($5'_{-91}$GCAAT$_{-87}3'$) and GC ($5'_{-123}$GGGCGGGCCCGCCCC$_{-109}3'$) boxes that correspond to the target sites for the transcription factors CTF/NF-1 and SP-1, respectively (Angel et al., 1988).

Partial sequences of the *c-jun* promoter were also determined for the rat (Kitabayashi et al., 1990), mouse (de Groot et al., 1991), and the chicken genes (Nishimura and Vogt, 1988). When aligned with the human *c-jun* promoter, a high degree of homology can be observed (Figure 9-2). The general organization of these promoters appears to be similar. Interestingly, the exact sequences and positions of the transcription factor binding sites described above are highly conserved or even identical between the human, mouse, and rat sequences. The chicken sequence is more variable but still contains many sites discussed for the human *c-jun* promoter. Following identification, these highly conserved DNA binding sites were analyzed as potential targets of the various stimuli described previously.

TARGET SEQUENCES INVOLVED IN THE REGULATION OF THE *c-jun* PROMOTERS

Several recent studies have attempted to identify the DNA elements that regulate the *c-jun* response to distinct stimuli. As we shall see, the exact results depend on the length of the promoter fragment analyzed, the nature of the reporter sequences (e.g., free or integrated into cellular DNA), and the cell type studied. First, the role of the AP-1 binding site (−72/−63) in *c-jun* regulation was first investigated (Angel et al., 1988). The presence of this site in regulatory regions of *c-jun* genes first suggested that AP-1 might regulate both basal and serum/TPA-induced expression of *c-jun*. Transfection experiments with short, truncated promoter clones into HeLa cells showed that this site is a key element for induction of human *c-jun* transcription after stimulation of cells by serum or TPA (Angel et al., 1988). Two different mechanisms lead to the transcriptional activation of *c-jun* in TPA-stimulated human fibroblasts. First, after TPA treatment, the preexisting c-Jun protein, expressed at the low but significant basal level, is rapidly dephosphorylated on specific serine and threonine residues located directly upstream of the basic DNA binding domain (Boyle et al., 1991; see also Chapter 5). Since *in vitro* phosphorylation of these residues reduces the DNA binding activity of Jun, these dephosphorylations may lead to an increase in the DNA binding activity of c-Jun (Boyle et al., 1991). However, whereas this scenario may take place at the AP-1 binding site of the Jun-dependent collagenase gene (which shows enhanced AP-1 binding after TPA stimulation of cells), *in vivo* footprinting analysis reveals complete binding of factors to both the distal and proximal AP-1 binding sites (Rozek and Pfeifer, 1993; Herr et al., 1994) already in nonstimulated cells. The activity of these factors, which have been identified as predominantly heterodimers of c-Jun and ATF-2 (van Dam et al., 1993; Herr et al., 1994), is likely regulated by a second type of change in the phosphorylation pattern of c-Jun; in response to TPA and growth factors, the phosphorylation of Jun within the transactivation domain is enhanced, resulting in an enhancement of the transactivation function of c-Jun (Chapter 5). However, c-Jun homodimers, which are potent transactivators (Angel et al., 1988; Bohmann and Tjian, 1989), are able to bind to the *c-jun* promoter, resulting in autoinduction of *c-jun* transcription (Angel et al., 1988).

The positive autoregulatory loop of AP-1 on the *c-jun* promoter is an attractive model for signal amplification and conversion of transient early events, generated at the cell surface by receptor stimulation, into longer term affects on gene expression (Angel et al.,

1988). This positive autoregulation seems to play an important role in different pathways by amplifying magnitude and/or timing of responses to an initial signal. This mechanism may not occur in all cells, however. Experiments in our laboratory with NIH3T3 cells have shown that only a limited increase (less than twofold) in AP-1 DNA binding activity occurs in the absence of protein synthesis (Spyrou, unpublished results), while a 10 to 20-fold increase in c-jun mRNA is observed under these conditions. A further rapid 10 to 20-fold increase of AP-1 activity requires de novo protein synthesis. It is possible that in these cells other nuclear factors are essential for the initial induction of c-jun transcription.

The −72/−63 AP-1 site can also activate c-jun expression by cellular or viral oncogenes. The initial studies by Wasylyk et al. (1988) showed that transcriptional activity of the polyoma enhancer, which contains an AP-1 site (first called PEA1), is strongly activated after serum addition or transformation by c-Ha-ras. As mentioned previously, overexpression of oncogenes like v-mos in mouse fibroblasts leads to the transcriptional activation of c-jun. Overexpression of PymT antigen also stimulates transcription of c-jun, which, in turn, activates the transcription and replication of the polyoma virus enhancer (Wasylyk et al., 1988; Martin et al., 1988; Muramaki et al., 1991). Analysis of deleted human c-jun promoter constructs shows that positive regulation of c-jun expression by v-mos or PymT is mediated by transcriptional activation through the −72/−63 AP-1 binding site (Schönthal and Feramisco, 1990; Schönthal et al., 1992).

This proximal AP-1 site is also essential for c-jun activation by DNA damaging agents such as UV irradiation or H_2O_2. Induction of c-jun expression by these stresses on the cell is stronger than for any other genes, suggesting a central role for c-jun in mediating a DNA repair response. In fact, UV irradiation increases c-jun transcription even more than does TPA (Devary et al., 1991). The response of a human c-jun promoter construct containing only 132 bp of upstream sequences is similar to that of the endogenous c-jun gene. Mutational inactivation of the AP-1 binding site at position −72/−63 completely abolishes its response to UV (Devary et al., 1991). However, another group defined a second independent UV response element, which corresponds to the second potential AP-1 binding site at position −190/−183 in the human c-jun promoter (Stein et al., 1992). UV treatment of HeLa cells increases the binding of transcription factors to both elements in vitro. However, as in the case of TPA, in vivo footprinting analysis does not reveal differences in the binding activity of Jun–ATF-2 heterodimers of untreated and UV-irradiated cells (Rozek and Pfeifer, 1993; Herr et al., 1994).

The DNA damaging agent Ara-C also induces c-jun transcription in the human myeloid leukemic cell line, KG1. In both the presence and absence of de novo protein synthesis, the Ara-C target element corresponds to the −72/−63 AP-1 binding site. Moreover, as in TPA-stimulated fibroblasts, enhancement of AP-1 activity involves posttranslational modifications. The interaction of activated AP-1 molecules with their targets confers Ara-C inducibility of the human c-jun gene (Brach et al., 1992b).

The −72/−63 AP-1 site is not essential for EGF induction of c-jun in HeLa cells. Deletion analysis of the mouse c-jun promoter reveals that the related serum responsive factor (RSRF) (binding site, which overlaps the distal potential TATA box sequence (−52), is an essential element for EGF stimulation of the promoter. Thus, c-jun AP-1 site does not seems to be required for EGF induction, and mediates only a weak response of c-jun to this agent (Han et al., 1992).

The addition of phorbol esters such as TPA to U937 human leukemia cells induces their differentiation toward macrophages. Initial events in this differentiation process include activation of c-jun gene expression (Unlap et al., 1992). To understand the exact mechanism of phorbol ester-mediated c-jun stimulation in this cell line, specific DNA recognition sites within the c-jun promoter were mutated. Interestingly, the effects of

```
HUMAN   GCATCCCCCGCATCCCCGTC......CGCAGCCGTGAACTTGAGCCCCCCTC......CATCAGAGGTTGCGAGCGTCGCCG..CTCGGC..AGCCACCGTCACTAGACAGTCAAACCC  -334
        * ********** ** * ***      * ***** ************* ****      *  ****    ** **   *  *   * ****  **** ************** ******
MOUSE   GGATCCCCCGCGTCTCGGTCCCAGTCCCCAGCCCTGAACTTGAGCCCGCCTC......CGACAGACTCCGCAAGGTCCTGCCGACCCGGC..AGCCTCCGTCACTAGACAGCCAAACC.  -331
        ********  **********   ****************************      ******************** ** ********  ***************************
RAT     GGATCCCCG.CGTCTCGGTCTGGGTCCCCAGCCCTGAACTTGAGCCCGCCTC......CGACAGACTCCGCAAGGTCCAGCGGACCCGGC..AGCCTCCGTCACTAGACAGCCAAACC.  -331
           **  *   *   * * *  *  *******           *** * ****      ** ** **  *** * *   * * **** ***   ****  **  *     *     *  * *
CHICKEN ..GTCGACCGAGGAGCCGGCCCGC.CCCCAGC.........AGCGCACCTCGCGTGCGCGCCACACAACGC.ACGGGCGGGGACGCGGGGAGCCGGCGCGAGGGCAGCTGCGGAGCG  -395

HUMAN   CAAGACGTCAGCCC...ACAATGCACCGGGCGGGCCGGGAAAAACGGC..CCGGGGAGGGGACCGGG.AAGAGAGGGCCGAG..ACGTGCGGCAGGGGGGAGGGT..AGGAGAAAGAAGG  -218
         *************   ************************** **     ********** *** ** ** * * * *  **    * **** ******* ***   *************
MOUSE   .AAGACGTCAGCCC...ACAATGCACCGGGCGGGCCGGGAAAGACTCGTGCCGGGGAGGGAACCCGGGAACACAAGCCGAAGCTGAGCGCGGGAGGGGGGGGGAGGAGGAGAAAGAAGG  -215
         *************   **************************** ** ***************************** ***   ************* *** *** ************
RAT     .AAGACGTCAGCCC...ACAATGCACCGGGCGGGCCGGGAAAGACGCG..CCGGGGAGGGAACCCGGGAACACAAGCCGGAGCAACGCGCGGGAGGGGGAGGG.AGGTGGAGAAAGAAGG  -218
           * ** **** *   **   **** * *  ** ****    *****  ****  ***   ** * *   ** * **       * ** *    **  *    *  *  * *** **
CHICKEN CGGGGCGGCAGCGCTGCACCCGGCACGGCGGCGGGCGGGTGGTACGCGA.CCGGTAAGGAGGCCGGCGGCGACCAACCCTTCCCCGCTCGGGACGCGGTAACCAGATATTCAAAAACAATA  -276

                                        AP1                                                         NF-Jun
HUMAN   GC....CCGACTGTAGGAGGGCAGCGG.....AGCATTACCTCATCCCGTGAGCCTCCGCGGGCCCAGAGAAGAATCTT............CTAGGGTGGAGTCTCCATGGTGAC  -124
        **    ** ********** *****       ***************    ***** ***** ****************            ********  ************ ***
MOUSE   GC....CCAACTGTAGGAGCGCAGC.......AGCATTACCTCATCCGTG.AGCCTTCGCGG.CCCAGAGAAGAATCTT............CTAGGGTGAGGTCTCCATGGCGAC  -124
        **    *******************       ***************    *********** ****************            ***************************
RAT     GC....CCAACTGTAGGAGCGCAGCGG.....AGCATTACCTCATCCCGTGACCCTTCGCGGGCCCAGAGAAGAATCTT............CTAGGGTGAGGTCTCCATGGCGAC  -124
              ** **   *    * * ** *     *****  *********    ** *  *** *** * ******  ***            * * ** ** * ** *   ** **
CHICKEN CGTCAGCCCAC...AATGCCCCGGCCGCCGCCAGCATATCCTCATCCCACAAGGC.CCGCCGGCGCTGAGAAGCTTCTAGGGGCGGGCTGGCCATGGCGACGGCTGCTATGC.AC  -165

                                                                   RSRF
                            SP1                CTF           AP1          TATA                      TATA
HUMAN   GGGCGGGC.........CCGCCCCCCTGAGAGCGACGCGAGCCAATGGGAAGGCCTTGGGGTGACATCATGGGCTATTTTTAGGGGTTGACTGGTAGCAGATAAGTGT  -25
        *** ****         ************** ****** ******************* **** *********************************** *****************
MOUSE   GGGTGGGC.........CCGCCCCCCTGAGAACGACGCAAGCCAATGGGAAAGCCTCGGGGTGACATCATGGGCTATTTTTAGGGATTGACTGGTAGCAGATAAGTGT  -25
        ********         ********* ***** ****************************** **** ************************************************
RAT     GGGTGGGC.........CCGCCCCCTTGAGAGCGACGCAAGCCAATGGGAAGGCCTTGGGGTGACATCATGGGCTATTTTTAGGGATTGACTGGTAGCAGATAAGTGT  -25
         **  *  *        ** ******  *****   **  ************* ** **  ** ** ************************ *     * *** ** *********
CHICKEN CGGGCGCC AGC GGCCCGCCTCCCCCGGAGAGGCGCGGTAGCCAATGGGGAG.CCGCGGGGTGACATCATGGGCTATTTTTAGCGGGCTCCCGGTCGCTGATAAGTGA  -58
```

```
                    +1
HUMAN   TGAGCTCGGCTGGATAAGGGCTCAGAGTTGC..........ACTGAGTGTGG...CTGAACCACGCAGCCCGGAGTGGAGGTGCGCGGAGTCAGGCGAGCAGACGACAGA   +71
        ******* ***********  ******** ***       ***********   * ** **** *** ****  *   *** ** *** ***********

MOUSE   TGAGCTCAGGCTGGATAAGGACTAAGAGTTGC..........ACTGAGTGTGG...CAGAGACAGCCTGGCAGGAGAG.....CCCTCAGGCGAGCAGAGACAGA   +65
        ************************ ********        ************   ******* **** ** ******    *****************************  **

RAT     TGAGCTCAGGCTGATAAGGACTAGAGTTGC...........ACTGAGTGTG....CAGAGACTGCCTAGCTGGAGAG.....CCCTCAGGCAGACAGACAGACCGA   +65
        *  ** ** **  ** ********                 *** ** *     *  *  *** *** *  **       ***  *** *  **   ***

CHICKEN AGGCTGCACGCCCGACGGG.CTCAGA GGGCGGGCGGGCGCGTCCGCCACTCAGCCGCCCGGACTCCGCGATCAGGCGCGCGCAG......CCCGGACTCGGGGT.ACTCGCGGC   +52
                    +1
```

Figure 9-2 Sequence alignment of human, mouse, rat, and chicken *c-jun* promoters. Comparison of these sequences reveals a high degree of homology. The general structure of these 5′ flanking regions, the sequences, and the relative position of the identified binding sites (represented by bold letters) are accurately conserved. A star identifies each identical nucleotide between the 4 analyzed sequences.

these mutations are dependent on the total length of the *c-jun* untranslated region studied. The same mutations result in different effects of PMA on *c-jun* inducibility depending on whether a short (–132/+170) or long (–1639/+740) *c-jun* promoter is used. In experiments with the short segment (–132/+170), mutation of the CTF binding site has no effect, mutation of the AP-1 sites decreases only partially the TPA-mediated *c-jun* induction, while deletion of the SP-1 site markedly increases it. Double mutations of both the CTF and SP-1 binding sites render *c-jun* transcription insensitive to TPA. However, each of the single or double mutations has no effect on *c-jun* transcription from a longer promoter construct (–1639/740). In fact, several sequences between –711 and –142, upstream of the SP-1, CTF, and AP-1 sites in the human *c-jun* promoter, mediate TPA-stimulated *c-jun* transcription in human hematopoietic cells. Mutational analysis excludes the possibility that the second AP-1-like binding site (–190/–183) is responsible for TPA induction. These results suggest that an, as yet, unidentified factor(s), distinct from c-Jun–ATF-2 heterodimers binding to the distal and proximal AP-1 sites, plays an important role in PMA activation of *c-jun* transcription in human leukemic cells (Unlap et al., 1992).

A novel nuclear factor, termed nuclear factor-Jun (NF-Jun), that contributes to *c-jun* induction in human hematopoietic cells, was recently identified by DNase I footprinting analysis (Brach et al., 1992a). NF-Jun mediates TPA or TNFα *c-jun* activation and recognizes an 11 bp sequence ($5'_{-141}$TGGAGTCTCCA$_{-131}3'$) just upstream of the SP-1 site (–119/–107) in the human *c-jun* promoter (see Figure 9-1). This site is highly conserved in the mouse and rat promoter but is more divergent in chicken (see Figure 9-2). In unstimulated cells, NF-Jun is located in the cytoplasm in an inactive form; cell stimulation induces its modification, activation, and translocation from the cytoplasm to the nucleus. This activation mechanism is reminiscent of the induction process of transcription factor NF-κB (Lenardo et al., 1987; Bäuerle and Baltimore, 1988). NF-Jun expression appears to be restricted to rapidly proliferating cells, such as human myeloid hematopoietic cells, and is undetectable in lung fibroblasts, monocytes, granulocytes, and resting T lymphocytes. NF-Jun may mediate rapid induction of the *c-jun* gene in a cell type-dependent and stimulus-specific manner.

RA-induced differentiation of embryonal carcinoma stem cells, such as F9 or P19, leads to increased AP-1 activity that may be essential for the upregulation of two genes considered as differentiation markers, EndoB and tissue plasminogen activator. The regulatory region of both genes contains AP-1 sites. The increase in AP-1 activity results in part from transcriptional activation of the *c-jun* gene (de Groot et al., 1991). This induction, observed after 48 hours of RA treatment, appears as a secondary response to RA that may be mediated by RA receptor β (RARβ) induced in these cells (de Groot et al., 1990). Analysis of the human *c-jun* promoter elements reveals that this *c-jun* response to RA in F9 cells requires a functional –72/–63 AP-1 binding site (Yang-Yen et al., 1990). This site appears to cooperate with other upstream regions located between –329 and –293 bp, which binds at least five different but, as yet, uncharacterized proteins. These sites, however, do not resemble the RA responsive elements identified in other promoters, indicating a more indirect effect of RAR on *c-jun* transcription (de Groot et al., 1991).

All the previously described results were obtained by transient transfection experiments. However, the kinetics and amplitude of the endogenous RA-mediated *c-jun* induction are more accurately reproduced after stable transfection of a *c-jun* promoter/bacterial reporter gene construct into F9 cells. Indeed, RA and E1A, which both efficiently induce F9 differentiation, strongly increase expression of the stably integrated *c-jun* construct (Kitabayashi et al., 1992). Deletions, mutations and insertions of the rat *c-jun* promoter have identified a new 145-bp region between –190 and –46 that is an RA and E1A responsive element (RERE). These RA and E1A responsive sequences contain five variants of the same motif 5′–C/GC/GGTGAC/GNT–3′ (see Figure 9-1). The two upstream motifs are adjacent and divergent, creating an imperfect palindrome, one of which overlaps the distal −190/−183 AP-1 binding site; the three downstream motifs are in 35- or 36-bp intervals with the same polarity, the last site overlapping the proximal −72/−63 AP-1 binding site. These sequences are also highly conserved in the human and mouse *c-jun* promoters. Thus, RA- and E1A-mediated *c-jun* expression results from stringent sequence and spacing requirements. In fibroblasts and HeLa cells, both the proximal and distal AP-1 binding sites were found to be required for E1A-dependent *c-jun* transcription (van Dam et al., 1990, 1993), suggesting that the factors binding to these sites (c-Jun–ATF-2) are direct targets of E1A action while the other RERE elements are binding sites for factors that are affected during differention of F9 cells. The RERE sequences resemble the half site of the hormone responsive motif. Nevertheless, nuclear receptors like RARβ do not seem to be directly involved in activation of the RERE. Interestingly, these RERE sequences mediate a strong *c-jun* induction from a stably integrated *c-jun* promoter construct, but this effect is not observed from transiently transfected promoters (Kitabayashi et al., 1992). Thus, these repetitive motifs might function efficiently only in an intrachromosomal context. The same stable transfection experiments also lead to the detection of a GT repeat sequence located between −543 and −500 bp in the rat *c-jun* promoter. Deletion of this motif decreases the response of *c-jun* to RA. The GT motif gives us a potential Z-DNA structure, which could influence formation and localization of nucleosomes and thereby enhance the RERE-mediated activation of *c-jun* gene (Kitabayashi et al., 1992).

DOWNREGULATION OF *c-jun* EXPRESSION

It may be important to attenuate the initial induction of the *c-jun* gene by mitogenic or differentiation signals, since a constitutive high *c-jun* synthesis could lead to abnormal cellular proliferation. In most systems, *c-jun* induction at the transcriptional level is transient and falls off rapidly in less than 1 to 2 hours. It is obvious that the positive autoregulatory loop postulated above has to be blocked rapidly in the nucleus.

The AP-1 binding sites are potential targets for negative *c-jun* regulation. While c-Jun is an efficient transcriptional activator of its own expression, a fourfold excess of JunB over c-Jun represses c-Jun activation (Chiu et al., 1989). This suggests a potential competition between the members of the Jun family. The same mechanism may also explain the inhibitory effect of another family of transcription factors that binds to the cAMP responsive element. CREB (cAMP responsive element binding protein) and its homologues (Ruppert et al., 1992) bind to the sequence (5′-TGAC$\underline{G}$TCA-3′), which is similar to the consensus AP-1 binding site. The affinity of CREB for imperfect CRE sequences is increased upon phosphorylation by the activated catalytic subunit of protein kinase A (PKA) (Yamamoto et al., 1988; Nichols et al., 1992). *In vitro*, CREB recognizes also the −72/−63 AP-1 binding site located in the *c-jun* promoter, but it does not recognize the canonical AP-1 site (TGACTCA). Transient transfection experiments reveal that an excess of CREB protein represses serum- or TPA-induced *c-jun* transcription. However, this repression is alleviated if CREB is phosphorylated by PKA (Lamph et al., 1990). The physiological relevance of overexpression of CREB homodimers (competing for binding of the physiological regulator c-Jun–ATF-2) remains unclear since this mode of regulation contrasts the effect of elevated cAMP levels on the endogenous *c-jun* gene. Increased cAMP levels in NIH3T3 (Mechta et al., 1990), HeLa (Angel et al., 1988), or thyrocytes (Reuse et al., 1991) stimulate PKA activity and repress both basal and TPA-induced *c-jun* expression. This inhibitory effect occurs at the transcriptional level, at least in NIH3T3 (Mechta, unpublished results), but the target elements in the *c-jun* promoter have not yet been characterized.

Other elements of the *c-jun* promoter may be important for the attenuation of its activity. Deletion of the CTF and SP-1 sites in a short *c-jun* promoter increases the basal and TPA-mediated human *c-jun* expression (Angel et al., 1988). Through an unknown mechanism, CTF and SP-1 sequences would then negatively influence *c-jun* transcription.

CONCLUSIONS

c-jun gene transcription is induced in response to various stimuli, in agreement with its putative role as a switch element in the processes of cell proliferation, transformation, and differentiation. The cloning of distinct mammalian *c-jun* genomic sequences has revealed that the general organization of human, rat, mouse, and chicken *c-jun* promoters is strongly conserved. The precise localization and sequence of specific protein recognition sites are similar or even identical in the rat, mouse, and human. This high degree of homology between the *c-jun* promoters of different mammalian species helps to explain why many different stimuli result in identical *c-jun* responses in different organisms. Finally, while it is clear that the c-Jun protein is essential for fibroblast growth and that its overexpression can cause cell transformation, the exact role of c-Jun activation during cell differentiation remains to be determined.

REFERENCES

Alcorn, J. A., Feitelberg, S. P., and Brenner, D. A., Transient induction of *c-jun* during hepatic regeneration, *Hepatology*, 11(6), 909, 1990.

Almendral, J. M., Sommer, D., Bravo, M. C., Burckhardt, H., Perera, J., and Bravo, R., Complexity of the early genetic response to growth factors in mouse fibroblasts, *Mol. Cell. Biol.*, 8, 2140, 1988.

Angel, P. and Karin, M., The role of Jun, Fos and the AP-1 complex in cell-proliferation and transformation, *Biochim. Biophys. Acta*, 1072, 129, 1991.

Angel, P., Hattori, K., Smeal, T., and Karin, M., The *jun* proto-oncogene is positively auto-regulated by its product, Jun/AP-1, *Cell*, 55, 875, 1988.

Bäuerle, P. A., and Baltimore, D., IKB: A specific inhibitor of the transcriptional factor NF-κB, *Science*, 242, 540, 1988.

Bohmann, D. and Tjian, R., Biochemical analysis of transcriptional activation by Jun: Differential activity of c- and v-Jun, *Cell*, 59, 709, 1989.

Boyle, W. J., Smeal, T., Defize, L. H. K., Angel, P., Woodgett, J. R., Karin, M., and Hunter, T., Activation of protein kinase C decreases phosphorylation of c-Jun at sites that negatively regulate its DNA-binding activity, *Cell*, 64, 573, 1991.

Brach, M. A., Herrmann, F., and Kufe, D. W., Activation of the AP-1 transcription factor by arabinofuranosylcytosine, *Blood*, 79(3), 728, 1992a.

Brach, M. A., Hermann, F., Yamada, H., Bäuerle, P. A., and Kufe, D. W., Identification of NF-Jun, a novel inducible transcription factor that regulates *c-jun* gene transcription. *EMBO J.*, 11(4), 1479, 1992b.

Brenner, D. A., Koch, K., and Leffert, H. L., Transforming growth factor a stimulates proto-oncogene *c-jun* expression and a mitogenic program in primary cultures of adult rat haptocytes, *DNA*, 8(4), 279, 1989a.

Brenner, D. A., O'Hara, M., Angel, P., Chojkier, M., and Karin, M., Prolonged activation of *jun* and *collagenase* genes by tumor necrosis factor a, *Nature*, 337, 661, 1989b.

Castellazzi, M., Spyrou, G., La Vista, N., Dangy, J.-P., Piu, F., Yaniv, M., and Brun, G., Overexpression of c-jun, junB, or JunD affects cell growth differently, *Proc. Natl. Acad. Sci. U.S.A.*, 88, 8890, 1991.

Chiu, R., Angel, P., and Karin, M., Jun-B differs in its biological properties from, and is a negative regulator of, c-jun, *Cell*, 59, 979, 1989.

de Groot, R. P., Kruyt, F. A. E., Van der Saag, P. T., and Kruijer, W., Ectopic expression of *c-jun* leads to differentiation of P19 embryonal carcinoma cells, *EMBO J.*, 9(6), 1831, 1990a.

de Groot, R. P., Schoorlemmer, J., Van Genesen, S. T., and Kruijer, W., Differential expression of *jun* and *fos* genes during differentiation of mouse P19 embryonal carcinoma cells, *Nucleic Acids Res.*, 18(11), 3195, 1990.

de Groot, R. P., Pals, C., and Kruijer, W., Transcriptional control of *c-jun* by retinoic acid, *Nucleic Acids Res.*, 19(17), 1585, 1991.

Devary, Y., Gottlieb, R. A., Lau, E. F., and Karin, M., Rapid and preferential activation of the *c-jun* gene during mammalian UV response, *Mol. Cell. Biol.*, 11(5), 2804, 1991.

Han, T.-H., Lamph, W. W., and Prywes, R., Mapping of epidermal growth factor-, serum-, and phorbol Ester-Responsive sequence elements in the *c-jun* promoter, *Mol. Cell. Biol.*, 12(10), 4472, 1992.

Hass, R., Brach, M., Kharbanda, S., Giese, G., Traub, P., and Kufe, D., Inhibition of phorbol ester-induced monocytic differentiation by dexamethasone is associated with down-regulation of *c-fos* and *c-jun*, *J. Cell. Physiol.*, 149(1), 125, 1991.

Hattori, K., Angel, P., Le Beau, M. M., and Karin, M., Structure and chromosomal localization of the functional intronless human *jun* proto-oncogene, *Proc. Natl. Acad. Sci. U.S.A.*, 85, 9148, 1988.

Herr, I., vanDam, H., and Angel, P., Binding of promoter associated AP-1 is not altered during induction and subsequent repression of the *c-jun* promoter by TPA and UV irradiation. *Carcinogenesis*, 15, 1105, 1994.

Kitabayashi, I., Saka, F., Gachelin, G., and Yokoyama, K., Nucleotide sequence of rat *c-jun* protooncogene, *Nucleic Acids Res.*, 18(11), 3400, 1990.

Kitabayashi, I., Kawakami, Z., Chiu, R., Ozawa, K., Matsuoka, T., Toyoshima, S., Umesono, K., Evans, R. M., Gachelin, G., and Yokoyama, K., Transcriptional regulation of the c-jun gene by retinoic acid and E1A during differentiation of F9 cells, *EMBO J.*, 11(1), 167, 1992.

Kovary, K. and Bravo, R., Expression of different Jun and Fos proteins during the G0-to-G1 transition in mouse fibroblasts: In vitro and in vivo associations, *Mol. Cell. Biol.*, 11(5), 2451, 1991.

Lamph, W., Dwarki, V. J., Ofir, R., Montminy, M., and Verma, I. M., Negative and positive regulation by transcription factor cAMP response element-binding protein is modulated by phosphorylation. *Proc. Natl. Acad. Sci. U.S.A.*, 87, 4320, 1990.

Lau, L. F. and Nathans, D., Expression of a set of growth-related immediate early genes in Balb/c 3T3 cells: Coordinate regulation with *c-fos* or *c-myc*, *Proc. Natl. Acad. Sci. U.S.A.*, 84, 1182, 1987.

Lenardo, M., Pierce, J. W., and Baltimore, D., Protein-binding sites in Ig gene enhancers determine transcriptional activity and inducibility, *Science*, 236, 1573, 1987.

Martin, M. E., Piette, J., Yaniv, M., Tang, W. J., and Folk, W. R., Activation of the polyomavirus enhancer by a murine activator protein 1 (AP-1) homolog and two contiguous proteins, *Proc. Natl. Acad. Sci. U.S.A.*, 85, 5839, 1988.

Mechta, F., Piette, J., Hirai, S. H., and Yaniv, M., Stimulation of protein kinase C or protein kinase A mediated signal transduction pathways shows three modes of response among serum inducible genes, *New Biol.*, 1(3), 297, 1990.

Morello, D., Lavenu, A., and Babinet, C., Differential regulation and expression of *jun*, *c-fos* and *c-myc* proto-oncogenes during mouse liver regeneration and after inhibition of protein synthesis, *Oncogene*, 5, 1511, 1990.

128

Muramaki, Y., Satake, M., Yamaguchi-Iwai, Y., Sakai, M., Muramatsu, M., and Ito, Y., The nuclear proto-oncogenes *c-jun* and *c-fos* as regulators of DNA replication, *Proc. Natl. Acad. Sci. U.S.A.*, 88, 3947, 1991.

Nichols, M., Weih, F., Schmid, W., DeVack, C., Kowenz-Leutz, E., Luckow, B., Boshart, M., and Schütz, G., Phosphorylation of CREB affects its binding to high and low affinity sites: Implications for cAMP induced gene transcription, *EMBO J.*, 11(9), 3337, 1992.

Nishimura, T. and Vogt, P. K., The avian cellular homologue of the oncogene *jun*, *Oncogene*, 3, 659, 1988.

Offringa, R., Gebel, S., van Dam, H., Timmers, M., Smits, A., Zwart, R., Stein, B., Bos, J. L., van der Eb, A., and Herrlich, P., A novel function of the transforming domain of E1a: Repression of AP-1 activity, *Cell*, 62, 527, 1990.

Pertovaara, L., Sistonen, L., Bos, T. J., Vogt, P. K., and Keski-Oja, J., Enhanced *jun* gene expression is an Early Genomic Response to transforming growth factor b stimulation, *Mol. Cell. Biol.*, 9(3), 1255, 1989.

Quantin, B. and Breathnach, R., Epidermal growth factor stimulates transcription of the *c-jun* proto-oncogene in rat fibroblasts, *Nature*, 334, 538, 1988.

Reuse, S., Pirson, I., and Dumont, J. E., Differential regulation of proto-oncogenes *c-jun* and *junD* expressions by protein tyrosine kinase, protein kinase C and cAMP mitogenic pathways in dog primary thyrocytes: TSH and cAMP induce proliferation but down regulate *c-jun* expression, *Exp. Cell Res.*, 196, 210, 1991.

Rozek, D. and Pfeifer, G. P., *In vivo* protein-DNA interactions at the *c-jun* promoter: preformed complexes mediate the UV response, *Mol. Cell Biol.*, 13, 9490, 1993.

Ruppert, S., Cole, T. J., Boshart, M., Scmid, E., and Schütz, G., Multiple mRNA isoforms of the transcription activator protein CREB: Generation by alternative splicing and specific expression in primary spermatocytes, *EMBO J.*, 11(4), 1503, 1992.

Ryseck, R. P., Hirai, S. I., Yaniv, M., and Bravo, R., Transcriptional activation of *c-jun* during the G0/G1 transition in mouse fibroblasts, *Nature*, 334, 535, 1988.

Schönthal, A. and Feramisco, J. R., Different promoter elements are required for the induced expression of *c-fos* and *c-jun* proto-oncogenes by the v-mos product, *New Biol.*, 2(2), 143, 1990.

Schönthal, A., Srinivas, S., and Eckhart, W., Induction of *c-jun* proto-oncogene expression and transcription factor AP-1 activity by the polyoma virus middle-sized tumor antigen, *Proc. Natl. Acad. Sci. U.S.A.*, 89, 4972, 1992.

Sistonen, L., Höltta, E., Mäkela, T. P., Keski-Oja, J., and Atilato, K., The cellular response to induction of the p21c-Ha-ras oncoprotein includes stimulation of *Jun* gene expression, *EMBO J.*, 8(3), 815, 1989.

Sobczack, J., Mechti, N., Tournier, M. F., Blachard, J. M., and Duguet, M., *c-myc* and *c-fos* gene regulation during mouse liver regeneration, *Oncogene*, 4, 1503, 1989.

Stein, B., Angel, P., van Dam H., Ponta, H., Herrlich, P., and van der Eb, A., Ultraviolet-radiation induced *c-jun* gene transcription: Two AP-1 like binding sites mediate the response, *Photochem. Photobiol.*, 55(3), 409, 1992.

Stiles, C. D., The biological role of oncogenes — insights from platelet derived growth factor, *Cancer Res.*, 45, 5215, 1985.

Trejo, J., Chambard, J.-C., Karin, M., and Heller Brown, J., Biphasic increase in *c-jun* mRNA is required for induction of AP-1-mediated gene transcription: Differential effects of muscarinic and thrombin receptor activation, *Mol. Cell. Biol.*, 12(10), 4742, 1992.

Unlap, T., Franklin, C. C., Wagner, F., and Kraft, A. D., Upstream regions of the *c-jun* promoter regulate phorbol ester-induced transcription in U937 leukemic cells, *Nucleic Acids Res.*, 20(4), 897, 1992.

van Dam, H., Offringa, R., Meijer, I., Stein, B., Smits, A. M., Herrlich, P., Bos, J. L., and van der Eb, A. J., Differential effects of the adenovirus E1A oncogene on members of the AP-1 transcription factor family, *Mol. Cell. Biol.*, 10(11), 5857, 1990.

van Dam. H., Duyndam. M., Rottier. R.. Bosch. A., de Vries-Smits. L., Herrlich. P., Zantema. A.. Angel. P., and van der Eb. A. J.. Heterodimer formation of cJun and ATF-2 is responsible for induction of c-jun by the 243 amino acid adenovirus E1A protein. *EMBO J.*, 12, 479, 1993.

Wasylyk. C., Imler. J. L., and Wasylyk. B., Transforming but not immortalising oncogenes activate the transcription factor PEA1. *EMBO J.*, 7(8), 2475, 1988.

Wu. B.. Fodor. E. J. B.. Edwards. R. H.. and Rutter. W.. Nerve growth factor induces the proto-oncogene *c-jun* in PC12 cells, *J. Biol. Chem.*, 264(15), 9000, 1989.

Yamaguchi-Iwai. Y.. Satake. M., Murakami. Y.. Sakai. M.. Muramatsu. M.. and Ito. Y.. Differentiation of F9 embryonal carcinoma cells induced by the *c-jun* and activated *c-Ha-ras* oncogenes. *Proc. Natl. Acad. Sci. U.S.A.*, 87, 8670, 1990.

Yamamoto. K. K.. Gonzales. G. A.. Biggs, W. H.. III. and Montminy. M. R.. Phosphorylation-induced binding and transcriptional efficacy of nuclear factor CREB. *Nature*, 334. 494, 1988.

Yang-Yen. H.. Chiu. R.. and Karin. M.. Elevation of AP-1 activity during F9 cell differentiation is due to increased *c-jun* transcription. *New Biol.*, 2(4), 351, 1990.

Zwiller. J.. Sassone-Corsi. P.. Kakazu. K.. and Boynton. A. L.. Inhibition of PDGF-induced *c-jun* and *c-fos* expression by a tyrosine protein kinase inhibitor. *Oncogene*, 6, 219, 1991.

Section III

FUNCTIONS IN
PHYSIOLOGICAL PROCESSES

Current estimates of the number of putative subunits that can dimerize and participate in AP-1 activity, range between 100 and 200. In view of this enormous diversity, it is not surprising that target genes of individual AP-1 factors are largely unknown. Many questions are unresolved: What are the dimerization rules? Does dimer formation in solution automatically yield a good DNA binding AP-1 factor? It is difficult to assess the actual mix of AP-1 factors *in vivo*. From overexpression studies with truncated Jun and subsequent coprecipitation with Jun-specific antibodies, Jun homodimers seem to be the minority of stable complexes formed, at least in cultured cells. F9 embryonal carcinoma cells may represent the only exception in that few AP-1 subunits are expressed, and, in particular, no Jun or Fos. Expressing Jun upon transfection of a Jun coding expression vector leads to substantial homodimerization.

Because of these unknowns and uncertainties, a discussion of target genes and physiological function of individual AP-1 members must be fragmentary. So, too, must any discussion about the question of signal transduction to the different AP-1 factors and the process of posttranslational activation. The chapters that follow attempt to shed some light on the roles of AP-1 under both physiological and pathological conditions.

Chapter 10

Identification of AP-1-regulated genes

Meinrad Busslinger and Gabriele Bergers

CONTENTS

INTRODUCTION

As introduced in Chapter 1, the transcription factor AP-1 is composed of dimeric complexes formed between three Jun family members (c-Jun, JunB, and JunD) and four Fos family members (c-Fos, FosB, Fra-1, and Fra-2). These proteins share a homologous region containing the basic DNA binding domain and an adjacent diverization motif known as the leucine zipper. Jun proteins are able to form homodimers, while Fos proteins are only capable of heterodimeric complexes with Jun proteins. Both Jun and Fos contribute to the transactivation function of the AP-1 complex, which stimulates transcription by binding to AP-1 elements (TGA G/C TCA) in enhancer and promoter regions. AP-1 can also repress gene activity through negative protein–protein interaction with other transcription factors (Ransone and Verma, 1990; Angel and Karin, 1991; Kerppola and Curran, 1991).

The transcription factor AP-1 has been implicated in diverse cellular processes including cell proliferation, differentiation, and neuronal function (Angel and Karin, 1991). The activity of this transcription factor rapidly and transiently increases in response to extracellular signals in most cells. AP-1 is therefore thought to play a central role in signal transduction by reprogramming gene expression. Consistent with this idea is the fact that the different members of the AP-1 gene family belong to the class of immediate early genes, whose transcription is also rapidly and transiently induced in response to external signals. In particular, expression of the fos genes has been considered to be important for cell cycle entry (c-fos and fosB) and progression (fra-1 and fra-2), since transfection of fos antisense RNA constructs or injection of specific antibodies inhibited, at least partially, these processes (Holt et al., 1986; Nishikura and Murray, 1987; Riabowol et al., 1988; Kovary and Bravo, 1991; Kovary and Bravo, 1992). However, recent analysis of c-fos-deficient mice has indicated that c-fos is not required for the proliferation of most cell types (Wang et al., 1992; Johnson et al., 1992). Instead, c-fos is important for the differentiation of several distinct tissues, as mice lacking c-fos suffer from bone defects, reduced numbers of lymphoid cells, abnormal gametogenesis, and altered behavior

0-8493-4573-1/94/$0.00+$.50
© 1994 by CRC Press, Inc.

133

(Wang et al., 1992; Johnson et al., 1992). In agreement with this, overexpression of c-fos in transgenic animals results in dysregulation of bone and cartilage differentiation (Rüther et al., 1987; Wang et al., 1991) and in altered maturation of T lymphocytes (Rüther et al., 1988).

Some insight into the regulatory function of AP-1 at the molecular level has been obtained by identifying and characterizing genes that are controlled by this transcription factor. Here we briefly review the literature dealing with this topic and discuss the utility of selective Fos induction systems for identifying novel AP-1-regulated genes.

IDENTIFICATION OF AP-1-REGULATED GENES BY PROMOTER ANALYSES

The most obvious approach for identifying genes that are regulated by the transcription factor AP-1 is to locate AP-1 elements in transcriptional control regions and determine their role for gene regulation. The latter is usually achieved by mutagenizing the potential AP-1 binding site *in vitro* and analyzing the effect of the mutation on transcriptional regulation in transient transfection experiments. This approach has led to the identification of functional AP-1 sites in the regulatory regions of more than 20 different genes (Table 10-1). Note that almost half of these genes code for growth factors and hormones, while others code for metalloproteases, transcription factors, or cell type-specific gene products. Moreover, most of the identified AP-1 sites are located in the proximal promoter region of these genes or, in the case of the nerve growth factor and cytokeratin 18 genes, even in the first intron.

Apart from identifying AP-1 target genes, these promoter analyses have also uncovered different regulatory functions of AP-1. For example, expression of each gene listed in Table 10-1 is induced by specific growth factors or hormones, and in most cases the AP-1 site indicated was shown to be involved in the transcriptional response to extracellular signals. These promoter analyses, therefore, support the notion that AP-1 plays an important role as nuclear mediator of signal transduction. In fact, AP-1 binding sites are also known as TPA-responsive elements (TRE), which mediate gene activation in response to signaling through protein kinase C-dependent pathways (Angel and Karin, 1991).

In addition to mediating transcriptional induction, the AP-1 site also plays an important role in basal level expression of several of the genes shown in Table 10-1 (Comb et al., 1988; Timmers et al., 1990; Gizang and Ziff, 1990; Lee et al., 1991; Buttice et al., 1991; Kovacic and Gardner, 1992). This is not surprising, since AP-1 was originally identified as a DNA-binding activity that interacts with a basal level element (BSE1) of the human metallothionein IIA gene promoter (Lee et al., 1987a, 1987b).

One AP-1 site upstream of a minimal promoter is usually not sufficient for transcriptional stimulation (Schönthal et al., 1988; Chiu et al., 1989; Buttice et al., 1991). Efficient regulation is only seen with either artificial promoter constructs containing multiple AP-1 sites or natural promoters harboring additional cooperating elements. Cooperation between AP-1 and other transcription factors has been documented for many of the inducible genes listed in Table 10-1 (Comb et al., 1988; Wasylyk et al., 1989; Gutman and Wasylyk, 1990; Kislaukis and Dobner, 1990; Auble and Brinckerhoff, 1991; Wasylyk et al., 1991; Jain et al., 1992; Nerlov et al., 1992; Sterneck et al., 1992; listed in Table 10-1). In particular, induction of the collagenase, stromelysin, urokinase, and interleukin-2 genes, and of polyoma enhancer activity depends on cooperation between AP-1 and transcription factors of the Ets protein family (Wasylyk et al., 1989; Gutman and Wasylyk, 1990; Wasylyk et al., 1991; Auble et al., 1991; Jain et al., 1992; Nerlov et al., 1992).

The promoter of the mouse proliferin gene contains a 25-bp "composite" glucocorticoid response element that is simultaneously bound by the glucocorticoid receptor and the transcription factor AP-1. In this case, c-Jun and c-Fos serve as selectors of hormone responsiveness, since they influence the activity of the glucocorticoid receptor on this element in an opposite manner. c-Jun mediates a positive glucocorticoid response, while high levels of c-Fos repress hormone responsiveness (Diamond et al., 1990). c-Jun and c-Fos have similar opposing affects on the regulation of phosphoenolpyruvate carboxykinase (Gurney et al., 1992) and atrial natriuretic peptide genes (Kovacic and Gardner, 1992). Moreover, binding of Jun–Fos complexes to upstream sequences of the osteocalcin gene has been shown to prevent its transcriptional stimulation by retinoic acid receptors (Schüle et al., 1990; Owen et al., 1990; see also Chapter 4, this volume).

c-Jun stimulates transcription of its own gene by binding to an AP-1 site in the proximal promoter (Angel et al., 1988). This positive autoregulatory loop appears to be responsible for prolonged c-jun expression following stimulation by extracellular signals such as tumor necrosis factor-α (Brenner et al., 1989) or transforming growth factor-$\beta1$ (TGF-$\beta1$) (Kim et al., 1990). TGF-$\beta1$ is also able to activate expression of its own gene. This autoinduction is mediated by Jun/AP-1, which binds to multiple AP-1 sites in the TGF-$\beta1$ promoter (Kim et al., 1990).

SELECTIVE Fos INDUCTION SYSTEMS

Identification of AP-1-regulated genes by promoter analyses is not only time consuming, but also biased, as initial information is required to suggest transcriptional regulation by AP-1. Furthermore, the assumption that a functional AP-1 site identified in transient transfection experiments is equally important for regulation of the endogenous gene is not necessarily valid. A more general approach would be to use differential cDNA cloning strategies to directly identify endogenous genes that are regulated by AP-1 within a given cell. AP-1 is, however, only one of several transcription factors that act as nuclear mediators of signal transduction and so it is difficult, if not impossible, to directly relate changes in gene expression to any one transcription factor. For this reason it would be desirable to uncouple activation of AP-1 from other events of signal transduction.

Most cycling cells express basal levels of Jun proteins, but lack detectable c-Fos and FosB (Kovary and Bravo, 1991, 1992), indicating that the latter two proteins are limiting components of AP-1 activity under these conditions. Introducing conditional Fos activity into cells could, therefore, provide a greatly simplified system that would facilitate identification of endogenous AP-1-regulated genes. Conventional induction systems consist of promoters that are regulated by endogenous transcription factors in response to glucocorticoid hormones (Hynes et al., 1981), heat shock (Topol et al., 1985), heavy metal ions (Mayo et al., 1982), or interferon (Goodbourn et al., 1985). However, all these inducers trigger an endogenous program of gene expression and, thus, elicit pleiotropic effects within the cell. In order to avoid these complications, we have established two novel induction systems for selective activation of the transcription factor Fos/AP-1, both of which are based on the inducible properties of the estrogen receptor (ER) Figure 10-1.

Most mammalian cells do not express endogenous ER and are thus unable to respond to estrogen. Therefore, if these cells were transfected with genes encoding ER proteins, estrogen would exclusively activate those exogenous proteins. For this reason, we established a transcriptional and a posttranslational induction system based on some of the inducible properties of the hormone binding domain of the human estrogen receptor (Figure 10-1). This domain contains ligand-inducible transactivation (Webster et al., 1988) and dimerization activities (Kumar and Chambon, 1988) as well as a general

Table 10-1 Genes with Functional AP-1 Sites

Gene	Position of AP-1 Site[a]	Cell Type Tested	Ref.
Collagenase	$-72/-66$	HeLa cells, fibroblasts	Angel et al., 1987a, 1987b; Auble and Brinckerhoff, 1991
Stromelysin	$-71/-65$	PyT21, NIH 3T3, F9 cells HeLa, HepG2 cells	Kerr et al., 1988; Nicholson et al., 1990; Buttice et al., 1991
Urokinase	$-1967/-1960; -1885/-1879$	HepG2, HeLa, HT1080 cells	Nerlov et al., 1991, 1992
TGF-β1	$-371/-365$[b]	A-549 cells, F9 cells	Kim et al., 1990
Interleukin-2	$-187/-181; -153/-147$	T lymphoma LBRM, EL4 cells	Muegge et al., 1989; Serfling et al., 1989
Endothelin-1	$-108/-102$	Bovine aortic endothelial cells	Lee et al., 1991
Atrial natriuretic peptide	$-241/-235$	Atrial, ventricular cardiocytes	Kovacic and Gardner, 1992
Nerve growth factor	$+35/+41$	Mouse embryo fibroblasts	Hengerer et al., 1990
Prodynorphin	$-257/-249$	NCB20 neuroblastoma cells	Naranjo et al., 1991
Neurotensin/ neuromedin N	$-188/-182$	PC12 cells	Kislaukis and Dobner, 1990
Proenkephalin	$-92/-86$	CV-1 cells, F9 cells	Comb et al., 1988; Sonnenberg et al., 1989
Myelomonocytic growth factor	$-91/-85$	HD11 macrophages	Sterneck et al., 1992
Proliferin	$-231/-225$[b]	L cells, HeLa cells, F9 cells	Diamond et al., 1990; Mordacq and Linzer, 1989
JE	$-52/-46$	Rat-1, HeLa cells	Timmers et al., 1990
Tyrosine hydroxylase	$-205/-199$	PC12 cells, PC8b cells	Gizang and Ziff, 1990; Yoon and Chikaraishi, 1992
Phosphoenolpyruvate carboxykinase	$-285/-268; -258/-252; -90/-82$	HepG2 cells	Gurney et al., 1992
Adipocyte P2	$-120/-114$	3T3-F442A, 3T3-L1 adipocytes	Herrera et al., 1989; Yang et al., 1989

Ovalbumin	−49/−43	Chicken embryo fibroblasts	Gaub et al., 1990
Osteocalcin	−508/−502	ROS cells	Schüle et al., 1990
Keratin 18 (Endo B)	+813/+819	F9 cells	Oshima et al., 1990
β-globin LCR	Two tandem AP-1 sites[c]	Erythroid cells	Ney et al., 1990; Talbot and Grosveld, 1991
ets-1	−455/−449[b]	HeLa cells	Jorcyk et al., 1991; Majerus et al., 1992
c-jun	−71/−64	F9 cells	Angel et al., 1988
SV40 enhancer	24/30; 114/120[J]	HeLa cells, HepG2 cells	Lee et al., 1987a; Chiu et al., 1987; Mermod et al., 1988
Polyoma enhancer	5114/5120[d]	3T3 cells, Ltk⁻ cells	Piette and Yaniv, 1987; Wasylyk et al., 1989
Metallothionein IIA	−104/−98	HeLa cells	Lee et al., 1987a, 1987b; Angel et al., 1987b

[a] The numbering is relative to the transcription initiation site; [b] Additional AP-1-like elements in the promoter region; [c] An erythroid member of the AP-1 family, NF-E2, binds to the two AP-1 sites in the hypersensitive region HS II of the β-globin locus control region (LCR) (Ney et al., 1990; Talbot and Grosveld, 1991); [d] The numbering corresponds to that of the viral DNA sequences.

138

Induction systems

A transcriptional   ### B posttranslational

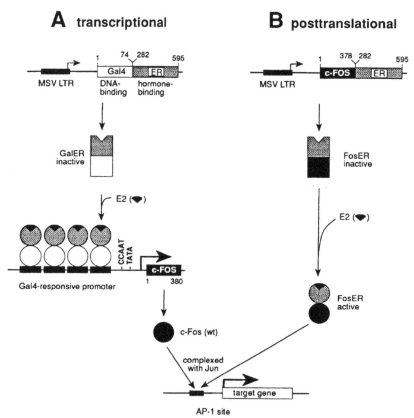

Figure 10-1 Schematic diagram of the Gal-ER and Fos-ER induction systems. The chimeric genes coding for (A) Gal-ER and (B) Fos-ER have been described by Webster et al. (1988) and Superti-Furga et al. (1991), respectively. ER, estrogen receptor; MSV LTR, Moloney murine sarcoma virus long terminal repeat; E2, β-estradiol.

"protein inactivation" function (Picard et al., 1988). These properties were used to develop a transcriptional induction system based on the hybrid transcription factor Gal-ER. This transcription factor (Webster et al., 1988) consists of the DNA binding domain (first 74 amino acids) of the yeast Gal4 protein fused to the hormone binding region of the estrogen receptor (amino acids 282–595, Figure 10-1A). Gal-ER was shown to regulate a transfected c-fos gene under the control of a Gal4-responsive promoter in a strictly estrogen-dependent manner (Braselmann et al., 1993). Moreover, rat fibroblast cell lines containing both the Gal-ER gene and the Gal4-responsive fos gene were morphologically transformed and grew in soft agar only in the presence of estrogen (Braselmann et al., 1993).

The "protein inactivation" function located in the hormone binding domain of the estrogen receptor is able to repress other activities present on the same polypeptide chain.

Table 10-2 **Fos-Regulated Genes Identified with the Fos-ER Induction System**

| Gene | Protein | Inducibility by Fos-ER in | | Ref. |
		Fibroblasts	PC12 Cells	
Fit-1	Secreted protein	Inducible	Not expressed	Superti-Furga et al., 1991; Bergers et al., 1994a
fra-1	Transcription factor	Inducible	Inducible	Braselmann et al., 1992
ODC	Biosynthetic enzyme	Not inducible	Inducible	Wrighton and Busslinger, 1993
TH	Biosynthetic enzyme	Not expressed	Inducible	Wrighton and Busslinger, 1993
Annexin II	Membrane-associated	Inducible	n.d.[a]	Braselmann et al., 1992
Annexin V	Membrane-associated	Inducible	Inducible	Braselmann et al., 1992
K18	Cytokeratin	Not expressed	Inducible	Wrighton et al., 1993

[a] n.d. = no data.

This repression results from interaction of the abundant hsp90 protein with the ligand-free hormone binding domain and is relieved by estrogen binding (Picard et al., 1988, 1990). This protein inactivation function has been successfully used to subject several transcription factors to hormonal regulation by fusing them to the ligand binding domain of the human estrogen receptor (Picard et al., 1988; Superti-Furga et al., 1991; Umek et al., 1991; Burk and Klempnauer, 1991; Boehmelt et al., 1992). As shown in Figure 10-1B, we linked the c-Fos protein at its *C* terminus to the hormone binding domain of the estrogen receptor; this Fos-ER fusion protein was then constitutively expressed from the MSV LTR promoter in transfected cells. In this way three activities of c-Fos, i.e., AP-1-dependent transactivation, repression of the c-fos promoter, and transformation of fibroblasts, were brought under hormonal control (Superti-Furga et al., 1991). In conclusion, both the transcriptional Gal-ER/Fos and the posttranslational Fos-ER induction systems are well suited for regulating Fos activity in fibroblasts (Superti-Furga et al., 1991; Braselmann et al., 1993).

IDENTIFICATION OF Fos-REGULATED GENES BY THE USE OF Fos INDUCTION SYSTEMS

THE Fos TARGET GENE Fit-1 CODES FOR A SECRETED PROTEIN

We used rat fibroblasts and pheochromocytoma (PC12) cells expressing Fos-ER to identify endogenous genes that can be regulated by Fos activity in these two cell types. The seven genes that we have so far identified in this way are listed in Table 10-2. The Fos-induced transcript 1 (Fit-1) was isolated by differential cDNA cloning from Fos-ER expressing fibroblasts (Superti-Furga et al., 1991). Fit-1 was the only cDNA that was independently isolated many times, suggesting that it represents the most abundant Fos-induced transcript in rat fibroblasts. The level of Fit-1 mRNA rapidly increases with estrogen treatment (Figure 10-2) due to transcriptional stimulation of the Fit-1 gene by Fos-ER (Superti-Furga et al., 1991; Bergers et al., 1994a). This increase in transcription rate even occurred in the presence of protein synthesis inhibitors, thus indicating a direct involvement of Fos-ER in the regulation of the Fit-1 gene (Bergers et al., 1994a).

Fit-1 is the rat homologue of the mouse T1 gene. Both genes are highly homologous not only in the gene but also in the 5′ flanking sequences, and code for proteins that are 80% identical (Klemenz et al., 1989; Tominaga et al., 1991; Bergers et al., 1994a). The

Rat1A - FosER

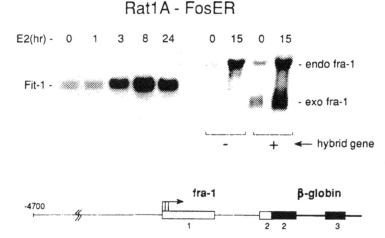

Figure 10-2 Fit-1 and fra-1 gene induction by Fos-ER. Fos-ER expressing Rat-1A fibroblasts were grown in full serum and then treated with estrogen (E2) for the indicated time, followed by northern analysis of total RNA with Fit-1 and fra-1 DNA probes. Equal RNA loading in all lanes was shown by rehybridization with a glyceraldehyde phosphate dehydrogenase gene probe. The chimeric gene shown below was transfected into Fos-ER expressing Rat-1A cells and its expression was analyzed in a cell pool of ~100 hygromycin-resistant clones (RNA signal marked as "exo fra-1").

T1 gene (also known as ST2) was shown to be inducible by serum or conditional expression of the activated Ha-ras or viral mos oncogenes (Klemenz et al., 1989; Tominaga et al., 1989). Serum stimulation of the Fit-1 gene was blocked by protein synthesis inhibitors, indicating that immediate early gene products such as c-Fos are involved in the normal regulation of this gene (Bergers et al., 1994). Fit-1 belongs therefore to the family of delayed early genes although weak expression of the Fit-1 and (T1/ST2) gene is seen during embryogenesis and in hemotopoietic tissues of the adult animal, abundant expression has so far been detected in cultured fibroblasts (Tominaga et al., 1989; Bergers et al., 1994a).

The Fit-1 and T1 proteins are members of the immunoglobulin superfamily. They are most closely related to the extracellular domain of the mouse interleukin-1 receptor and to the immunoglobulin-like repeats of the human carcinoembryonic antigen (Tominaga et al., 1989; Klemenz et al., 1989). The Fit-1 and T1 proteins are heavily glycosylated and contain a signal peptide at their N terminus, but lack a membrane anchor (Klemenz et al., 1989; Bergers et al., 1994a). In agreement with this, the Fit-1 protein is secreted by fibroblasts (Bergers et al., 1994a). Recently we have discovered that the Fit-1 gene also codes for a membrane-based protein which is homologous to the entire protein of the type I interleukin-1 receptor (see note added in proof).

THE fra-1 GENE IS POSITIVELY AUTOREGULATED BY AP-1 ACTIVITY

Two members of the AP-1 gene family, c-jun and c-fos, are known to be regulated by AP-1 activity itself (Angel et al., 1988; Schönthal et al., 1989; Sassone-Corsi et al., 1988; König et al., 1989; Gius et al., 1990). Therefore, we investigated whether induction of Fos activity in Fos-ER expressing fibroblasts affects expression of other members of the jun and fos gene families. Of all the genes tested, only the fra-1 gene was clearly upregulated by estrogen activation of Fos-ER (Figure 10-2). An increase in fra-1 mRNA synthesis was also observed in fibroblasts expressing native c-Fos protein under the control of the transcriptional Gal-ER induction system (Braselmann et al., 1993). In addition, fra-1 expression was strongly induced by estrogen in Fos-ER expressing PC12 cells, indicating that upregulation of the fra-1 gene by Fos/AP-1 appears to be a more general phenomenon (Bergers et al., submitted).

Fra-1 gene induction by Fos/AP-1 occurred at the level of transcription initiation and was strictly dependent on the presence of a functional DNA binding domain in the Fos-ER protein (Bergers et al., submitted). This evidence suggests that Fos/AP-1 directly regulates the fra-1 gene by binding to cis-regulatory sequences. To test this hypothesis, we cloned the rat fra-1 gene and linked its 5′ region in the second exon to the rabbit β-globin gene (Figure 10-2). Such a chimeric gene, containing 4700 bp of fra-1 5′ flanking sequences, was stably transfected into Rat-1A cells expressing Fos-ER, and a representative cell pool was analyzed by northern blot hybridization for estrogen-dependent stimulation of the transfected gene ("exo fra-1" in Figure 10-2B). These experiments demonstrated that sequences in the 5′ region of the fra-1 gene mediate regulation by Fos/AP-1.

Elevated levels of endogenous fra-1 mRNA were also observed in fibroblast cells overexpressing c-Jun, FosB, or Fra-1, thus indicating that the fra-1 gene is upregulated by other members of the AP-1 family as well (Bergers et al., submitted). In particular, this evidence suggests that the Fra-1 protein may be able to positively autoregulate transcription from its own promoter. The kinetics of fra-1 induction following serum or phorbol ester stimulation are delayed and prolonged in comparison to those of the c-fos gene (Cohen et al., 1989; Matsui et al., 1990; Kovary and Bravo, 1992). This protracted expression pattern may reflect transcriptional upregulation of the fra-1 gene by increased levels of other members of the AP-1 family, most likely by increased c-Fos and FosB activity in stimulated cells. The recent analysis of c-fos$^-$/c-fos$^-$ fibroblasts instead of ES cells supports this hypothesis, as the fra-1 gene could be induced by serum to only low levels in these cells compared to wild-type fibroblasts (Z.-Q. Wang and E. Wagner, personal communication).

At present we can only speculate about the significance of fra-1 induction by c-Fos/AP-1 activity. The dimerization and DNA binding activities of Fra-1 are similar, if not identical, to those of other members of the Fos family (c-Fos, FosB, and Fra-2) (Cohen et al., 1989; Ryseck and Bravo, 1991). However, the transcriptional properties of Fra-1 differ from those of c-Fos. Chimeric proteins consisting of the Gal4 DNA binding domain linked to the entire Fra-1 protein are unable to transactivate a Gal4-responsive reporter gene in transfected NIH 3T3 cells, in contrast to the corresponding Gal4–cFos fusion proteins (Bergers et al., submitted). Moreover, the Fra-1 protein stimulates AP-1-dependent transcription as a complex with JunD, but fails to do so in the presence of c-Jun and JunB (Okuno et al., 1991). More important, Fra-1 suppresses transactivation by c-Fos efficiently and in a dose-dependent manner (Okuno et al., 1991). Induction of fra-1 expression by c-Fos may, therefore, be a mechanism to down-modulate the transcriptional activity of c-Fos, analogous to the naturally occurring truncated form of FosB, which also inhibits AP-1 activity (Nakabeppu and Nathans, 1991; Wisdom et al., 1992). However, in contrast to c-Fos, the Fra-1 protein is expressed at significant levels in cycling cells and is apparently essential for cell cycle progression (Kovary and Bravo, 1992). It is, therefore, conceivable that c-Fos and Fra-1 regulate different sets of AP-1-responsive genes within the same cell.

OTHER Fos-RESPONSIVE GENES

By differential cDNA screening we have identified two additional Fos-inducible transcripts in rat fibroblasts that code for annexins II and V (Braselmann et al., 1992). We have furthermore shown that AP-1 binds with high affinity to a consensus AP-1 recognition sequence (Braselmann et al., 1992), which is present at position −128/−122 in the promoter of the human annexin II gene (Spano et al., 1990). A similar AP-1 site may also be present upstream of the rat annexin II gene, thus mediating transcriptional stimulation by Fos/AP-1. Annexins are a family of eight related proteins that bind to cellular membranes and phospholipids in a Ca^{2+}-dependent manner. Although no clear biological function has yet been assigned to any of these proteins, they have been implicated in the regulation of membrane traffic and exocytosis, mediation of cytoskeleton–membrane interactions and mitogenic signal transduction (Crompton et al., 1988; Klee, 1988; Haigler et al., 1989).

We have recently demonstrated that stimulation of Fos-ER activity induces neuronal PC12 cells to differentiate to an epitheloid cell type expressing cytokeratin filaments. Indeed, Fos-ER rapidly induces the synthesis of keratin 18 (K18) mRNA in these cells (Wrighton et al., 1993). Oshima et al. (1990) previously identified a functional AP-1 site in the first intron of the K18 gene that is responsible for transcriptional stimulation by c-Jun and c-Fos in undifferentiated F9 embryonal carcinoma cells. Our finding that an increase in Fos activity is sufficient to activate the endogenous K18 gene in PC12 cells thus confirms and extends the finding of Oshima et al. (1990) that K18 is a Fos target gene.

Fos/AP-1 has been implicated as a transcriptional regulator of the tyrosine hydroxylase (TH) gene in PC12 cells by transient transfection experiments and protein–DNA binding studies that demonstrated the requirement of a functional AP-1 site in the −200 region for basal and NGF-induced expression of the TH gene (Gizang and Ziff, 1990; Yoon and Chikaraishi, 1992). In agreement with these data, we demonstrated that induction of Fos-ER activity in PC12 cells results in a strong and rapid increase in the transcription rate of the endogenous TH gene, thus establishing a direct role for Fos in TH regulation (Wrighton and Busslinger, 1993).

Transcription of the ornithine decarboxylase (ODC) gene is induced by NGF in PC12 cells with delayed early kinetics relative to c-fos expression, suggesting that Fos/AP-1 mediates NGF induction of this gene (Greenberg et al., 1985). In agreement with this hypothesis, activation of Fos-ER induces ODC mRNA synthesis with similar kinetics and to the same maximal level in PC12 cells as NGF treatment. ODC induction occurs even in the absence of protein synthesis and is entirely dependent on an intact DNA binding function of the Fos-ER protein (Wrighton and Busslinger, 1993). These data demonstrate that Fos/AP-1 activity directly regulates the endogenous ODC gene in PC12 cells. Interestingly, ODC expression could not be induced by Fos-ER in rat fibroblasts, although transcription of the ODC gene is fully stimulated by serum in these cells. To account for this cell-specific action of Fos-ER, we have proposed that stimulation of the ODC gene by Fos-ER requires either a specific modification of the protein or, more likely, cooperation with another transcription factor(s) that is constitutively expressed in PC12 cells but is absent in unstimulated fibroblasts (Wrighton and Busslinger, 1993).

CONCLUSION

During the last decade we have experienced the discovery of a wealth of transcription factors involved in the control of different aspects of development, differentiation, or signal transduction. Paradoxically, it has proven difficult to identify downstream target genes responsible for the biological effects elicited by these transcription factors. Different strategies have been employed for this purpose. The so-called enhancer detection method was first developed in *Drosophila* (Wagner et al., 1991) and is now also applied in transgenic mice (Joyner, 1991) to search for genes with an expression pattern similar to that of important developmental regulators. An alternative procedure for identifying targets relies on specific antibodies raised against the transcription factor under study. These antibodies can be used for immunopurification and subsequent cloning of chromatin fragments that are bound by the transcription factor (Gould et al., 1990). Here we have described the utility of selective induction systems to identify Fos target genes. We have characterized seven genes, to date, that are regulated by Fos/AP-1 activity (Table 10-2). Many more AP-1-regulated genes have been analyzed in the past by the more classical approach of identifying functional AP-1 sites in promoters of known genes by protein–DNA binding studies, *in vitro* mutagenesis, and transient transfection experiments (Table 10-1). However, the importance of AP-1 for *in vivo* regulation of the endogenous allele remains to be determined for many of these genes (Table 10-1).

The use of Fos induction systems for identifying AP-1-regulated genes has several advantages. Most important, the effect of AP-1 is directly studied on the endogenous gene rather than on a cloned and transfected copy of the same gene. The endogenous gene is embedded in its proper regulatory environment and chromatin structure and is linked to all its control sequences, some of which may be located at a great distance from the gene. In contrast, only a limited amount of gene-flanking sequences can be analyzed in transfection experiments, which may explain why the majority of functional AP-1 sites identified by this method are located in proximal promoter regions (Table 10-1). Preliminary transfection experiments with the Fit-1 and fra-1 genes have indeed indicated that the sequences mediating regulation by Fos/AP-1 are distant from the transcription start sites of these genes (unpublished data). Another advantage of the posttranslational Fos-ER induction system is that the Fos-ER protein is synthesized within the cell in an inactive form. Fos-ER can, therefore, be activated in the absence of ongoing protein synthesis, which allowed us to demonstrate a direct role of Fos/AP-1 in the regulation of the Fit-1 and ODC genes (Bergers et al., 1994; Wrighton and Busslinger, 1993). Furthermore, mutant Fos-ER proteins lacking a functional DNA binding domain can be used to show that Fos/AP-1 exerts its regulatory affect by binding to cis-acting sequences, as demonstrated for the fra-1 and ODC genes (Bergers et al., submitted; Wrighton and Busslinger, 1993).

Fos induction systems are highly specific and greatly simplified compared to normal signaling mechanisms, as the activity of a single immediate early gene product, Fos, is stimulated in the absence of other induced immediate early gene products. These induction systems, therefore, facilitate the identification of genes that depend solely on increased AP-1 activity for their transcriptional stimulation. Under physiological conditions, however, activation of the transcription factor AP-1 is only one aspect of the complex intracellular response triggered by extracellular signals. In particular, a variety of other transcription factors are simultaneously co-activated at either the transcriptional or posttranslational level (Hunter and Karin, 1992). Hence, it is not surprising that Fos-ER alone failed to induce genes such as the proenkephalin and neurotensin/neuromedin N genes (Wrighton and Busslinger, 1993), which are known to be activated by AP-1 only in cooperation with other induced factors (Comb et al., 1988; Kislaukis and Dobner, 1990). The ubiquitously expressed ODC gene may exemplify yet a third class of genes that can be stimulated by Fos-ER in one cell type but not in another, depending on the cell-specific regulatory environment (Bergers et al., submitted; Wrighton and Busslinger, 1993).

In most cells c-fos is only transiently induced and may, thus, elicit short-term effects. It is important to note that the Fos-ER induction system allows one to mimic and study this behavior of the endogenous c-fos gene. This is best illustrated by the ODC and fra-1 genes that are transiently rather than permanently induced by Fos-ER in PC12 cells (Wrighton and Busslinger, 1993). Moreover, the Fos-ER protein can be activated by estrogen for defined time periods to distinguish short-term and long-term affects of Fos on cell differentiation. Reichmann et al. (1992) thus demonstrated that short-term stimulation of Fos-ER results in reversible loss of the polarized phenotype of epithelial cells. In contrast, long-term stimulation of Fos-ER causes these cells to depolarize irreversibly and to undergo conversion to fibroblastoid cells, suggesting that members of the AP-1 family are important for both transient and permanent changes of the epithelial cell phenotype (Reichmann et al., 1992).

Identifying AP-1 target genes by the use of Fos induction systems represents a "classical" gain-of-function approach. Recently, an alternative strategy has become feasible, as cell lines and transgenic mice lacking a functional c-jun (Hilberg and Wagner, 1992; Hilberg et al., 1993) or c-fos (Wang et al., 1992; Johnson et al., 1992) gene have been produced by targeted disruption of these loci through homologous recombination. These experiments demonstrated that c-jun is essential for early mouse development (Hilberg et al., 1993), while lack of c-fos creates a pleiotropic phenotype characterized by bone and hematopoietic defects (Wang et al., 1992; Johnson et al., 1992). Matched cell lines derived from wild-type and fos- or jun-deficient mice may prove to be important tools for identifying those genes that are controlled by c-Fos or c-Jun activity in the affected tissues. This loss-of-function approach is likely to complement the strategies described above for identifying AP-1-regulated genes.

NOTE ADDED DURING PROOF

Since the submission of this review article, we have discovered that expression of the rat Fit-1 gene gives rise to two different mRNA isoforms. The Fit-1M mRNA isolated from spleen codes for a membrane-bound protein which is most closely related in its extracellular, transmembrane, and intracellular domains to the type I interleukin-1 (IL-1) receptor. The Fit-1S mRNA of fibroblasts directs, instead, the synthesis of a secreted protein consisting of only the extracellular domain (Fit-1S was referred to as Fit-1 in this article). Analysis of the exon-intron structure of the Fit-1 gene indicated that the Fit-1S and Fit-1M mRNAs are transcribed from two different promoters and that the sequence differences at their 3′ ends result from alternative 3′ processing. Northern blot analysis with specific 5′ and 3′ probes directly demonstrated tight coupling between alternative promoter usage and 3′ processing of the Fit-1 transcripts. During ontogeny the mouse Fit-1 gene (known as T1 or ST2) is first expressed in the fetal liver of the embryo and later in the lung and hematopoietic tissues of the adult. The mRNA coding for the membrane-bound protein is more abundantly expressed in all of these tissues, while the transcript for the secreted form predominates in fibroblasts and mammary epithelial cells. Differential regulation of two distinct promoters is thus used to determine the ratio between secreted and membrane-bound forms of Fit-1 (T1/ST2) which may modulate signaling in response to IL-1 (Bergers, et al., submitted).

ACKNOWLEDGMENTS

We are grateful to H. Beug and L. Ballou for their critical reading of the manuscript.

REFERENCES

Angel, P. and Karin, M., The role of Jun, Fos and the AP-1 complex in cell-proliferation and transformation, *Biochim. Biophys. Acta*, 1072, 129, 1991.

Angel, P., et al., 12-O-tetradecanoyl-phorbol-13-acetate induction of the human collagenase gene is mediated by an inducible enhancer element located in the 5′-flanking region, *Mol. Cell. Biol.*, 7, 2256, 1987a.

Angel, P., Imagawa, M., Chiu, R., Stein, B., Imbra, R. J., Rahmsdorf, H. J., Jonat, C., Herrlich, P., and Karin, M., Phorbol ester-inducible genes contain a common *cis* element recognized by a TPA-modulated trans-acting factor, *Cell*, 49, 729, 1987.

Angel, P., Hattori, K., Smeal, T. and Karin, M., The jun proto-oncogene is positively autoregulated by its product, Jun/AP-1, *Cell*, 55, 875, 1988.

Auble, D. T. and Brinckerhoff, C. E., The AP-1 sequence is necessary but not sufficient for phorbol induction of collagenase in fibroblasts, *Biochemistry*, 30, 4629, 1991

Bergers, G., Reikerstorfer, A., Braselmann, S., Graninger, P., and Busslinger, M., Alternative promoter usage of the Fos-responsive gene *Fit-1* generates mRNA isoforms coding for either secreted or membrane-bound proteins related to the IL-1 receptor, *EMBO J.*, 13, 1176, 1994.

Bergers, G., Braselmann, S., Graninger, P., Wrighton, C., and Busslinger, M., Sequences in the 5' end of the *fra-1* gene mediate transcriptional stimulation by Fos/AP-1 activity, *Oncogene*, submitted.

Boehmelt, G., Walker, A., Kabrun, N., Mellitzer, G., Beug, H., Zenke, M., and Enrietto, P. J., Hormone-regulated v-*rel* estrogen receptor fusion protein: reversible induction of cell transformation and cellular gene expression, *EMBO J.*, 11, 4641, 1992.

Braselmann, S., Bergers, G., Wrighton, C., Graninger, P., Superti-Furga, G., and Busslinger, M., Identification of Fos target genes by the use of selective induction systems, *J. Cell. Sci.*, 16 (Suppl.), 97, 1992.

Braselmann, S., Graninger, P., and Busslinger, M., A selective transcriptional induction system for mammalian cells based on Gal4-estrogen receptor fusion proteins, *Proc. Natl. Acad. Sci. U.S.A.*, 90, 1657-1661, 1993.

Brenner, D. A., O'Hara, M., Angel, P., Chojkier, M., and Karin, M., Prolonged activation of jun and collagenase genes by tumour necrosis factor-alpha, *Nature*, 337, 661, 1989.

Burk, O. and Klempnauer, K.-H., Estrogen-dependent alterations in the differentiation state of myeloid cells caused by a v-myb/estrogen receptor fusion protein, *EMBO J.*, 10, 3713, 1991.

Buttice, G., Quinones, S., and Kurkinen, M., The AP-1 site is required for basal expression but is not necessary for TPA-response of the human stromelysin gene, *Nucleic Acids Res.*, 19, 3723, 1991.

Chiu, R., Imagawa, M., Imbra, R. J., Bockoven, J. R., and Karin, M., Multiple cis- and trans-acting elements mediate the transcriptional response to phorbol esters, *Nature*, 329, 648, 1987.

Chiu, R., Angel, P., and Karin, M., Jun-B differs in its biological properties from, and is a negative regulator of, c-Jun, *Cell*, 59, 979, 1989.

Cohen, D. R., Ferreira, P. C., Gentz, R., Franza, B. J., and Curran, T., The product of a fos-related gene, fra-1, binds cooperatively to the AP-1 site with Jun: Transcription factor AP-1 is comprised of multiple protein complexes, *Genes Dev.*, 3, 173, 1989.

Comb, M., Mermod, N., Hyman, S. E., Pearlberg, J., Ross, M. E., and Goodman, H. M., Proteins bound at adjacent DNA elements act synergistically to regulate human proenkephalin cAMP inducible transcription, *EMBO J.*, 7, 3793, 1988.

Crompton, M. R., Moss, S. E., and Crumpton, M. J., Diversity in the lipocortin/calpactin family, *Cell*, 55, 1, 1988.

Diamond, M. I., Miner, J. N., Yoshinaga, S. K., and Yamamoto, K. R., Transcription factor interactions: Selectors of positive or negative regulation from a single DNA element, *Science*, 249, 1266, 1990.

Gaub, M. P., Bellard, M., Scheuer, I., Chambon, P., and Sassone-Corsi, P., Activation of the ovalbumin gene by the estrogen receptor involves the fos-jun complex, *Cell*, 63, 1267, 1990.

Gius, D., Cao, X. M., Rauscher, F. E., Cohen, D. R., Curran, T., and Sukhatme, V. P., Transcriptional activation and repression by Fos are independent functions: the C terminus represses immediate-early gene expression via CArG elements, *Mol. Cell Biol.*, 10, 4243, 1990.

Gizang, G. E. and Ziff, E. B., Nerve growth factor regulates tyrosine hydroxylase gene transcription through a nucleoprotein complex that contains c-Fos, *Genes Dev.*, 4, 477, 1990.

Goodbourn, S., Zinn, K., and Maniatis, T., Human beta-interferon gene expression is regulated by an inducible enhancer element. *Cell*, 41, 509, 1985.

Gould, A. P., Brookman, J. J., Strutt, D. I., and White, R. A., Targets of homeotic gene control in Drosophila, *Nature*, 348, 308, 1990.

Greenberg, M. E., Greene, L. A., and Ziff, E. B., Nerve growth factor and epidermal growth factor induce rapid transient changes in proto-oncogene transcription in PC12 cells. *J. Biol. Chem.*, 260, 14101, 1985.

Gurney, A. L., Park, E. A., Giralt, M., Liu, J., and Hanson, R. W., Opposing actions of Fos and Jun on transcription of the phosphoenolpyruvate carboxykinase (GTP) gene. Dominant negative regulation by Fos, *J. Biol. Chem.*, 267, 18133, 1992.

Gutman, A. and Wasylyk, B., The collagenase gene promoter contains a TPA and oncogene-responsive unit encompassing the PEA3 and AP-1 binding sites, *EMBO J.*, 9, 2241, 1990.

Haigler, H. T., Fitch, J. M., Jones, J. M., and Schlaepfer, D. D., Two lipocortin-like proteins, endonexin II and anchorin CII, may be alternate splices of the same gene. *Trends Biochem. Sci.*, 14, 48, 1989.

Hengerer, B., Lindholm, D., Heumann, R., Rüther, U., Wagner, E. F., and Thoenen, H., Lesion-induced increase in nerve growth factor mRNA is mediated by c-fos, *Proc. Natl. Acad. Sci. U.S.A.*, 87, 3899, 1990.

Herrera, R., Ro, H. S., Robinson, G. S., Xanthopoulos, K. G., and Spiegelman, B. M., A direct role for C/EBP and the AP-1-binding site in gene expression linked to adipocyte differentiation. *Mol. Cell. Biol.*, 9, 5331, 1989.

Hilberg, F. and Wagner, E. F., Embryonic stem (ES) cells lacking functional c-*jun:* Consequences for growth and differentiation. AP-1 activity and tumorigenicity, *Oncogene*, 7, 2371, 1992.

Hilberg, F., Aguzzi, A., Howells, N., and Wagner, E. F., c-Jun is essential for normal mouse development and hepatogenesis, *Nature*, 365, 179, 1993.

Holt, J. T., Gopal, T. V., Moulton, A. D., and Nienhuis, A. W., Inducible production of c-fos antisense RNA inhibits 3T3 cell proliferation, *Proc. Natl. Acad. Sci. U.S.A.*, 83, 4794, 1986.

Hunter, T. and Karin, M., The regulation of transcription by phosphorylation, *Cell*, 70, 375, 1992.

Hynes, N. E., Kennedy, N., Rahmsdorf, U., and Groner, B., Hormone-responsive expression of an endogenous proviral gene of mouse mammary tumor virus after molecular cloning and gene transfer into cultured cells, *Proc. Natl. Acad. Sci. U.S.A.*, 78, 2038, 1981.

Jain, J., McCaffrey, P. G., Valge, A. V., and Rao, A., Nuclear factor of activated T cells contains Fos and Jun, *Nature*, 356, 801, 1992.

Johnson, R. S., Spiegelmann, B. M., and Papaioannou, V., Pleiotropic effects of a null mutation of the c-*fos* proto-oncogene, *Cell*, 71, 577, 1992.

Jorcyk, C. L., Watson, D. K., Mavrothalassitsi, G. J., and Papas, T. S., The human ETS1 gene: Genomic structure, promoter characterization and alternative splicing, *Oncogene*, 6, 523, 1991.

Joyner, A. L., Gene targeting and gene trap screens using embryonic stem cells: New approaches to mammalian development, *Bioessays*, 13, 649, 1991.

Kerppola, T. K. and Curran, T., Transcription factor interactions: Basics on zippers, *Curr. Opinion Struct. Biol.*, 1, 71, 1991.

Kerr, L. D., Holt, J. T., and Matrisian, L. M., Growth factors regulate transin gene expression by c-fos-dependent and c-fos-independent pathways, *Science*, 242, 1424, 1988.

Kim, S. J., Angel, P., Lafyatis, R., Hattori, K., Kim, K. Y., Sporn, M. B., Karin, M., and Roberts, A. B., Autoinduction of transforming growth factor beta 1 is mediated by the AP-1 complex, *Mol. Cell. Biol.*, 10, 1492, 1990.

Kislaukis, E. and Dobner, P. R., Mutually dependent response elements in the cis-regulatory region of the neurotensin/neuromedin N gene integrate environmental stimuli in PC12 cells, *Neuron*, 4, 783, 1990.

Klee, C. B., Ca2+-dependent phospholipid- (and membrane-) binding proteins, *Biochemistry*, 27, 6645, 1988.

Klemenz, R., Hoffmann, S. and Werenskiold, A. K., Serum- and oncoprotein-mediated induction of a gene with sequence similarity to the gene encoding carcinoembryonic antigen, *Proc. Natl. Acad. Sci. U.S.A.*, 86, 5708, 1989.

König, H., Ponta, H., Rahmsdorf, U., Buscher, M., Schönthal, A., Rahmsdorf, H. J., and Herrlich, P., Autoregulation of fos: The dyad symmetry element as the major target of repression. *EMBO J.*, 8, 2559, 1989.

Kovacic, M. B. and Gardner, D. G., Divergent regulation of the human atrial natriuretic peptide gene by c-jun and c-fos, *Mol. Cell. Biol.*, 12, 292, 1992.

Kovary, K. and Bravo, R., The jun and fos protein families are both required for cell cycle progression in fibroblasts, *Mol. Cell. Biol.*, 11, 4466, 1991.

Kovary, K. and Bravo, R., Existence of different Fos/Jun complexes during the G_0-to-G_1 transition and during exponential growth in mouse fibroblasts: Differential role of Fos proteins, *Mol. Cell. Biol.*, 12, 5015, 1992.

Kumar, V. and Chambon, P., The estrogen receptor binds tightly to its responsive element as a ligand-induced homodimer, *Cell*, 55, 145, 1988.

Lee, W., Haslinger, A., Karin, M., and Tjian, R., Activation of transcription by two factors that bind promoter and enhancer sequences of the human metallothionein gene and SV40, *Nature*, 325, 368, 1987a.

Lee, W., Mitchell, P., and Tjian, R., Purified transcription factor AP-1 interacts with TPA-inducible enhancer elements, *Cell*, 49, 741, 1987b.

Lee, M. E., Dhadly, M. S., Temizer, D. H., Clifford, J. A., Yoshizumi, M., and Quertermous, T., Regulation of endothelin-1 gene expression by Fos and Jun. *J. Biol. Chem.*, 266, 19034, 1991.

Majerus, M. A., Bibollet, R. F., Telliez, J. B., Wasylyk, B., and Bailleul, B., Serum, AP-1 and Ets-1 stimulate the human ets-1 promoter, *Nucleic Acids Res.*, 20, 2699, 1992.

Matsui, M., Tokuhara, M., Konuma, Y., Nomura, N., and Ishizaki, R., Isolation of human fos-related genes and their expression during monocyte-macrophage differentiation, *Oncogene*, 5, 249, 1990.

Mayo, K. E., Warren, R., and Palmiter, R. D., The mouse metallothionein-I gene is transcriptionally regulated by cadmium following transfection into human or mouse cells, *Cell*, 29, 99, 1982.

Mermod, N., Williams, T. J., and Tjian, R., Enhancer binding factors AP-4 and AP-1 act in concert to activate SV40 late transcription in vitro, *Nature*, 332, 557, 1988.

Mordacq, J. C. and Linzer, D. I., Co-localization of elements required for phorbol ester stimulation and glucocorticoid repression of proliferin gene expression. *Genes Dev.*, 3, 760, 1989.

Muegge, K., Williams, T. M., Kant, J., Karin, M., Chiu, R., Schmidt, A., Siebenlist, U., Young, H. A., and Durum, S. K., Interleukin-1 costimulatory activity on the interleukin-2 promoter via AP-1, *Science*, 246, 249, 1989.

Nakabeppu, Y. and Nathans, D., A naturally occurring truncated form of FosB that inhibits Fos/Jun transcriptional activity, *Cell*, 64, 751, 1991.

Naranjo, J. R., Mellström, B., Achaval, M., and Sassone, C. P., Molecular pathways of pain: Fos/Jun-mediated activation of a noncanonical AP-1 site in the prodynorphin gene. *Neuron,* 6, 607, 1991.

Nerlov, C., Rorth, P., Blasi, F., and Johnsen, M., Essential AP-1 and PEA3 binding elements in the human urokinase enhancer display cell type-specific activity, *Oncogene,* 6, 1583, 1991.

Nerlov, C., De Cesare, D., Pergola, F., Caracciolo, A., Blasi, F., Johnson, M., and Verde, P., A regulatory element that mediates co-operation between a PEA3-AP-1 element and an AP-1 site is required for phorbol ester induction of urokinase enhancer activity in HepG2 hepatoma cells. *EMBO J.,* 11, 4573, 1992.

Ney, P. A., Sorrentino, B. P., McDonagh, K. T., and Nienhuis, A. W., Tandem AP-1-binding sites within the human beta-globin dominant control region function as an inducible enhancer in erythroid cells, *Genes Dev.,* 4, 993, 1990.

Nicholson, R. C., Mader, S., Nagpal, S., Leid, M., Rochette, E. C., and Chambon, P., Negative regulation of the rat stromelysin gene promoter by retinoic acid is mediated by an AP1 binding site. *EMBO J.,* 9, 4443, 1990.

Nishikura, K. and Murray, J. M., Antisense RNA of proto-oncogene c-fos blocks renewed growth of quiescent 3T3 cells, *Mol. Cell. Biol.,* 7, 639, 1987.

Okuno, H., Suzuki, T., Yoshida, T., Hashimoto, Y., Curran, T., and Iba, H., Inhibition of jun transformation by a mutated fos gene: Design of an anti-oncogene, *Oncogene,* 6, 1491, 1991.

Oshima, R. G., Abrams, L., and Kulesh, D., Activation of an intron enhancer within the keratin 18 gene by expression of c-fos and c-jun in undifferentiated F9 embryonal carcinoma cells, *Genes Dev.,* 4, 835, 1990.

Owen, T. A., Bortell, R., Yocum, S. A., Smock, S. L., Zhang, M., Abate, C., Shalhoub, V., Aronin, N., and Wright, K. L., Coordinate occupancy of AP-1 sites in the vitamin D-responsive and CCAAT box elements by Fos-Jun in the osteocalcin gene: Model for phenotype suppression of transcription, *Proc. Natl. Acad. Sci. U.S.A.,* 87, 9990, 1990.

Picard, D., Salser, S. J., and Yamamoto, K. R., A movable and regulatable inactivation function within the steroid binding domain of the glucocorticoid receptor, *Cell,* 54, 1073, 1988.

Picard, D., Khursheed, B., Garabedian, M. J., Fortin, M. G., Lindquist, S., and Yamamoto, K. R., Reduced levels of hsp90 compromise steroid receptor action in vivo, *Nature,* 348, 166, 1990.

Piette, J. and Yaniv, M., Two different factors bind to the alpha-domain of the polyoma virus enhancer, one of which also interacts with the SV40 and c-fos enhancers, *EMBO J.,* 6, 1331, 1987.

Ransone, L. J. and Verma, I. M., Nuclear proto-oncogenes fos and jun, *Annu. Rev. Cell. Biol.,* 6, 539, 1990.

Reichmann, E., Schwarz, H., Deiner, E. M., Leitner, I., Eilers, M., Berger, J., Busslinger, M., and Beug, H., Activation of an inducible c-FosER fusion protein causes loss of epithelial polarity and triggers epithelial-fibroblastoid cell conversion, *Cell,* 71, 1103, 1992.

Riabowol, K. T., Vosatka, R. J., Ziff, E. B., Lamb, N. J., and Feramisco, J. R., Microinjection of fos-specific antibodies blocks DNA synthesis in fibroblast cells, *Mol. Cell. Biol.,* 8, 1670, 1988.

Rüther, U., Garber, C., Komitowski, D., Müller, R., and Wagner, E. F., Deregulated c-fos expression interferes with normal bone development in transgenic mice, *Nature,* 325, 412, 1987.

Rüther, U., et al., c-fos expression interferes with thymus development in transgenic mice, *Cell,* 53, 847, 1988.

Ryseck, R. P. and Bravo, R., c-JUN, JUN B, and JUN D differ in their binding affinities to AP-1 and CRE consensus sequences: Effect of FOS proteins, *Oncogene*, 6, 533, 1991.

Sassone-Corsi, P., Sisson, J. C., and Verma, I. M., Transcriptional autoregulation of the proto-oncogene fos, *Nature*, 334, 314, 1988.

Schönthal, A., Herrlich, P., Rahmsdorf, H. J., and Ponta, H., Requirement for fos gene expression in the transcriptional activation of collagenase by other oncogenes and phorbol esters, *Cell*, 54, 325, 1988a.

Schönthal, A., Buscher, M., Angel, P., Rahmsdorf, H. J., Ponta, H., Hattori, K., Chiu, R., Karin, M., and Herrlich, P., The Fos and Jun/AP-1 proteins are involved in the downregulation of Fos transcription, *Oncogene*, 4, 629, 1989.

Schüle, R., Umesono, K., Mangelsdorf, D. J., Bolado, J., Pike, J. W., and Evans, R. M., Jun-Fos and receptors for vitamins A and D recognize a common response element in the human osteocalcin gene, *Cell*, 61, 497, 1990.

Serfling, E., Barthelmäs, R., Pfeuffer, I., Schenk, B., Zarius, S., Swoboda, R., Mercurio, F., and Karin, M., Ubiquitous and lymphocyte-specific factors are involved in the induction of the mouse interleukin 2 gene in T lymphocytes, *EMBO J.*, 8, 465, 1989.

Sonnenberg, J. L., Rauscher, F., III, Morgan, J. I., and Curran, T., Regulation of proenkephalin by Fos and Jun, *Science*, 246, 1622, 1989.

Spano, F., Raugei, G., Palla, E., Colella, C., and Melli, M., Characterization of the human lipocortin-2-encoding multigene family: Its structure suggests the existence of a short amino acid unit undergoing duplication, *Gene*, 95, 243, 1990.

Sterneck, E., Müller, C., Katz, S., and Leutz, A., Autocrine growth induced by kinase type oncogenes in myeloid cells requires AP-1 and NF-M, a myeloid specific, C/EBP-like factor, *EMBO J.*, 11, 115, 1992.

Superti-Furga, F. G., Bergers, G., Picard, D., and Busslinger, M., Hormone-dependent transcriptional regulation and cellular transformation by Fos-steroid receptor fusion proteins, *Proc. Natl. Acad. Sci. U.S.A.*, 88, 5114, 1991.

Talbot, D. and Grosveld, F., The 5'HS2 of the globin locus control region enhances transcription through the interaction of a multimeric complex binding at two functionally distinct NF-E2 binding sites, *EMBO J.*, 10, 1391, 1991.

Timmers, H. T., Pronk, G. J., Bos, J. L., and van der Eb, A., Analysis of the rat JE gene promoter identifies an AP-1 binding site essential for basal expression but not for TPA induction, *Nucleic Acids Res.*, 18, 23, 1990.

Tominaga, S., A putative protein of a growth specific cDNA from BALB/c-3T3 cells is highly similar to the extracellular portion of mouse interleukin 1 receptor, *FEBS Lett.*, 258, 301, 1989.

Tominaga, S., Jenkins, N. A., Gilbert, D. J., Copeland, N. G., and Tetsuka, T., Molecular cloning of the murine ST2 gene. Characterization and chromosomal mapping, *Biochim. Biophys. Acta*, 1090, 1, 1991.

Topol, J., Ruden, D. M., and Parker, C. S., Sequences required for in vitro transcriptional activation of a Drosophila hsp 70 gene, *Cell*, 42, 527, 1985.

Umek, R. M., Friedman, A. D., and McKnight, S. L., CCAAT-enhancer binding protein: A component of a differentiation switch, *Science*, 251, 288, 1991.

Wagner, B. J. T., Wilson, C., Gibson, G., Schuh, R., and Gehring, W. J., Identification of target genes of the homeotic gene Antennapedia by enhancer detection, *Genes Dev.*, 5, 2467, 1991.

Wang, Z. Q., Grigoriadis, A. E., Mohle, S. U., and Wagner, E. F., A novel target cell for c-fos-induced oncogenesis: Development of chondrogenic tumours in embryonic stem cell chimeras, *EMBO J.*, 10, 2437, 1991.

Wang, Z.-Q., et al., Bone and haematopoietic defects in mice lacking c-*fos*, *Nature*, 360, 741-744, 1992.

Wasylyk. C., Flores. P., Gutman. A., and Wasylyk. B., PEA3 is a nuclear target for transcription activation by non-nuclear oncogenes. *EMBO J.*, 8, 3371, 1989.

Wasylyk. C., Gutman. A., Nicholson, R., and Wasylyk, B., The c-Ets oncoprotein activates the stromelysin promoter through the same elements as several non-nuclear oncoproteins. *EMBO J.*, 10, 1127, 1991.

Webster. N. J., Green, S., Jin, J. R., and Chambon, P., The hormone-binding domains of the estrogen and glucocorticoid receptors contain an inducible transcription activation function. *Cell*, 54, 199, 1988.

Wisdom. R., Yen. J., Rashid. D., and Verma. I. M., Transformation by FosB requires a trans-activation domain missing in FosB2 that can be substituted by heterologous activation domains, *Genes Dev.*, 6, 667, 1992.

Wrighton. C. and Busslinger. M., Direct transcriptional stimulation of the ornithine decarboxylase gene by Fos in PC12 cells, but not in fibroblasts, *Mol. Cell. Biol.*, 13, 4657-4669, 1993.

Wrighton. C., Ully. H., and Busslinger. M., Reversal of NGF-induced differentiation and induction of a novel phenotype in PC12 cells by conditional Fos activity, *J. Cell Biol.*, 1994: submitted.

Yang. V. W., Christy. R. J., Cook. J. S., Kelly, T. J., and Lane. M. D., Mechanism of regulation of the 422(aP2) gene by cAMP during preadipocyte differentiation. *Proc. Natl. Acad. Sci. U.S.A.*, 86, 3629, 1989.

Yoon. S. O. and Chikaraishi. D. M., Tissue-specific transcription of the rat tyrosine hydroxylase gene requires synergy between an AP-1 motif and an overlapping E box-containing dyad. *Neuron*, 9, 55, 1992.

Chapter 11

Fos/Jun in Brain Function

José R. Naranjo, Britt Mellström, and Paolo Sassone-Corsi

CONTENTS

INTRODUCTION

Since the first report showing the increase in c-*fos* mRNA levels after membrane depolarization (Morgan and Curran, 1986), many neurobiologists have studied the physiological relevance of immediate early nuclear proto-oncogenes in brain function. As a result, several scientific reports have reviewed the descriptive aspects of this problem, showing changes in immediate early genes (IEGs) in different models of physiological, pharmacological, and pathological conditions (Morgan and Curran, 1991). While functional correlations have been suggested in several cases, there are only few examples in which early response genes can be clearly implicated.

As a general principle, early response genes have been considered as nuclear messengers, coupling changes at the membrane level with modifications in gene expression. These changes may determine short- to long-term responses of the neuron. A direct application of this concept has been to use the inducibility of early response genes as a marker of neuronal activity. Most early response genes are known to encode transcriptional regulators. Thus, one goal has been to identify the putative target genes of the early response. However, in most cases the target genes have not been uncovered. Furthermore, the rapid discovery of new members of the immediate early gene class and the high degree of homology among some of them, questions the specificity of the tools used in early studies. For this reason, new strategies using immediate early lacZ fusion genes are now successfully applied, either *in vitro* or *in vivo*, to specifically define the patterns of basal expression and inducibility (Smeyne et al., 1992a, 1992b). By site-directed mutagenesis of these fusion genes, the relative function of operative cis-acting elements could possibly be unmasked *in vivo*.

In this chapter, our goal is not to summarize all the available information (see previous reviews from Sheng and Greenberg, 1990; Wisden et al., 1990; Morgan and Curran, 1991), but rather to focus on the few emerging aspects of the functional significance of early response genes of the leucine zipper class in brain function.

NEURONAL GENES AS TARGETS

An important step required to link c-*fos* induction to neuronal function is the identification of targets for the gene transactivation elicited by the products of the immediate early genes. The target gene expression will constitute the specific response of the system, and the nature of the response may validate the physiological meaning of the early induction.

Several strategies have been used for the identification of putative target genes. One approach involves the analysis of the regulatory region of candidate target genes. This may allow the identification of promoter sequences recognized by proteins encoded by specific immediate early genes (Sonnenberg et al., 1989; Naranjo et al., 1991). We will discuss the factors containing the leucine zipper dimerization motif that bind to cis-acting elements such as the TPA-responsive element (TRE) and the cAMP-responsive element (CRE) (Chiu et al., 1988; Sassone-Corsi et al., 1988a, 1988b). Another strategy consists in the identification of the components of the second wave of gene expression (intermediate genes) by differential or subtractive screening of two cDNA libraries obtained under basal and induced conditions (Lanahan et al., 1991; Mohn et al., 1991; Mellström et al., 1992).

The identification of the prodynorphin (Dyn) gene as a potential target for nuclear complexes containing Fos, could well serve as an example of the first approach. Early studies defined an induction of Dyn mRNA in lumbar spinal cord after pain stimulation (Iadarola et al., 1986; Hollt et al., 1987) in areas exhibiting increased Fos immunoreactivity after the same stimulation (Draisci and Iadarola, 1989). By double labeling technique it was shown that intense nuclear Fos immunostaining is present in the same neurons that showed a strong hybridization signal for Dyn mRNA (Naranjo et al., 1991b), suggesting a molecular relationship between the events. Subsequent analysis of the prodynorphin promoter revealed the presence of the sequence TGACAAACA, centered at position −257, which was required for pain-induced transcription. Although this sequence is not a consensus AP-1 site, this element could accommodate the binding of the AP-1/ATF families. In fact, the sequence did specifically interact with nuclear complexes containing the Fos protein. This same site was shown to be a target of Jun-mediated transactivation in transient transfection experiments (Naranjo et al., 1991b). However, the subunit partners of Jun and Fos binding to this site are unknown. So far this approach has allowed the identification of several target genes encoding neurotransmitters/neuromodulators (Gizeng-Ginsberg and Ziff, 1990; Haun and Dixon, 1990; Hengerer et al., 1990). Since long-term adaptive neuronal responses to adverse stimuli often imply deep structural changes, such as synaptic membrane remodeling and radical variations in the connectivity of the neuron, it is conceivable that some Fos–Jun target genes would encode products directing these processes.

IMMEDIATE EARLY GENES AND HIGHLY SPECIALIZED NEURONAL FUNCTIONS

SENSORY INTEGRATION

Since the initial work by Hunt et al. (1987), in which the accumulation of Fos immunoreactivity in neurons of the dorsal spinal cord after noxious stimulation was shown, additional studies have extended this observation to other models of sensory information. The purpose has been to establish the physiological meaning of the Fos induction.

The molecular, temporal, and anatomical features of the activation of IEGs after peripheral or visceral noxious stimulation have been studied (Ceccatelli et al., 1989; Draisci and Iadarola, 1989; Wisden et al., 1990; Naranjo et al., 1991a). A major conclusion emerging from these analyses is that c-*fos* induction in external laminae of the dorsal horn tends to be intense, short lasting, and possibly related to relatively restricted and

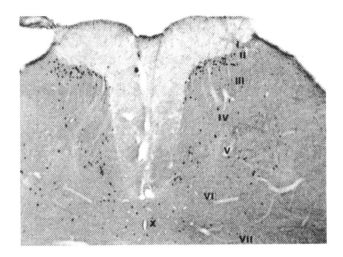

Figure 11-1 c-*fos* induction in external laminae of the dorsal horn.

tightly controlled transcriptional modifications (see Figure 11-1). Conversely, IEG induction in the inner laminae is slower and more sustained, which might correlate with prolonged transcriptional effects that regulate adaptive responses to the noxious stimulation. Previous analyses of the changes in gene expression after noxious stimulation offered several candidates as potential target genes in this model. In addition to the prodynorphin gene, two other potential target genes for transactivation by Fos–Jun in the spinal cord are the proenkephalin and the protachykinin genes. As with the prodynorphin gene, induced levels of the proenkephalin and protachykinin transcripts and peptides have been described after pain stimulation (Iadarola et al., 1986; Noguchi et al., 1992), and molecular analysis of their promoters revealed that they contain AP-1 and CRE sequences (Comb et al., 1988; Sonnenberg et al., 1989). While the nuclear protein(s) participating with Fos in the activation of these genes have not been identified, co-induction in the same cell of several IEGs has been demonstrated after noxious stimulation (Naranjo et al., 1991). Olfactory input has been shown to drive high basal levels of Fos immunoreactivity in projecting areas of the cerebral cortex of the lizard (Blasco-Ibanez et al., 1992). Induced levels of Fos after olfactory stimulation are also found in the median preoptic area of the male rat. This region is associated with sexual behavior (Robertson et al., 1991).

Finally, induction of IEGs in the visual cortex (but not in the frontal cortex) of the cat occurs after a brief visual experience and is important for the development of adult-like response properties at this level (Rosen et al., 1992). Furthermore, the induction is specific for c-*fos* and *jun*B, while levels of c-*jun* remain unaffected. As in the olfactory system, no target genes have been identified following the early induction of IEGs. It is important to stress that all inducibility studies should take into account that the various members of the same family of IEGs may be expressed differently, depending on the specific cell type of a given tissue. This is, for instance, true of the *jun* genes in the rat brain (Mellström et al., 1991) (see Figure 11-2).

CIRCADIAN RHYTHM

The internal pacemaker responsible for circadian rhythms in various physiological processes is postulated to reside in the hypothalamic suprachiasmatic nucleus (SCN).

c–jun

junB

junD

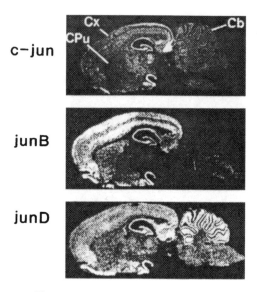

Figure 11-2 *jun* genes in the rat brain.

Synchronization of the circadian clock to the daily cycle of light is an endogenous function of the SCN, as has been shown with isolated neurons *in vitro*, which are intrinsically capable of generating circadian rhythms such as vasopressin secretion or electrical activity (Inouye and Kawamura, 1982; Schwartz and Reppert, 1985; Earnest et al., 1992). The effect of light is mediated via visual pathways that terminate in the SCN. Light has a modulatory action on the circadian periodicity of the pacemaker, whose sensitivity oscillates during the cycle. Induced levels of c-*fos* and *jun*B mRNA, and induced amounts of nuclear complexes able to recognize AP-1 sequences, occur in the ventrolateral SCN by brief light exposure during times when light is capable of modulating circadian rhythmicity (Aronin et al., 1990; Rusak et al., 1990; Kornhauser et al., 1990, 1992). The photic threshold of the c-*fos* induction is comparable to the threshold for the phase shifting effect of light on the pacemaker. The photic phase shifting of the behavioral activity rhythm also depends upon the circadian time at which the light occurs, linking the c-*fos* induction to the pathway for entrainment of the circadian oscillator (Aronin et al., 1990; Kornhauser et al., 1990; Rusak et al., 1990). The functional meaning of c-*fos* expression in the SCN has not been defined, but an increase in c-*fos* expression may reflect the activation of discrete neuronal subpopulations by primary glutamatergic afferents involved in transduction of the light stimulus (Abe et al., 1992; Vindlacheruvu et al., 1992). Fine regulation of gene expression is likely to be crucial for both the generation and control of circadian rhythm and for the modulation of biochemical rhythms by the circadian system. Thus, generation of the AP-1 complex is both gated by and, in turn, likely to influence the circadian pacemaker (Kornhauser et al., 1992).

The proposed relationship between AP-1 and the circadian clock may be analogous to the role of the Drosophila *per* gene. Levels of *per* transcript and Per protein cycle in a circadian manner (Siwicki et al., 1988; Hardin et al., 1990), and expression of the wild-type Per protein is required for normal circadian rhythms. Mutant alleles of the Drosophila *per* gene produce altered circadian periods or abolish rhythmicity. Thus, *per* is one example of a molecule that both acts on (or is a component of) the pacemaker and is, in turn, regulated by it. The induction of c-*fos* by light is also controlled by the circadian oscillator, and this phenomenon may by parallel to the *per* situation. Furthermore, Per is homologous to characterized transcription factors and has a nuclear localization. The fact that a component of a transcription complex is regulated by both light and circadian phase may have significant implications for understanding the links between photic information/transduction and gene expression (Kornhauser et al., 1992). It will be interesting to examine whether expression of Per and transcription of the genes, related to c-*fos* and c-*jun* (dFRA and dJRA, respectively) (Perkins et al., 1990), are regulated similarly.

Three genes have been proposed as putative targets for transactivation by the IEGs in the SCN of rodents. The vasoactive intestinal peptide (VIP) and the neuropeptide Y (NPY) genes are expressed in the same subdivision of the SCN (the ventrolateral) where induction of Fos occurs, and they colocalize with Fos in the same neurons (Card and Moore, 1989; Earnest et al., 1992). The vasopressin gene (AVP) undergoes circadian fluctuations in the SCN but not in other hypothalamic nuclei. The fluctuations include changes in vasopressin mRNA levels (Uhl and Reppert, 1986) and the length of the poly(A) tail of the transcript (Robinson et al., 1988).

OSMOTIC STIMULATION

It has been demonstrated that the neurosecretory neurons of the paraventricular and supraoptic hypothalamic areas respond to osmotic stimulation with increase in the firing rate, neuronal hypertrophy, and increase in the expression of vasopressin, oxytocin, and prodynorphin genes (Swanson and Sawchenko, 1983; van Toll et al., 1987; Sherman et al., 1988). Sagar et al. (1988) noticed that sham-operated as well as electrically stimulated animals exhibited increased Fos immunostaining in the hypothalamic paraventricular and supraoptic nuclei. They investigated whether the increased Fos staining was due to failure of the rats to drink normally after anesthesia, and found that the induction correlated with water deprivation. Further studies have shown that the expression of Fos related antigens (FRAs) is more prolonged than c-Fos in different experimental models of osmotic stimulation (Lafarga et al., 1992). In addition, induction of c-*jun* in supraoptic neurons after osmotic stimulation has also been reported (Carter and Murphy, 1990). More recently, it has been shown that c-*fos* induction is followed by the specific increase of the antagonistic isoforms of CRE modulator (CREM) in supraoptic neurons (Mellström et al., 1993), suggesting the possibility that CREM could be responsible for the well-described decrease of c-*fos* expression that follows induction (Foulkes et al., 1991b).

Changes in nuclear and nucleolar size, parameters related to transcription rate and metabolic status, occur in supraoptic neurons after osmotic stimulation (Sagar et al., 1988). The increased nuclear size is linked with the overexpression of FRAs in many SON neurons and can be related to chromatin reorganization. Experimental evidence in nonneuronal cells indicated that the activation of transcriptionally competent genes is induced by partial depletion of histone H1, with consequent relaxation of the higher order structure of chromatin domains containing these genes (Weintraub, 1984). The "open" state of these chromatin domains generates "nuclease hypersensitive sites", which allows enhanced access of diffusible transcriptional factors, such as c-Fos and FRAs, to regulatory regions of inducible genes. Since the large nuclei of supraoptic neurons have a euchromatic configuration, this could facilitate the accessibility of c-Fos and FRA proteins to target sequences following osmotic stimulation. This model could be suitable for electron microscopic analysis of nuclear loci where c-Fos and FRA proteins are located, and could reveal the nature of other proteins participating in the transcriptional activation after osmotic stimulation.

In the paraventricular nucleus, c-Fos immunoreactivity colocalizes with corticotropin-releasing factor immunoreactivity (Ceccatelli et al. 1989), while in the supraoptic nucleus, induced levels of Fos colocalize with vasopressin-, oxytocin-, and prodynorphin-expressing neurons (Giovannelli et al., 1990; Naranjo et al., manuscript in preparation), supporting a physiological role for activation by IEGs after osmotic stimulation.

IMMEDIATE EARLY GENES AND NEURONAL CELL DEATH

A number of studies have identified the induction of immediate early genes as a signal leading to changes in the cell cycle. These changes include progression in the cycle, cell division, differentiation, maturation, terminal differentiation, and apoptosis (He and Rosenfeld, 1991; Oppenheim, 1991). Several groups have reported high levels of Fos immunoreactivity in neural progenitor cells during their differentiation (Smeyne et al., 1992b; Gonzalez-Martin et al., 1992) and in naturally occurring neuronal death in the cerebral cortex or in the dentate gyrus during development (Oppenheim, 1991; Gould et al., 1991; Gonzalez-Martin et al., 1992). Furthermore, induction of IEGs follows shortly after the exposure to neurotoxic agents (Murphy et al., 1991), and precedes neuronal death induced in various ways, such as the exogenous application of toxins, neuronal loss after infarction, ischemia or epileptic discharges (Dragunow and Robertson, 1987; Morgan et al., 1987; Jorgensen et al., 1989). In an *in vitro* model of neuronal degeneration induced by overstimulation of NMDA receptors (a subclass of the glutamate receptor superfamily), the induction of c-*fos* correlates with the amount of Ca^{2+} entry into the cytoplasm and with the fate of the neuron (Szekely et al., 1987). Since the process of neuronal cell death requires new protein synthesis (Oppenheim, 1991; Martin et al., 1988), the search for Fos/Jun target genes, the so-called "death genes," is now in progress in many laboratories. So far, several genes have been shown to be upregulated in neurodegenerative diseases or after brain ischemia (Griffin et al., 1989; Hogquist et al., 1991). Some of them belong to the heat shock family of proteins (HSP), which are produced by cells in response to stress, and are usually considered to be protective (Chiang et al., 1989). AP-1-like sites have been described in the promoter region of the 72-kDa HSP and in the amyloid precursor protein (Sassone-Corsi et al., 1988b; Wirak et al., 1991), although proof of a functional relationship is still missing. In this respect it should be noted that transactivation of the nerve growth factor gene by Fos–Jun has been shown to be mediated through a functional AP-1 site located in the promoter region of this gene (Hengerer et al., 1990).

Naturally occurring neuronal degeneration in the nematode *Caenorhabditis elegans* is due to the uncoordinate expression of the *ced-3, ced-4,* and *ced-9* genes (Hengartner et al., 1992). According to the model, *ced-3* and *ced-4* function as death genes leading to degeneration of the neuron in the absence of *ced-9* gene expression, the cell-survival gene (Driscoll and Chalfie, 1992). Activation of these three genes occurs soon after birth, though it is not known whether the induction is autonomous or whether these genes need external signals (Hengartner et al., 1992). Recently, prevention of programmed cell death in sympathetic neurons was described after overexpression of the *bcl-2* proto-oncogene, the human counterpart of the *ced-9* gene (Garcia et al., 1992; Hengartner et al., 1992). The ultimate relationship between activation of immediate early genes and the induction of the *bcl-2* gene remains to be elucidated.

THE SHORT LOOP

If one accepts that immediate early activation of IEGs in the CNS is important for the rapid setting of alert mechanisms to respond to environmental stimuli, an essential property expected from the system would be to undergo a fast termination process in order to be prepared for the arrival of successive stimulatory inputs. One basic idea is that the early induction activates its own termination process. Indeed, the negative feedback regulation of the c-Fos protein on the expression of the c-*fos* gene has been known for several years (Sassone-Corsi et al., 1988b; Jorgensen et al., 1989). In addition, it has been recently demonstrated that antagonistic isoforms of the CREM gene (Foulkes et al., 1991a) participate in the downregulation of c-*fos* (Foulkes et al., 1991b), and are induced early after sensory stimulation in the supraoptic nucleus (Mellström et al., 1993). This activation is specific, and does not occur for a splice variant that functions as transcriptional activator, CREMτ (Mellström et al., 1993; Naranjo et al, manuscript in preparation). Similarly, we have observed induction of CREM antagonists after pain stimulation in the external laminae of the dorsal horn, an area where transcriptional changes are known to be tightly controlled, while in the inner laminae no modification in CREM expression was observed (Naranjo et al., manuscript in preparation).

ACKNOWLEDGMENTS

B. M. is supported by Glaxo S. A. Work in our laboratories is supported by the Ministerio Educación y Ciencia, Communidad Autónoma de Madrid, Fundación Ramón Arces, CNRS, INSERM, Association pour la Recherche contre le Cancer and Rhône-Poulenc Rorer.

REFERENCES

Abe, H., Rusak, B., and Robertson, H. A., NMDA and nonNMDA receptor antagonists inhibit photic induction of Fos protein in the Hamster suprachiasmatic nucleus, *Brain Res. Bull.*, 28, 813, 1992.

Aronin, N., Sagar, S. M., Sharp, F. R., and Schwartz, W. J., Light regulates expression of a Fos-related protein in rat suprachiasmatic nuclei, *Proc. Natl. Acad. Sci. U.S.A.*, 87, 5959, 1990.

Blasco-Ibanez, J. M., Martinez-Guijarro, J. M., Mellström, B., Lopez-Garcia, C., and Naranjo, J. R., Olfactory input regulates basal levels of Fos proteins in the cerebral cortex of the lizard Podarcis hispanica, *Neuroscience*, 50, 647, 1992.

Card, J. P. and Moore, R. Y., Organization of lateral geniculate-hypothalamic connections in the rat, *J. Comp. Neurol.*, 284, 135, 1989.

Carter, D. A. and Murphy, D., Regulation of c-*fos* and c*jun* expression in the rat supraoptic nucleus, *Cell. Mol. Neurobiol.*, 10, 435, 1990.

Ceccatelli, S., Villar, M. J., Goldstein, M. J., and Hokfelt, T., Expression of c-Fos immunoreactivity in transmitter characterized neurons after stress, *Proc. Natl. Acad. Sci. U.S.A.*, 86, 9569, 1989.

Chiang, H.-L., Terlecky, S. R., Plant, C. P., and Dice, J. F., A role for a 70-kilodalton heat shock protein in lysosomal degradation of intracellular proteins, *Science*, 246, 382, 1989.

Chiu, R., Boyle, W. J., Meek, J., Smeal, T., Hunter, T., and Karin, M., The c-fos protein interacts with c-jun/AP-1 to stimulate transcription of AP-1 responsive genes, *Cell*, 54, 541, 1988.

Comb, M., Mermod, N., Hyman, S. E., Pearlberg, J., Ross, E., and Goodman, H. M., Proteins bound at adjacent DNA elements act synergistically to regulate human proenkephalin cAMP-inducible transcription, *EMBO J.*, 7, 3793, 1988.

Dragunow, M. and Robertson, H. A., Kindling stimulation induces c-Fos protein(s) in granule cells of the rat dentate gyrus, *Nature*, 329, 441, 1987.

Draisci, G. and Iadarola, M., Temporal analysis of increases in c-*fos*, preprodynorphin and preproenkephalin mRNAs in rat spinal cord, *Mol. Brain Res.*, 6, 31, 1989.

Driscoll, M. and Chalfie, M., Developmental and abnormal cell death in *C. elegans*, *Trends Neurosci.*, 15, 15, 1992.

Earnest, D. J., Ouyang, S., and Olschowa, J. A., Rhythmic expression of Fos-related proteins within rat suprachiasmatic nucleus during constant retinal illumination, *Neurosci. Lett.*, 140, 19, 1992.

Foulkes, N. S., Borrelli, E., and Sassone-Corsi, P., CREM gene: Use of alternative DNA-binding domains generates multiple antagonists of cAMP-induced transcription, *Cell*, 64, 739, 1991a.

Foulkes, N. S., Laoide, B. M., Schlotter, F., and Sassone-Corsi, P., Transcriptional antagonist cAMP-responsive element modulator (CREM) down-regulates c-*fos* cAMP-induced expression, *Proc. Natl. Acad. Sci. U.S.A.*, 88, 5448, 1991b.

Garcia, I., Martinou, I., Tsujimoto, Y., and Martinou, J. C., Prevention of programmed cell death of sympathetic neurons by the *bcl-2* proto-oncogene, *Science*, 258, 302, 1992.

Giovannelli, L., Shiromani, P. J., Jirikowski, G. F., and Bloom, F. E., Oxytocin neurons in the rat hypothalamus exhibit c-Fos immunoreactivity upon osmotic stress, *Brain Res.*, 531, 299, 1990.

Gizeng-Ginsberg, E. and Ziff, E. B., Nerve growth factor regulates tyrosine hydroxylase gene transcription through a nucleoprotein complex that contains c-Fos, *Genes Dev.*, 4, 477, 1990.

Gonzalez-Martin, C., de Diego, I., Crespo, D., and Fairn, A., Transient c-*fos* expression accompanies naturally occurring cell death in the developing interhemispheric cortex of the rat, *Dev. Brain Res.*, 68, 83, 1992.

Gould, E., Woolley, C. S., and McEwen, B. S., Naturally occurring cell death in the developing dentate gyrus of the rat, *J. Comp. Neurol.*, 304, 408, 1991.

Griffin, W. S., Stanley, L. C., Ling, C., White, L., McLeod, V., Perrot, L. J., White, C. L., and Araoz, C., Brain interleukin 1 and S-100 immunoreactivity are elevated in Down syndrome and Alzheimer disease, *Proc. Natl. Acad. Sci. U.S.A.*, 81, 7611, 1989.

Hardin, P. E., Hall, J. C., and Rosbash, M., Feedback of the Drosophila *period* gene product on circadian cycling of its messenger RNA levels, *Nature*, 343, 536, 1990.

Haun, R. S. and Dixon, J. E., A transcriptional enhancer essential for the expression of the rat cholecystokinin gene contains a sequence identical to the -296 element of the human c-*fos* gene. *J. Biol. Chem.*, 265, 15455, 1990.

He, X. and Rosenfeld, M. G., Mechanisms of complex transcriptional regulation: Implications for brain development, *Neuron*, 7, 183, 1991.

Hengartner, M. O., Ellis, R. E., and Horvitz, H. R., *Caenorhabditis elegans* gene ced-9 protects cells from programmed cell death, *Nature*, 356, 494, 1992.

Hengerer, B., Lindholm, D., Heumann, R., Rather, U., Wagner, E. F., and Thoenen, H., Lesion-induced increase in nerve growth factor mRNA is mediated by c-*fos*, *Proc. Natl. Acad. Sci. U.S.A.*, 87, 3899, 1990.

Hoffman, E. C., Reyes, H., Chu, F.-F., Sander, F., Conley, L. H., Brooks, B. A., and Hankinson, O., Cloning of a factor required for activity of the Ah (Dioxin) receptor, *Science*, 252, 954, 1991.

Hogquist, K. A., Nett, M. A., Unanue, E. R., and Chaplin, D. D., Interleukin 1 is processed and released during apoptosis, *Proc. Natl. Acad. Sci. U.S.A.*, 88, 8485, 1991.

Hollt, V., Haarmann, I., Millan, M. J., and Herz, A., Prodynorphin gene expression is enhanced in the spinal cord of chronic arthritic rats, *Neurosci. Lett.*, 73, 90, 1987.

Hunt, S. P., Pini, A., and Evan, G., Induction of c-*fos*-like protein in spinal cord neurons following sensory stimulation, *Nature*, 328, 632, 1987.

Iadarola, M. J., Douglass, J., Civelli, O., and Naranjo, J. R., Increased spinal cord dynorphin mRNA during peripheral inflammation, in *Progress in Opioid Research NIDA Research Monograph*, vol. 75, Holaday, J. W., Law, P.-Y., and Herz, A., Eds., Department of Health and Human Services, Rockville, MD, p. 406, 1986.

Inouye, S. T. and Kawamura, H., Characteristics of a circadian pacemaker in the suprachiasmatic nucleus, *J. Comp. Physiol.*, 146, 153, 1982.

Jorgensen, M. B., Deckert, J., Wright, D. C., and Gehler, D. R., Delayed c-*fos* proto-oncogene expression in the rat hippocampus induced by transient global cerebral ischemia: An *in situ* hybridization study, *Brain Res.*, 484, 393, 1989.

Kornhauser, J. M., Nelson, D. E., Mayo, K. E., and Takahashi, J. S., Photic and circadian regulation of c-*fos* gene expression in the hamster suprachiasmatic nucleus, *Neuron*, 5, 127, 1990.

Kornhauser, J. M., Nelson, D. E., Mayo, K. E., and Takahashi, J. S., Regulation of *jun*B messenger RNA and AP-1 activity by light and a circadian clock, *Science*, 255, 1581, 1992.

Lafarga, M., Berciano, M. T., Martinez-Guijarro, F. J., Andres, M. A., Mellström, B., Lopez-Garcia, C., and Naranjo, J. R., Fos expression and nuclear size in osmotically stimulated supraoptic nucleus neurons, *Neuroscience*, 50, 867, 1992.

Lanahan, A., Williams, J. B., Sanders, L. K., and Nathans, D., Growth factor-induced delayed early response genes, *Mol. Cell. Biol.*, 12, 3919, 1991.

Martin, D. P., Schmidt, R. E., DiStefano, P. S., Lowry, O. H., Carter, J. G., and Johnson, E. M., Jr., Inhibitors of protein synthesis and RNA synthesis prevent neuronal death caused by nerve growth factor deprivation, *J. Cell Biol.*, 106, 829, 1988.

Mellström, B., Achaval, M., Montero, D., Naranjo, J. R., and Sassone-Corsi, P., Differential expression of the *jun* family members in rat brain, *Oncogene*, 6, 1959, 1991.

Mellström, B., Barrio, L. C., and Naranjo, J. R., Differential screening for NMDA-induced genes in primary neuronal cultures, in Second NIMH Conference on Molecular Neurobiology, 1992.

Mellström, B., Naranjo, J. R., Foulkes, N. S., Lafarga, M., and Sassone-Corsi, P., Transcriptional response to cAMP in brain: Basal and induced expression of CREM antagonists, *Neuron*, 10, 655, 1993.

Mohn, K. L., Laz, T. M., Hsu, J.-C., Melby, A. E., Bravo, R., and Taub, R., The immediate early growth response in regenerating liver and insuline-stimulated H-35 cells: Comparison with serum-stimulated 3T3 cells and identification of 41 novel immediate early genes, *Mol. Cell. Biol.*, 11, 381, 1991.

Mohr, E. and Richter, D., Sequence analysis of the promoter region of the rat vasopressin gene, *FEBS Lett.*, 2, 305, 1990.

Morgan J. I. and Curran T., Role of ion flux in the control of c-*fos* expression, *Nature*, 322, 552, 1986.

Morgan, J. I. and Curran, T., Stimulus-transcription coupling in the nervous system: Involvement of the inducible proto-oncogenes *fos* and *jun*, *Annu. Rev. Neurosci.*, 14, 421, 1991.

Morgan, J. I., Cohen, D. R., Hempstead, J. L., and Curran T., Mapping patterns of c-*fos* expression in the central nervous system after seizure, *Science*, 237, 192, 1987.

Murphy, T. H., Worley, P. F., Nakabeppu, Y., Christy, B., Gastel, J., and Baraban, J. M., Synaptic regulation of immediate early gene expression in primary cultures of cortical neurons, *J. Neurochem.*, 57, 1862, 1991.

Naranjo, J. R., Mellström, B., Achaval, M., Lucas, J. J., Del Rio, J., and Sassone-Corsi, P., Co-induction of *jun*B and c-*fos* in a subset of neurons in the spinal cord, *Oncogene*, 6, 223, 1991a.

Naranjo, J. R., Mellström, B., Achaval, M., and Sassone-Corsi, P., Molecular pathways of pain: Fos/Jun-mediated activation of the prodynorphin gene through a non-canonical AP-1 site, *Neuron*, 6, 607, 1991b.

Noguchi, K., Dubner, R., and Ruda, M. A., Preproenkephalin mRNA in spinal dorsal horn neurons is induced by peripheral inflammation and is colocalized with Fos and Fos-related proteins, *Neuroscience*, 46, 561, 1992.

Oppenheim, R. W., Cell death during development of the nervous system, *Annu. Rev. Neurosci.*, 14, 453, 1991.

Perkins, K. K., Admon, A., Patel, N., and Tjian, R., The Drosophila Fos-related AP-1 protein is a developmentally regulated transcription factor, *Genes Dev.*, 4, 822, 1990.

Robertson, G. S., Pfaus, J. G., Atkinson, L. J., Matsumura, H., Phillips, A.G., and Fibiger, H. C., Sexual behaviour increases c-*fos* expression in the forebrain of male rat, *Brain Res.*, 564, 352, 1991.

Robinson, B. G., Frim, D. M., Schwartz, W. J., and Majzoub, J. A., Vasopressin mRNA in the suprachiasmatic nucleus: Daily regulation of polyadenylate tail length, *Science*, 241, 342, 1988.

Rosen, K. M., McKormack, M. A., Villa-Komaroff, L., and Mower, G. D., Brief visual experience induces immediate early gene expression in the cat visual cortex, *Proc. Natl. Acad. Sci. U.S.A.*, 89, 5437, 1992.

Rusak, B., Robertson, H. A., Wisden, W., and Hunt, S. P., Light pulses that shift rhythms induce gene expression in the suprachiasmatic nucleus, *Science*, 248, 1237, 1990.

Sagar, S. M., Sharp, F. R., and Curran, T., Expression of c-Fos protein in the brain: Metabolic mapping at the cellular level, *Science*, 240, 1328, 1988.

Sassone-Corsi, P., Lamph, W. W., Kamps, M., and Verma, I. M., *fos*-associated cellular p39 is related to nuclear transcription factor AP-1, *Cell*, 54, 553, 1988a.

Sassone-Corsi, P., Sisson, J. C., and Verma, I. M., Transcriptional autoregulation of the proto-oncogene *fos*, *Nature*, 334, 314, 1988b.

Sassone-Corsi, P., Visvader, J., Ferland, L., Mellon, P. L., and Verma, I. M., Induction of proto-oncogene *fos* transcription through the adenylate cyclase pathway: Characterization of a cAMP-responsive element, *Genes Dev.*, 2, 1529, 1988c.

Schwartz, W. J., and Reppert, S. M., Neural regulation of the circadian vasopressin rhythm in cerebrospinal fluid: A preeminent role for the suprachiasmatic nuclei, *J. Neurosci.* 5, 2771, 1985.

Sheng, M. and Greenberg, M. E., The regulation and function of c-*fos* and other immediately early genes in the nervous system, *Neuron*, 4, 477, 1990.

Sherman, T. G., Day, R., Civelli, O., Douglass, J., Herbert, E., Akil, H., and Watson, S. J., Regulation of hypothalamic magnocellular neuropeptides and their mRNAs in the Brattleboro rat: Coordinate responses to further osmotic challenge, *J. Neurosci.*, 8, 3785, 1988.

Siwicki, K. K., Eastman, C., Petersen, G., Rosbash, M., and Hall, J. C., Antibodies to the *period* product of Drosophila reveal tissue distribution and rithmic changes in the visual system, *Neuron*, 1, 141, 1988.

Smeyne, R. J., Curran, T., and Morgan, J. I., Temporal and spatial expression of a *fos*-lacZ transgene in the developing nervous system, *Mol. Brain Res.*, 16, 158, 1992a.

Smeyne, R. J., Schilling, K., Robertson, L., Luk, Oberdick, J., Curran, T., and Morgan, J., Fos-lacZ transgenic mice: Mapping sites of gene induction in the central nervous system, *Neuron*, 8, 13, 1992b.

Sonnenberg, J. L., Rauscher, F. J., III, Morgan, J. I., and Curran, T., Regulation of proenkephalin by Fos and Jun, *Science*, 246, 1622, 1989.

Swanson, L. W. and Sawchenko, P. E., Hypothalamic integration: Organization of the paraventricular and supraoptic nuclei, *Annu. Rev. Neurosci.*, 6, 269, 1983.

Szekely, A. M., Barbaccia, M. L., and Costa, E., Activation of specific glutamate receptor subtypes increases c-*fos* proto-oncogene expression in primary cultures of neonatal rat cerebellar granule cells, *Neuropharmacology*, 26, 1779, 1987.

Uhl, G. R. and Reppert, S. M., Suprachiasmatic nucleus vasopressin messenger RNA: Circadian variation in normal and Brattleboro rats, *Science*, 232, 390, 1986.

van Toll, H. H. M., Vorhuis, D. Th. A. M., and Burbach, J. P. H., Oxitocyn gene expression in discrete hypothalamic magnocellular cell groups is stimulated by prolonged salt loading, *Endocrinology*, 120, 71, 1987.

Vindlacheruvu. R. R., Ebling, F. J. P., Maywood. E. S., and Hastings, M. H., Blockade of glutamatergic neurotransmission in the suprachiasmatic nucleus prevents cellular and behavioural responses of the circadian system to light. *Eur. J. Neurosci.,* 4. 673, 1992.

Weintraub. H., Histone H1-dependent chromatin superstructures and the suppression of gene activity, *Cell,* 38. 17, 1984.

Wirak. D. O., Bayney, R., Kundel, C. A., Lee, A., Scangos, G. A., Trapp, B. D., and Unterbeck, A. J., Regulatory region of the human amyloid precursor protein (APP) gene promotes neuron-specific gene expression in the CNS of transgenic mice, *EMBO J.,* 10, 289, 1991.

Wisden, W., Errington. M. L., Williams, S., Dunnett. S. B., Waters, C., Hitchcock, D., Evan, G., Bliss. T. V. P., and Hunt. S. P., Differential expression of immediate early genes in the hippocampus and spinal cord. *Neuron,* 4. 603, 1990.

Chapter 12

c-Fos in Differentiation and Development

Ulrich Rüther

CONTENTS

INTRODUCTION

About ten years have passed since the first report (Müller et al., 1982) was published in which the expression of c-Fos was analyzed in different tissues in mice during development. Since that time several dozens of publications have addressed the expression of endogenous c-Fos as well as the analyses of gene transfer experiments in either tissue culture systems or mice.

IN VITRO: CELL CULTURE EXPERIENCES

Relatively few cell lines have been used to investigate the involvement of c-Fos in differentiation. Furthermore, some data have been derived from organ culture experiments. Most of the cell lines only differentiate *in vitro* upon induction. Since almost every external stimulus can induce expression of the *c-fos* gene (see Chapter 8, this volume), it was quite a complicated task to separate cause and consequence in the differentiating process. For example, the monomyelocytic cell line HL60 differentiates into macrophages upon treatment with the phorbol ester TPA (Rovera et al., 1979). This differentiation is accompanied by high expression of c-Fos (Müller et al., 1984; Mitchell et al., 1985). However, variants of the HL60 line can differentiate in the absence of c-Fos expression (Mitchell et al., 1986).

Another example are PC12 cells, which can be induced to differentiate by either nerve growth factor (NGF) or dexamethasone (Kruijer et al., 1985). However, only induction by NGF results in activation of the *c-fos* gene (Kruijer et al., 1985). Remarkably, c-Fos can activate expression of the NGF gene (Hengerer et al., 1990), which suggests the initiation of an autocrine mechanism. This, however, seemingly contradicts the result that constitutive overexpression of c-Fos blocks the differentiation of PC12 cells when induced by NGF (Ito et al., 1989). It is tempting to speculate that different expression levels of c-Fos elicit opposite effects in differentiation or that c-Fos has different functions during the course of differentiation.

Another cell line system widely used to study differentiation are embryonal carcinoma (EC) cells. In undifferentiated EC cells, such as F9 or P19, c-Fos is expressed at very low levels, but is elevated in the course of differentiation to endoderm (Müller, 1983). The potential involvement of c-Fos in the differentiation of EC cells was tested by transfection of different *c-fos* gene constructs into F9 EC cells (Müller and Wagner, 1984; Rüther et al., 1985). Here, the spontaneous differentiation frequency of these cells was clearly increased as a consequence of c-Fos overexpression. These data are supported by *fos*–antisense experiments in F9 cells where blocking fos expression led to inhibition of the differentiation to endoderm (Edwards et al., 1988).

Further linkage of fos function to differentiation was investigated in B cells. One consequence of elevated expression of c-Fos in transgenic mice (Rüther et al., 1988) is that it appears to interfere with B cell function. This was analyzed in primary cultures of B cells isolated from different transgenic mouse lines. First, constitutive c-Fos expression blocked the differentiation of B cells to IgG_1-producing cells (Koizumi et al., 1993). Second, by using inducible *c-fos* constructs, differentiation of B cells to IgG_2b-producing cells was augmented when Fos was expressed only during the first 2 days after induction of differentiation. However, IgG_2b production was suppressed when c-Fos was further expressed at day 3 of differentiation (Takada et al., 1993). Thus, c-Fos might have different functions at various times during the differentiation of B cells.

Finally, organ culture systems have been used to investigate the pattern of c-Fos expression in the course of differentiation of osteogenic progenitors. In mouse mandibular condyles, cells of the progenitor zone differentiate and form new bone during *in vitro* cultivation. There is evidence that these cells express high levels of c-Fos prior to activation of genes characteristic of osteoblasts (Closs et al., 1990).

All these *in vitro* data suggest that c-Fos is a gene product that can either initiate or block certain differentiation processes. However, whether c-Fos is essential in differentiation by itself or just a component in one of several pathways could not be investigated in any of these systems.

IN VIVO: c-fos EXPRESSION IN MICE

PROFILE OF c-fos EXPRESSION

The first report about the expression of c-Fos in mice (Müller et al., 1982) described it as restricted to the extraembryonic tissues and placenta in mouse development and to bone and skin in adult mice. Later studies defined, by means of *in situ* analyses, the temporal and spatial pattern of c-Fos expression more precisely. Following ontogeny, c-Fos is first expressed in the trophectoderm of the preimplantation blastocyst (Whyte and Stewart, 1989). In the next stage analyzed, namely, late midgestation (day 13.5 to 14.5 of mouse development), c-Fos is expressed in the mesodermal web tissue of the digits, the growth regions of developing bones, and cartilage (Sandberg et al., 1988; Heckl and Wagner, 1989). In late gestation (day 17 of mouse development), high levels of c-Fos are found again in the mesodermal web tissue and the growth regions of long bones (Dony and Gruss, 1987; Togni et al., 1988; Heckl and Wagner, 1989). In addition, c-Fos is expressed in the intestine, developing cartilage, and the spinal cord as well as in certain structures in the peripheral nervous system (Caubet, 1989; Heckl and Wagner, 1989).

Just before birth, there is a marked expression of c-Fos in almost every organ (e.g., heart, liver, thymus, skin, lung, and gut) that declines one day after birth (Kasik et al., 1987). In healthy adult mice, c-Fos is only weakly expressed. Thus, c-Fos can be considered as a developmentally regulated gene with a precise spatial and temporal pattern. This suggests a specific function in certain developmental processes.

MICE EXPRESSING ADDITIONAL c-fos

If c-Fos is a key regulator in development, one would expect its activity to be dominant. Therefore, alteration of c-Fos expression, e.g., ectopically, should lead to consequences in development. Furthermore, ectopic c-Fos expression might help to unravel the function of c-Fos. Based on this idea, different transgenic and chimeric mouse lines have been generated that overexpress c-Fos in several organs at different levels. First, when c-Fos was expressed using the human metallothionein promoter in either transgenic or chimeric mice, chondro- and osteogenic hyperplasias and tumors developed (Rüther et al., 1987, 1989; Wang et al., 1991). The development of bone-associated tumors, however, was specific for transgenic constructs in which the proto-oncogene c-Fos had been converted into the transforming version. Second, when the nontransforming c-fos proto-oncogene was linked to the murine MHC class I promoter H-2 k^b, and, thereby, overexpressed in almost every organ, mice displayed a marked alteration of the thymus architecture and B cell function was impaired (Rüther et al., 1988; Takao et al., 1991).

Thus, only certain cell types are susceptible to a dominant action of c-Fos. They belong either to the chondro–osteo lineage or are part of the hematopoietic system. Since these cells are known to express endogenous c-Fos at certain stages in development, one can speculate that the level of c-Fos is crucial for their normal development.

MICE LACKING c-fos

The ultimate proof for the function of c-Fos in differentiation and development is the analysis of mice lacking c-Fos. This is performed by the inactivation of the c-fos gene via homologous recombination in embryonic stem (ES) cells. The ES cells carrying one mutant c-fos allele are then used to generate chimeric mice that can transmit the inactivated allele to offspring. These heterozygous F$_1$ mice will produce, by brother–sister mating, mice without functional c-Fos.

Using this approach, two groups recently published their findings about c-Fos-negative mice (Johnson et al., 1992; Wang et al., 1992). The predominant phenotype in mice without c-Fos was a disturbance of bone remodeling, called osteopetrosis. In almost every bone the bone marrow cavity was reduced because of massive production of new bone. The growth plates in bone were also affected, being reduced and highly irregular in the zone of proliferating chondrocytes. However, the zone of hypertropic chondrocytes was found to be increased.

The bone changes likely lead to other phenotypes. Teeth were present, but their eruption was apparently blocked by an abnormal amount of bone in the jaw. In the thymus, the total number of thymocytes was reduced about 10-fold, whereas the relative number of mature thymocytes was increased. In the spleen, B cells were found to be 75% reduced. However, myeloid cells showed a fourfold increase. All these findings can be interpreted to result from the drastic changes in bone, which might interfere with hematopoiesis. Bone marrow transfer studies now underway will explore the direct role of c-Fos in these phenotypes.

In addition, gametogenesis in both female and male mice was affected (Johnson et al., 1992). However, this finding was not consistent for all the homozygous animals analyzed. For spermatogenesis a disturbance might be expected, since c-Fos was found to be expressed throughout sperm development (Pelto-Huikko et al., 1991). Mating of heterozygous animals revealed a non-mendelian ratio of the different genotypes that likely represents a transmission distortion in the female germline (Wang et al., 1992).

Finally, both studies indicate that Fos-negative mice display abnormal behavior, such as no reaction to stress (Johnson et al., 1992; Wang et al., 1992). However, this behavior might be the consequence of systemic bone alterations. Furthermore, histological analysis did not show any gross changes in the brain.

CONCLUSIONS

The tissue culture experiments had previously indicated that c-Fos has a distinct function in differentiation. However, because of the restricted potential of the few *in vitro* differentiation systems, several of the results were inconsistent and could even be interpreted as resulting from artificial conditions. Nevertheless, these findings, as well as the c-Fos expression profile in development, have initiated several *in vivo* experiments.

The *in vivo* studies turned out to be more consistent. First, they documented that c-Fos is essential for normal bone development. Second, the correct amount of c-Fos is important for normal bone development, otherwise c-Fos exerts a dominant activity. Third, c-Fos seems not to have an essential role in proliferation and growth control, as was believed for several years. This also agrees with the growth and differentiation behavior of Fos-negative ES cells (Field et al., 1992). Fourth, although c-Fos is essential for normal development of certain structures, it is dispensable for embryonic development, since Fos-negative mice are viable and can even mate. Thus, c-Fos can be considered as a key regulator in specific tissues, such as bone and hematopoietic cells, were it can exert a dominant function.

ACKNOWLEDGMENTS

I am very grateful to Robert Hipskind for critical reading of the manuscript.

REFERENCES

Caubet, J.-F., c-fos proto-oncogene expression in the nervous system during mouse development, *Mol .Cell. Biol.*, 9, 2269, 1989.

Closs, E. I., Murray, A. B., Schmidt, J., Schön, A., Erfle, V., and Strauss, P. G., c-fos expression precedes osteogenic differentiation of cartilage cells in vitro, *J. Cell Biol.*, 111, 1313, 1990.

Dony, C. and Gruss, P., Proto-oncogene c-fos expression in growth regions of fetal bone and mesodermal web tissue, *Nature*, 328, 711, 1987.

Edwards, S. A., Rundell, A. Y., and Adamson, E. D., Expression of c-fos antisense RNA inhibits the differentiation of F9 cells to parietal endoderm, *Dev. Biol.*, 129, 91, 1988.

Field, S., Johnson, R. S., Mortenson, R., Papaioannou, V. E., Spiegelman, B. M., and Greenberg, M. E., Fos is not required for the growth or differentiation of ES cells, *Proc. Natl. Acad. Sci. U.S.A.*, 89, 9306, 1992.

Heckl, K. and Wagner, E. F., In situ analysis of c-fos expression in transgenic mice, in *Molecular Genetics of Early Drosophila and Mouse Development*, Capecchi, M. R., Ed., Cold Spring Harbor Laboratory Press, Cold Spring Harbor, 1989.

Hengerer, B., Lindholm, D., Heumann, R., Rüther, U., Wagner, E. F., and Thoenen, H., Lesion-induced increase in nerve growth factor mRNA is mediated by c-fos, *Proc. Natl. Acad. Sci. U.S.A.*, 87, 3899, 1990.

Ito, E., Sonnenberg, J. L., and Narayanan, R., Nerve growth factor-induced differentiation in PC-12 cells is blocked by fos oncogene, *Oncogene*, 4, 1193, 1989.

Johnson, R. S., Spiegelman, B. M., and Papaioannou, V., Pleiotropic effects of a null mutation in the c-fos proto-oncogene, *Cell*, 71, 577, 1992.

Kasik, J. W., Wan, Y. Y., and Ozato, K., A burst of c-fos gene expression in the mouse occurs at birth, *Mol. Cell. Biol.*, 7, 3349, 1987.

Koizumi, T., Ochi, Y., Imoto, S., Sakai, N., Takao, S., Kobayashi, S., Matsuoka, M., Sakano, S., Rüther, U., and Tokuhisa, T., Deregulated c-fos modulates B cell responses to switch mediators, *Cell. Immunol.*, 149, 82, 1993.

Kruijer, W., Schubert, D., and Verma, I. M., Induction of the proto-oncogene fos by nerve growth factor, *Proc. Natl. Acad. Sci. U.S.A.,* 82, 7330, 1985.

Mitchell, R. L., Zokas, L., Schreiber, R. D., and Verma, I. M., Rapid induction of the expression of proto-oncogene fos during human monocytic differentiation, *Cell,* 40, 209, 1985.

Mitchell, R. L., Henning-Chubb, C., Huberman, E., and Verma, I. M., c-fos expression is neither sufficient nor obligatory for differentiation of monomyelocytes to macrophages, *Cell,* 45, 497, 1986.

Müller, R., Differential expression of cellular oncogenes during murine development and in teratocarcinoma cell lines, *Cold Spring Harbor Conferences in Cell Proliferation,* Vol. 10, p. 451, 1983.

Müller, R. and Wagner, E. F., Differentiation of F9 teratocarcinoma stem cells after transfer of c-fos proto-oncogenes, *Nature,* 311, 438, 1984.

Müller, R., Slamon, D. J., Tremblay, J. M., Cline, M. J., and Verma, I. M., Differential expression of cellular oncogenes during pre- and postnatal development of the mouse, *Nature,* 299, 640, 1982.

Müller, R., Müller, D., and Guilbert, L., Differential expression of c-fos in hematopoietic cells: Correlation with differentiation of monomyelocytic cells in vitro, *EMBO J.,* 3, 1887, 1984.

Pelto-Huikko, M., Schultz, R., Koistinaho, J., and Hökfelt, T., Immunocytochemical demonstration of c-fos protein in sertoli cells and germ cells in rat testis, *Acta Physiol. Scand.,* 141, 283, 1991.

Rovera, G., O'Brien, T. G., and Diamond, L., Induction of differentiation in human promyelocytic leukemia cells by tumor promoters, *Science,* 204, 868, 1979.

Rüther, U., Wagner, E. F., and Müller, R., Analysis of the differentiation-promoting potential of inducible c-fos genes introduced into embryonal carcinoma cells, *EMBO J.,* 4, 1775, 1985.

Rüther, U., Komitowski, D., Schubert, F. R., and Wagner, E. F., c-fos expression induces bone tumors in transgenic mice, *Oncogene,* 4, 861, 1989.

Rüther, U., Garber, C., Komitowski, D., Müller, R., and Wagner, E. F., Deregulated c-fos expression interferes with normal bone development in transgenic mice, *Nature,* 325, 412, 1987.

Rüther, U., Müller, W., Sumida, T., Tokuhisa, T., Rajewsky, K., and Wagner, E. F., c-fos expression interferes with thymus development in transgenic mice, *Cell,* 53, 847, 1988.

Sandberg, M., Vuorio, T., Hirvonen, H., Alitalo, K., and Vuorio, E., Enhanced expression of TGF-(beta) and c-fos mRNAs in the growth plates of developing human long bones, *Development,* 102, 461, 1988.

Takada, M., Koizumi, T., Bachiller, D., Rüther, U., and Tokuhisa, T., Deregulated c-fos modulates IgG2b production of B cells mediated by lipopolysaccharide, *Immunobiology,* 188, 233, 1993.

Takao, S., Sakai, N., Hatano, M., Hanioka, K., Rüther, U., and Tokuhisa, T., IgG response is impaired in H2-c-fos transgenic mice, *Int. Immunol.,* 3, 369, 1991.

Togni, P., Niman, H., Raymond, V., Sawchenko, P., and Verma, I. M., Detection of fos protein during osteogenesis by monoclonal antibodies, *Mol. Cell. Biol.,* 8, 2251, 1988.

Wang, Z.-Q., Grigoriadis, A. E., Möhle-Steinlein, U., and Wagner, E. F., A novel target cell for c-fos induced oncogenesis: Development of chondrogenic tumours in embryonic stem cell chimeras, *EMBO J.,* 10, 2437, 1991.

Wang, Z.-C., Ovitt, C., Grigoriadis, A. E., Möhle-Steinlein, U., Rüther, U., and Wagner, E. F., Bone and haematopoietic defects in mice lacking c-fos, *Nature,* 360, 741, 1992.

Whyte, A. and Stewart, H. J., Expression of the proto-oncogene fos (c-fos) by preimplantation blastocysts of the pig, *Development,* 105, 651, 1989.

Chapter 13

Modulation of Fos and Jun in Response to Adverse Environmental Agents

Hans Jobst Rahmsdorf

CONTENTS

INTRODUCTION

Adverse environmental influences such as heat, radiation, or high concentrations of heavy metals not only lead to the destruction of cellular integrity, but also induce several ordered processes that involve productive cellular reactions. These reactions include the activation of transcription factors, the transcription of RNAs, and the translation of proteins that are not synthesized in nonexposed cells; a general cessation of replication; or, alternatively, the overreplication of certain parts of the genome. Several endpoints of these productive cellular reactions are conceivable: the cells may return to their preaffected state; or die, in many cases in an ordered process (apoptosis); or, and this will happen only to a minority of cells, will acquire mutations in regulatory genes involved in the control of cell growth or differentiation, which may lead to the outgrowth of cell clones with a transformed phenotype. Which of these possible endpoints prevails may depend on the noxious agent involved and the dosage. Heat shock, for instance, and certain plant metabolites, such as the tumor promoter 12-*O*-tetradecanoyl-phorbol-13-acetate (TPA), do not transform cells efficiently, while certain chemicals and radiation qualities do.

Common to all environmental adversities, however, is that cells react immediately with the activation of transcription factors and the subsequent enhanced transcription of certain genes. This chapter will provide examples of this response by discussing activation and synthesis of members of the AP-1 family. In cells encountering environmental noxes, the basic equipment with Jun and Fos that most cells maintain is subject to activation by posttranslational modification. *De novo* synthesis of both proteins also contributes to the higher activity of AP-1 in cells encountering environmental adversities. First, I will describe our present knowledge of stress-induced posttranslational activation of AP-1. Second, I will discuss the processes involved in enhanced Fos and Jun synthesis. The signal chain leading from the primary interaction of the noxious agent with the cell to the activation of AP-1 will be described in a third section. Finally, I will discuss the possible consequences of enhanced AP-1 activity for the fate of the cell.

POSTTRANSLATIONAL MODIFICATION OF AP-1

To date, induction of posttranslational modification of AP-1 by nonphysiologic agents has only been investigated with a restricted assortment of noxes. The tumor promoter TPA was the first to be shown to modulate AP-1 activity (Angel et al., 1987). TPA activates directly a second messenger pathway that may also be addressed by several growth factors, the protein kinase C pathway (Nishizuka, 1986). The other process that was investigated is irradiation with short-wavelength ultraviolet light (UV-C). UV induces, as does TPA, the transcription of the collagenase gene through the AP-1 binding site in the promoter of this gene (Stein et al., 1989b). Other agents from our environment that induce transcription of AP-1-dependent genes, such as gamma-rays, 4-nitroquinoline-oxide, hydroxy urea, radical inducing agents, and radical scavengers, have not yet been investigated for their effects on the posttranslational modification of AP-1. It is likely that they also induce modifications, but it is not known whether these modifications are similar or different from those induced by UV and TPA. The posttranslational modification of isolated AP-1 by redox reactions *in vitro* and the amino acid residues and accessory proteins possibly involved in this process (Abate et al., 1990; Xanthoudakis and Curran, 1992; Oehler et al., 1993) will not be discussed here; they may play a role in the activation of AP-1-dependent genes by antioxidants (see below).

PHORBOL ESTERS AND UV INDUCE DIFFERENT MODIFICATIONS OF c-FOS

Fos is a phosphoprotein (Curran et al., 1984; Barber and Verma, 1987; Müller et al., 1987; Lee et al., 1988). Thus, it can be labeled with either S-35-methionine or P-32. When precipitated with antibodies and run out by one-dimensional gel electrophoresis, Fos forms a smeary band of approximately 55 kDa, suggesting that the Fos protein from untreated cells is heterogeneously modified. Fos protein isolated from HeLa cells at 5 min after TPA treatment migrates more slowly. This slowly migrating form persists in the cells for at least 45 min, and even 2 hours after treatment some of the Fos protein still is in this form. The changed mobility of the c-Fos protein does not seem to be due to a change in phosphorylation, since phosphatase treatment of Fos immunoprecipitated from control and TPA-treated cells does not abolish the difference in migration of the two Fos species. UV, in contrast to TPA, does not induce the formation of the slowly migrating Fos form (Gebel, 1992).

Peptide analysis of ^{32}P-labeled Fos protein from untreated, TPA-treated, and UV-treated cells reveals similar induced changes in the phosphopeptide patterns. Upon trypsin digestion of the Fos protein from untreated cells, three phosphopeptides are seen. TPA or UV treatment of cells induces the dephosphorylation of two peptides and an enhanced phosphorylation of the third one (Gebel, 1992). Digestion of the Fos protein with chymotrypsin leads to the same result: TPA and UV induce a similar change in the phosphopeptide pattern. These findings demonstrate that two nonphysiologic agents, UV and TPA, send both similar and dissimilar signals to the Fos protein; the major modification seen in one-dimensional gel electrophoresis is only induced by treatment of the cells with TPA.

MODIFICATIONS OF c-JUN INDUCED BY PHORBOL ESTERS AND UV-C

A similar result as with the c-Fos protein is seen with the c-Jun protein; again, some of the modifications appear to be similar in UV- and TPA-treated cells, while others seem to differ. In one-dimensional resolutions, TPA induces a form of the c-Jun protein that migrates slightly more slowly than the c-Jun protein from untreated cells. This slight alteration in the migration of c-Jun with TPA treatment of cells seems to be due to TPA-induced phosphorylation, since phosphatase treatment of c-Jun from untreated and TPA-treated cells leads to a c-Jun form of identical migration in one-dimensional gel electrophoresis. With two-dimensional resolutions of tryptic digests of c-Jun precipitated from nontreated and TPA-treated HeLa cells, two differences in the phosphopeptide pattern appear. A peptide in the DNA binding domain, which in untreated cells is phosphorylated at three positions, loses one of its phosphate groups (Boyle et al., 1991). Moreover, a phosphopeptide (x) derived from the transactivation domain of c-Jun, whose phosphorylation depends on serine 73, is hyperphosphorylated (Pulverer et al., 1991; Radler-Pohl et al., 1993).

UV induces two forms of c-Jun by posttranslational modification: one form migrating at the same position as the TPA-induced form, and displaying an identical phosphopeptide pattern; and one with apparent larger molecular weight. This larger form also shows the loss of a phosphate group in the DNA binding domain, the hyperphosphorylation of phosphopeptide x and peptide y (which depends on serine 63), and the appearance of peptide v, which has not yet been mapped definitively (Devary et al., 1992; Radler-Pohl et al., 1993).

THE POSTTRANSLATIONAL MODIFICATIONS OF AP-1
APPEAR TO BE FUNCTIONALLY RELEVANT

The posttranslational modifications induced by TPA and UV could be fortuitous byproducts with no relevance for the function of the transcription factor AP-1, or they could mediate carcinogen-induced activation of AP-1. Evidence of carcinogen-induced activation of AP-1, is given by the finding that gene constructs driven by an AP-1 binding site are transcriptionally induced within minutes after treatment of cells with phorbol esters or UV, long before increased levels of AP-1 can be detected. Moreover, transcriptional activation also occurs in the absence of protein synthesis (Radler-Pohl et al., 1993). That this transcriptional activation is mediated by members of the AP-1 family follows from experiments showing that deprivation of cells from Fos and Jun by the antisense technique inhibits carcinogen-induced transcription of AP-1-dependent genes (Schönthal et al., 1988a, 1988b). The capability of overexpressed Jun to increase transcription from AP-1-dependent genes (Angel and Karin, 1991) has been used to determine whether the induced modifications described above may be relevant for the capability of the c-Jun protein to activate transcription. The serines shown to be dephosphorylated (serine 243 in the DNA binding domain) or hyperphosphorylated (serines 63 and 73 in the transactivation domain) were exchanged against amino acids that cannot be phosphorylated [phenylalanine in the case of serine 243; and leucine (Pulverer et al., 1991) or, more conservatively, alanine (Smeal et al., 1991; Baker et al., 1992) in the case of serines 63 and 73]. Whereas exchange of phenylalanine for serine 243 elevates the transactivating capability of c-Jun, exchange of serines 63 and 73 for leucines, which have a much bulkier side group as compared to serines, lowers the transactivating ability of the c-Jun protein. It is not clear whether this is because leucine cannot be phosphorylated or whether leucine changes the conformation of the protein. Substitution of the serines by alanine does not affect the ability of c-Jun to transactivate reporters. It does, however, inhibit the cooperation of ras and c-Jun in the induction of an AP-1-dependent reporter (Smeal et al., 1991); the replacement of serines 63 and 73 by leucines or alanines inhibits the UV-induced shift of c-Jun to the slow migration position and UV-induced c-Jun activation (Devary et al., 1992; Radler-Pohl et al., 1993). In summary, these experiments demonstrate that post-translational modifications of c-Jun activate its transcriptional activity.

Similar evidence has been collected for the c-Fos protein. As described above, the major structural modification induced by phorbol esters is a shift in migration. In order to determine the site in the protein responsible for this shift, carboxy terminal deletions of the protein were analyzed after transient transfection. A mutant lacking the 6 carboxy terminal amino acids shows the TPA-induced shift in migration; when a further 45 amino acids are deleted, the TPA-induced shift is lost; in addition, TPA loses its capability to transactivate a Gal4–fos hybrid protein directing the expression of a Gal4-dependent reporter (Gebel, 1992).

INDUCED TRANSCRIPTION OF THE *c-fos* AND THE *c-jun* GENES

In contrast to the situation with regard to posttranslational modification of AP-1, many more adverse agents have been examined for their ability to induce the transcription of the genes coding for AP-1 family members. Induced transcription is, of course, the result of posttranslational activation of other transcription factors. The adverse agents examined include UV irradiation, gamma-rays, neutrons, heat, magnetism, and chemicals such as phorbol esters, oxidants, antioxidants, DNA modifying chemicals, and heavy metals. All these agents have been shown to induce the transcription of either one or both proto-oncogenes (see Table 13-1).

INDUCED TRANSCRIPTION OF THE *c-fos* GENE

With regard to *c-fos* gene transcription, it may be easier to enumerate agents that do not induce the transcription of the gene than to name the agents that do induce. Practically all agents that disturb the homeostasis of the cell seem to induce the transcription (or at least mRNA accumulation) of the gene, albeit, it seems, with vastly different efficiencies. The fact that most agents interfering with cellular equilibrium induce *c-fos* gene transcription, should, however, not imply that this process is a "nonspecific" cellular reaction. The steps involved in induction and the sequences in the *c-fos* gene responsible for induction can be clearly defined.

Most agents induce *c-fos* RNA accumulation in a rapid and transient manner, and it has been shown for several agents, by nuclear "run on" analyses, that RNA accumulation is due to induced transcription. The first radiation quality that was shown to induce *c-fos* gene transcription was UV-C (Angel et al., 1985). Later UV-B irradiation (280 to 320 nm) was also shown to induce *c-fos* RNA accumulation (Hollander and Fornace, 1989; Shah et al., 1993). However, the ability of UV light to induce *c-fos* gene transcription (per quantum UV light absorbed) falls rapidly toward wavelengths around 300 nm, concomitantly with the reduced ability of such wavelengths to induce thymidine dimers and cell killing (Stein et al., 1989b). Whereas UV-C induces maximal *c-fos* RNA accumulation at doses (30 J/m^2) at which around 10% of the cells survive and in many cell lines examined, gamma irradiation induces appreciable *c-fos* RNA accumulation only at much higher doses (around 50 Gy). Among the cell lines examined, gamma induced accumulation only in human HL60 promyelocytic leukemia cells, but not in the human epithelial tumor cell line SQ-20B, HeLa S3 cells, or two normal human cell lines derived from fibroblasts (AG-1522) and kidney epithelium (cell line 293) (Sherman et al., 1990; Hallahan et al., 1991). The low and variable induction of *c-fos* by gamma irradiation as compared to UV suggests that the bulky DNA distortions induced by UV are more efficient signaling molecules than the products generated by gamma irradiation (see below). Modulation of *c-fos* gene expression has also been seen after irradiation of whole animals with long-wavelength UV, gamma-rays, or neutrons (Brunet and Giacomoni, 1990; Munson and Woloschak, 1990).

Two other physical agents have recently been shown to induce c-fos RNA accumulation: a static magnetic field (Hiraoka et al., 1992) and heat shock (Andrews et al., 1987; Hollander and Fornace, 1989; Colotta et al., 1990; Bukh et al., 1990). Heat shock induces fos RNA accumulation through both increased transcription of the gene and mRNA stabilization. Furthermore, altered gravity conditions appear to influence transcription from the c-fos (and the c-jun) gene (de Groot et al., 1990, 1991).

In addition to physical agents, a plethora of chemicals induces c-fos RNA accumulation: anorganic ions such as potassium, cadmium, zinc, barium, and arsenite (Morgan and Curran, 1986; Curran and Morgan, 1986; Andrews et al., 1987; Colotta et al., 1990); various metabolites derived from plants and animals, such as the tumor promoters 12-O-tetradecanoyl-phorbol-13-acetate, thapsigargin, and okadaic acid (Greenberg and Ziff, 1984; Prywes and Roeder, 1986; Ran et al., 1986; Schönthal et al., 1991b, 1991c); and DNA modifying chemicals, such as 4-nitroquinoline-oxide, methyl-methane-sulfonate, cis-Pt (II) diamminedichloride, adriamycin, N-acetoxy-2-acetylaminofluorene, or N-methyl-N-nitro-N-nitrosoguanidine (Hollander and Fornace, 1989). Hydrogen peroxide and other radical-inducing treatments have also been shown to induce fos RNA accumulation (Shibanuma et al., 1988, 1990; Crawford et al., 1988; Nose et al., 1991). Recent experiments show that not only oxidants but also antioxidants (such as pyrrolidone dithiocarbamate and N-acetyl-L-cysteine) activate c-fos (and c-jun) synthesis (Meyer et al., 1993).

Although in most cases the induction of c-fos RNA accumulation is fast but transient due to the rapid block of transcription through the newly synthesized c-Fos protein (Schönthal et al., 1988b, 1989; Sassone-Corsi et al., 1988; Lucibello et al., 1989; König et al., 1989; Shaw et al., 1989a) and the short half-life of its RNA (Cochran et al., 1983; Müller et al., 1984; Greenberg and Ziff, 1984; Kruijer et al., 1984; Rahmsdorf et al., 1987), exceptions have been noted. Both heat shock and inhibitors of protein synthesis induce a more long-lasting accumulation of c-fos RNA, possibly because these agents stabilize the RNA through the inhibition of protein synthesis. The tumor promoter okadaic acid has also been shown to induce prolonged c-fos RNA accumulation, but in this case it has been argued that the inhibition of protein synthesis by the tumor promoter does not cause the abnormal kinetics (Schönthal et al., 1991c). Instead, okadaic acid may interfere with the inactivation of the transcription factor controlling the expression of the c-fos gene (Zinck et al., 1993). In mouse 3T3 cells, UV, in contrast to phorbol esters, induces a second round of c-fos RNA accumulation after the first transient increase (Rahmsdorf et al., 1992), and in F9 teratocarcinoma stem cells UV induces a rapid increase in fos RNA levels, which continues for at least 6 hours (Auer et al., 1994). Moreover, in murine JB6 epidermal cells UV-B induces two rounds of c-fos RNA accumulation (Shah et al., 1993). Although the molecular details of these unusual kinetics have not yet been studied, two mechanisms come to mind: stabilization of c-fos RNA by the inducer, and the induced secretion of growth factors (Krämer et al., 1993), which may act back on the cells and induce a second round of c-fos RNA accumulation.

The elements in the c-fos gene responsible for transcriptional induction have been mapped by deletion analysis and by ligating portions of the gene to reporter genes, with subsequent analyses of these recombinant genes in stable and transient transfection assays (see Chapter 8). Induced transcription by all adverse agents [UV-C (Büscher et al., 1988), UV-B (Shah et al., 1993), phorbol esters (Fisch et al., 1987; Büscher et al., 1988), the tumor promoter thapsigargin (Schönthal et al., 1991b), and active oxygen species (Amstad et al., 1992)] is mediated through the same enhancer element that mediates serum induction of the gene: the serum response or dyad symmetry element located between positions −300 and −320 (Treisman, 1985). This element is addressed and always occu-

Table 13-1 Nonphysiological Agents Modulating the Expression of c-fos and c-jun

Agent	Dose Range	Cells/Tissue	c-fos	c-jun	Ref.
UV-C	$2-30 \text{ J/m}^2$	HeLa	+	+	Angel et al., 1985
		Human fibroblasts,			Ronai et al., 1988
		3T3, F9, CHO,			Büscher et al., 1988
		keratinocytes			Hollander and Fornace, 1989
					Devary et al., 1991
					Stein et al., 1989b
					Stein et al., 1992
UV-B	300 J/m^2	CHO	+	n.d.	Hollander and Fornace, 1989
	$2.000-12,000 \text{ J/m}^2$	JB6 epidermal cells	+	n.d.	Stein et al., 1989b
					Shah et al., 1991
UV-A		Hairless mouse epidermis	+	n.d.	Brunet and Giacomoni, 1990
Gamma-rays	$2-50 \text{ Gy}$	HL60	+	+	Sherman et al., 1990
		U-937, human diploid fibroblasts	+	+	Hallahan et al., 1991
Static magnetic field	$0.18-0.2 \text{ T}$	HeLa	+	n.d.	Hiraoka et al., 1992
Gravity		A431	–	–	de Groot et al., 1990, 1991
Heat shock		HeLa	+	n.d.	Andrews et al., 1987
		CHO	+	n.d.	Hollander and Fornace, 1989
		3T3	+	n.d.	Colotta et al., 1990
		Lymphoid cells	+	+	Bukh et al., 1990
Barium Chloride	$1 \text{ m}M$	PC12	+	n.d.	Curran and Morgan, 1986
Potassium chloride	$5 \text{ m}M$	PC12	+	n.d.	Morgan and Curran, 1986
Cadmium chloride + zinc chloride	$10^{-4} M$	HeLa	+	n.d.	Andrews et al., 1987
Sodium arsenite	$80 \text{ μ}M$	HeLa	+	n.d.	Andrews et al., 1987
		3T3	+	n.d.	Colotta et al., 1990

					References
12-O-Tetradecanoyl-phorbol-13-acetate	1–100 ng/ml	Many cell lines	+	+	Greenberg and Ziff, 1984; Prywes and Roeder, 1986; Ran et al., 1986; Lamph et al., 1988; Devary et al., 1991
Thapsigargin	10–1,000 nM	3T3	+	+	Schönthal et al., 1991b
Okadaic acid	250–1,000 nM	3T3	+	+	Schönthal et al., 1991a,c
4-Nitroquinoline-oxide	1 µg/ml	CHO	+	n.d.	Hollander and Fornace, 1989
Methyl-methane-sulfonate	200 µg/ml	CHO	+	n.d.	Hollander and Fornace, 1989
cis-Pt(II)diamminedichloride	45 µg/ml	CHO	+	n.d.	Hollander and Fornace, 1989
Adriamycin	2 µg/ml	CHO	+	n.d.	Hollander and Fornace, 1989
N-acetoxy-acetylaminofluorene	20 µM	CHO	+	n.d.	Hollander and Fornace, 1989
N-methyl-N-nitro-N-nitrosoguanidine	30 µM	CHO	+	n.d.	Hollander and Fornace, 1989
cis-Pt(II)diamminedichloride	10^{-4}–10^{-5} M	Human myeloid leukemia cells	+	+	Rubin et al., 1992
Bifunctional alkylating agents		Colo 320HSR	+	n.d.	Futscher and Erickson, 1990
Radical inducing treatment		3T3, JB6 mouse epidermal cells, mouse osteoblastic cells, HeLa	+	+	Shibanuma et al., 1988, 1990; Crawford et al., 1988; Nose et al., 1991; Devary et al., 1991; Amstad et al., 1992
Radical scavengers		HeLa	+	+	Meyer et al., 1993

Note: +, enhanced expression; –, reduced expression; n.d., not determined pied by a multicomponent transcription factor complex (Herrera et al., 1989; König, 1991). Several proteins bind to this element (see Chapter 6); between these proteins is the serum response factor (SRF) (Treisman, 1987; Schröter et al., 1987; Prywes and Roeder, 1987; Norman et al., 1988), which binds as a dimer. SRF complexes with a 62-kDa protein, the ternary complex factor (p62^{TCF}), that has been shown to belong to the family of ets proteins (Shaw et al., 1989b; Schröter et al., 1990). SRF, p62^{TCF}, and the serum response element form a ternary complex, which allows transcription from the c-fos promoter. It is not clear, which of the proteins receives the signal generated by agents inducing the transcription of the c-fos gene. It has been shown recently that the mitogen activated protein kinase 2 (MAP-2 kinase) modifies p62^{TCF} in vitro and that this modification may lead to a change in the electrophoretic mobility of the ternary complex (Gille et al., 1992; Zinck et al., 1993; Marais et al., 1993). Moreover, the kinetic of induced TCF modification corresponds exactly to the kinetic of induced c-fos transcription (Zinck et al., 1993; see also Chapters 6 and 8). Thus, since several of the environmental agents known to induce transcription of the c-fos gene activate MAP-2 kinase (see below), p62^{TCF} may well be at the receiving end of the signal chain.

INDUCED TRANSCRIPTION OF THE c-jun GENE

Because the c-jun gene was discovered several years after the c-fos gene, the list of agents influencing transcription of this gene is not as long as it is for the c-fos gene. However, the list of agents inducing the transcription of the two genes is largely overlapping (see Table 13-1), and in several investigations both genes have been studied in parallel. Coordinate induction of both genes by most treatments does not imply that the activation of one of the two genes depends on the product of the other gene, since both genes are also activated in the absence of ongoing protein synthesis. Rather, the primary signal reaching the cell seems to feed into a signal cascade that reaches both genes.

UV-C, gamma-rays, active oxygen species, and various tumor promoters have been investigated for their capacity to induce the transcription of the members of the jun family. While junD mRNA does not fluctuate appreciably, when cells are treated with UV, both junB and c-jun RNA accumulate with kinetics similar to those seen with c-fos (Krämer et al., 1990; Devary et al., 1991; Stein et al., 1992). In contrast to the situation with c-fos, which, in the cell lines examined, is induced by gamma-rays only in HL60 cells, c-jun is induced by (rather high doses of) gamma-rays in several cell lines, including HL60 cells (promyelocytic leukemia cells), U-937 cells (monocytic leukemia cells), AG-1522 cells (diploid foreskin fibroblasts), 293 cells (kidney epithelial cells), and SQ-20B cells (epithelial tumor cells) (Hallahan et al., 1991). In addition, H_2O_2 and the tumor promoters TPA, thapsigargin, and okadaic acid induce transcription of the c-jun and junB genes (Lamph et al., 1988; Devary et al., 1991; Schönthal et al., 1991a, 1991b).

As with the c-fos gene, most of the regulation of c-jun gene expression is at the level of transcription, although posttranscriptional components also may be involved (Sherman et al., 1990; Schönthal et al., 1991a). The first cis-acting element in the c-jun promoter that was shown to respond to external stimuli was a sequence between positions −71 and −64 that resembles an AP-1 binding site. Sequence similarity and the finding that this site responds to overexpressed c-jun protein were taken as an indication, that c-jun positively regulates its own promoter (Angel et al., 1988). A second AP-1-like binding site is located between positions −190 and −183. Both sites mediate additively the inducibility of the c-jun promoter by UV, interleukin-1α, the viral protein E1A, and phorbol esters (Devary et al., 1991; Stein et al., 1992; Mügge et al., 1993; van Dam et al., 1993; Herr et al., 1994). The factors binding to these sites are just being characterized. For both sequences heterodimers of c-Jun and ATF-2 (and perhaps other relatives of ATF-2) seem to be the important players and at the receiving end of signal transduction pathways (van Dam et al., 1993).

Interestingly, the AP-1 binding sites in the *c-jun* promoter are also occupied under noninduced conditions (Herr et al., 1994). Regulation of activity of the binding transcription factors appears to occur while bound to the DNA.

HOW DO ENVIRONMENTAL ADVERSITIES INDUCE ACTIVATION OF AP-1 AND ENHANCED SYNTHESIS OF FOS AND JUN?

In the bacterium *Escherichia coli* several proteins monitor the presence of environmental adversities and translate them into a change in the program of genes expressed. UV-induced DNA damage, or a consequence of it, is recognized by recA protein, whose activation initiates the SOS response (Walker, 1985). Alkylation damage to DNA induces the adaptive response, which involves increased transcription of the gene coding for O^6-methyl-guanine-DNA methyltransferase (Vaughan et al., 1991). Oxidative stress enhances, possibly directly, the activity of the two transcription factors oxyR and soxRS, which in turn activate the expression of dependent genes (Demple, 1991). In mammalian cells, the molecules recognizing the inducer and translating its presence into biochemical reactions have been only partially characterized. UV-induced collagenase gene transcription, which is detectable only hours after UV irradiation, depends critically on the persistence of DNA damage. This is suggested by the finding that *Xeroderma pigmentosum* cells, which cannot repair UV-induced DNA damage, are induced to transcribe the gene at much lower UV doses than those needed in wild-type cells (Schorpp et al., 1984; Stein et al., 1989b). Thus, UV-induced DNA damage may be the initiator of UV-induced collagenase (XHF1 in Schorpp et al., 1984) gene transcription. Since the repair process takes hours, the same sort of conclusion cannot be drawn for UV-induced endpoints, which occur minutes after UV irradiation, such as AP-1 activation or the transcription of immediate early genes. Other approaches have to be taken in the future to determine whether UV-induced DNA lesions are intermediates in this signal chain. Such experiments could involve the use of cells that overexpress the enzyme photolyase (which reverts immediately thymidine dimers); the injection of UV-damaged DNA into the nucleus (in order to investigate directly whether damaged DNA is the triggering signal for the UV-induced immediate reactions); and the use of enucleated cells (cytoplast). As our preliminary experiments with cytoplasts showed, in contrast to nucleated cells, no UV-induced MAP-2 kinase activation (see below), and for the sake of simplicity, I hypothesize that the immediate cellular reactions to UV are due to UV-induced DNA damage. Whether this assumption is also applicable for other DNA damaging agents, such as reactive oxygen species, gamma-rays, or alkylating agents, is yet unclear.

The situation is much better with most of the tumor promoters that activate AP-1 in that the primary interacting molecules have been defined. TPA binds to and activates the phospholipid- and Ca-dependent protein kinase C (Nishizuka, 1986); okadaic acid inhibits specifically the protein phosphatases 1 and 2A (Bialojan and Takai, 1988; Hescheler et al., 1988); and thapsigargin discharges intracellular calcium stores by inhibition of the Ca^{2+}-ATPase of the endoplasmatic reticulum (Thastrup et al., 1990). It is currently accepted that these tumor promoters feed into signal transduction pathways that are activated also by growth factors.

Thus, at least two signal transduction pathways, one initiated at the cell surface or in the cytoplasm, and the other one possibly initiated in the cell nucleus, lead to the activation of AP-1 and the transcription factors that control the expression of *c-fos* and *c-jun*. Where do these signal transduction pathways meet? Since it has been shown that UV-irradiated cells release growth factors that transmit the UV response to nonirradiated cells (Schorpp et al., 1984; Rotem et al., 1987; Stein et al., 1989a) and that UV irradiation induces the translocation of the cytoplasmic transcription factor NFκB into the nucleus (Stein et al., 1989b), UV apparently triggers stimulation of membranal and cytoplasmic pathways. The question is whether these are needed for the posttranslational modification of AP-1 and for the enhanced transcription of *c-fos* and *c-jun* (Devary et al., 1992; Herrlich et al., 1992; Radler-Pohl et al., 1993). Within minutes after UV irradiation the tyrosine kinases Src and Fyn are activated (Devary et al., 1992). Ras is converted from the GDP-binding form to the GTP-binding form, and the cytoplasmic protein kinases c-Raf and MAP-2 are activated (Devary et al., 1992; Radler-Pohl et al., 1993). Moreover, mutant forms of Ras, Src, Raf, and MAP-2, which interfere with the function of the respective endogenous proteins, inhibit UV-induced transcriptional activation of AP-1-dependent reporter gene constructs (Devary et al., 1992; Radler-Pohl et al., 1993; Sachsenmaier et al., 1994; and submitted). These findings suggest that the UV-induced signal transduction pathway feeds into the pathways activated by growth factors and tumor promoters and that Src, Ras, Raf, and MAP-kinases are obligatory intermediates in the UV-induced signal transduction chain.

Is it possible that the UV-induced signal chain involves the activation of growth factors and their receptors and elicits an autocrine growth factor cycle? As mentioned above, after treatment of HeLa cells or human primary fibroblasts with UV, growth factors are released from the irradiated cells. HeLa cells, for example, release interleukin-1α and basic fibroblast growth factor into the culture medium (Krämer et al., 1993). These growth factors act back on the irradiated cell and induce or enhance collagenase mRNA accumulation. This has been demonstrated by the inhibitory effect of suramin, a drug that inhibits growth factor–receptor interactions (Betsholtz et al., 1986), on UV-induced collagenase mRNA accumulation, and by the finding that antibodies directed against the two growth factors interfere severely with this process (Krämer et al., 1993). Two sets of data suggest that not only the late effects of UV, but also the early effects (activation of MAP-2 kinase, modification of c-Jun, induced transcription of *c-fos* and *c-jun*) are mediated by growth factor receptors: (1) suramin inhibits these early effects without interfering with the same processes induced by phorbol ester, and (2) pretreatment of cells with UV (which leads to transient activation of MAP-2 kinase disappearing 30 to 60 min after UV) (Radler-Pohl et al., 1993) inhibits strongly subsequent activation of MAP-2 kinase not only by UV, but also by interleukin-1α, basic fibroblast growth factor, or EGF, without interfering with phorbol ester-induced MAP-2 kinase activation. Also, pretreatment of cells with interleukin-1α or basic fibroblast growth factor inhibits a subsequent stimulation of MAP-2 kinase by UV, suggesting that the growth factor and UV-C-induced signal transduction pathways share common exhaustible components, possibly the growth factor receptors themselves (Sachsenmaier et al., submitted). Protein kinase C family members do not seem to be downregulated after UV irradiation (Büscher et al., 1988; Devary et al., 1991). This may be in contrast to the situation with gamma-rays, where experiments with cells that have been depleted of protein kinase C by treatment with phorbol esters, have suggested that depleted cells do not react any more to gamma-ray irradiation with enhanced *c-jun* gene transcription (Hallahan et al., 1991).

Are reactive oxygen intermediates involved in transmission of the ultraviolet response? Are they the primary elicitors of the response? Arguments have been made for and against this possibility. On one hand, radical-inducing treatments such as H_2O_2 or gamma-rays induce AP-1 and several AP-1-dependent genes; but a closer look reveals that efficient inducers of radical formation, such as H_2O_2 and gamma-rays, induce *c-fos* and *c-jun* mRNA much less efficiently than does UV-C (if applied at equitoxic doses) (Rahmsdorf, unpublished results). Furthermore, long-wavelength ultraviolet irradiation, which has been shown to induce a prooxidant state in the cell (Lautier et al., 1992), induces the ultraviolet response much less efficiently (per quantum absorbed) than does UV (Stein et al., 1989b), which has not been shown to induce a prooxidant state. In addition, the different dose dependence of collagenase mRNA accumulation in wild-type cells and *Xeroderma pigmentosum* cells could not be explained if ultraviolet acted

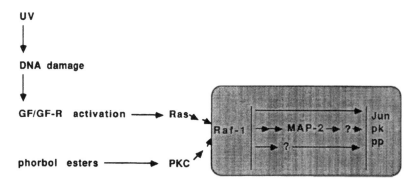

Figure 13-1 Hypothetical scheme of the chain of events leading to UV- and phorbol ester-induced AP-1 activation. (GFs growth factor, GF-R = growth factor receptor, pk = protein kinase, pp = protein phosphatase)

through the generation of oxygen radicals (wild-type and *Xeroderma pigmentosum* cells do not differ in the repair of DNA lesions induced by reactive oxygen intermediates). It remains to be explained why radical scavengers such as *N*-acetyl-cysteine (besides inducing fos and jun synthesis) (Meyer et al., 1993) interfere with UV-induced transcription (Devary et al., 1992). In view of the above arguments, a reasonable hypothesis is that UV does not induce gene transcription by generating reactive oxygen intermediates, but that it needs a high constitutive level of these intermediates in order to work. Treatment of cells with antioxidants may thus render cells unresponsive to UV. A similar conclusion was drawn from findings with various T cells and monocytic cell lines: in these cells the radical scavenger butylated hydroxyanisole inhibited TPA-induced NFκB activation. In the same cells, neither TPA generated reactive oxygen intermediates nor H_2O_2 (which generated these intermediates) activated NFκB (Israël et al., 1992).

In summary, the UV-induced signal transduction chain seems to be extraordinarily complicated (see Figure 13-1): Beginning with damage to DNA, it may involve a reverse signal flow from the nucleus into the periphery of the cell, which leads to activation of growth factors and their receptors. Activated receptors signal back into the nucleus through the activation of membrane standing tyrosine kinases, ras, raf, and MAP-2. As both UV and phorbol esters need raf kinase for signaling, but only UV needs ras (at least in Hela cells) (Radler-Pohl et al., 1993; Sachsenmaier et al., 1994), it seems that the UV- and phorbol ester-induced signal chains converge on the raf kinase.

POSSIBLE BIOLOGICAL CONSEQUENCES
OF CARCINOGEN INDUCED AP-1 ACTIVATION

The consequences, of course, depend on the actual mix of activated subunits of AP-1. Assessment of this mix is poor. Major activations, however, after UV and TPA concern c-fos and c-jun, and we may speculate what AP-1 (Fos/Jun) causes in cells. AP-1 is a family of transcription factors that may bind to numerous promoters and activate numerous genes. Induced AP-1 synthesis is generally seen as a means to transform short-lasting extracellular signals into longer lasting responses. Only a few AP-1 (Fos/Jun) responsive genes are known, e.g., genes coding for several metalloproteinases (Angel et al., 1987; Sirum-Conolly and Brinckerhoff, 1991), plasminogen activators (Nerlov et al., 1992), metallothionein II A (Angel et al., 1987; Lee et al., 1987), proliferin (Mordacq and Linzer, 1989), and the lymphocyte homing receptor CD44 (Hofmann et al., 1993). In addition, several viral promoters such as polyoma (Wasylyk et al., 1987) and SV40 (Angel et al., 1987; Lee et al., 1987) are controlled by AP-1 binding sites. Only in a few cases is there evidence to suggest which consequences the activation of these genes may have for the affected cell. Two complementary methods have been used to approach the problem: either the components of AP-1 or AP-1-responsive genes have been overexpressed independently of the inducing signal or the proteins have been eliminated by homologous recombination of truncated genes with the wild-type alleles, by the antisense technique or the injection of antibodies.

OVEREXPRESSION OF AP-1 OR AP-1-DEPENDENT GENES

c-fos and *c-jun* are the cellular counterparts of the viral transforming genes *v-fos* and *v-jun*, which, upon introduction into appropriate cells, alone or in combination with other oncogenes, induce the full spectrum of cellular transformation. This topic is reviewed in chapters 14 and 15. I will list here only a few examples in which the overexpression of the components of AP-1 or AP-1-dependent genes has been analyzed with respect to the generation of a cellular phenotype.

Indeed, a large spectrum of phenotypes can be induced by the overexpression of the components of AP-1 or AP-1-dependent genes independently of the inducers, namely environmental adversities. The phenotypes obtained resemble the phenotypes that are induced by environmental noxes directly. This suggests, but does not prove, that the proteins induced by environmental noxes induce the phenotype examined. The following examples illustrate this issue.

Some of the late consequences of treatment of cells with carcinogens are chromosomal aberrations, mutations, and gene amplification. Since most environmental noxes induce Fos expression, it was examined whether enhanced Fos expression may lead, independently of environmental signals, to genomic instability. This is indeed the case. The fos gene, driven by either the glucocorticoid responsive element of the mouse mammary tumor virus long terminal repeat or the heavy metal responsive element of the metallothionein II A gene, induces, upon appropriate activation of the enhancer unit, an at least twofold increase in chromosomal aberrations, gene mutations, and in the number of recombinants, as measured by reconstitution of an active gene from two inactive gene constructs (van den Berg et al., 1991, 1993). Moreover, the specific inhibition of Fos synthesis by the antisense technique inhibits substantially the UV-induced accumulation of chromosomal aberrations (van den Berg et al., 1991). These findings strongly suggest that c-Fos (or a c-Fos-dependent gene) is indeed involved in the generation of carcinogen-induced genomic instability.

Metallothioneins are induced by glucocorticoids and noxious agents from the environment, such as heavy metals, UV, and phorbol esters, and it is well established that overexpression of these proteins renders cells resistant to heavy metals. Basal and induced expression of metallothioneins depends largely on an AP-1 binding site in the promoter of the gene. Since metallothioneins are efficient radical scavengers in cell extracts (Thornalley and Vasak, 1985), it was examined whether cells overexpressing these proteins and resistant against heavy metals are also resistant against gamma irradiation, which is thought to exert much of its toxicity through the generation of radicals. The hypothesis turned out to be wrong; cells overexpressing metallothioneins show the same radiosensitivity as control cells. However, these cells are significantly more resistant against several other DNA-damaging agents, such as N-methyl-N-nitrosourea, N-methyl-N-nitro-N-nitrosoguanidine, cisplatin, morphalan, and chlorambucil (Kelley et al., 1988; Kaina et al., 1990). It is not known how metallothioneins increase the resistance of cells against these agents, some of which are used in chemotherapy.

Several genes coding for extracellular proteases are controlled by AP-1 binding sites. These proteases have long been suspected to be involved in tumor progression, invasion, and metastasis, processes that are enhanced by many carcinogens. Several attempts have been made to support this notion, either by overexpressing the genes or by interfering negatively with the function of the gene products. It was shown, for example, that the metastatic potential of several transformed rat cell lines correlates positively with the expression of the metalloproteinases stromelysin 1 and 2, but not with the expression of two type IV collagenases (Sreenath et al., 1992). Mouse cells transfected with gene constructs expressing plasminogen activators show, in an experimental metastasis assay (the cells are injected into the bloodstream), enhanced metastatic potential (Cajot et al., 1989; Axelrod et al., 1989). Inhibitors of expression or function of plasminogen activators and metalloproteases, such as E1A, and the tissue inhibitor of metalloproteases (TIMP) inhibit invasion in experimental metastasis assays (Reich et al., 1988; Frisch et al., 1990; Offringa et al., 1990). Downregulation of the metalloproteinase inhibitor TIMP with the antisense technique led to tumorigenic transformation of Swiss 3T3 cells (Khokha et al., 1989). However, overexpression of either a human collagenase gene or a human stromelysin gene in nonmetastatic rat tumor cells (harboring an activated ras oncogene and growing as subcutaneous solid tumors; this is called a spontaneous metastasis assay), did not lead to the metastatic spread of the cells. The locally growing tumors, however, contained an enhanced number of blood vessels (Giles and Ponta, unpublished). Taken together, these results suggest that extracellular proteases are indeed involved in tumorigenesis, but that the overexpression of a certain protease may not suffice to transform a nonmetastatic into a metastatic cell in a spontaneous metastasis assay.

A variant form of the lymphocyte homing receptor CD44 was shown to confer metastatic potential to nonmetastatic cells (Günthert et al., 1991). Since the gene coding for this protein critically depends on an AP-1 binding site and is positively regulated by phorbol esters and ras (Hofmann et al., 1993), it may also be a target addressed by carcinogen-induced AP-1 activation.

INHIBITION OF AP-1

In contrast to the approach discussed above, the knockout approach is by far more convincing: if a protein supposed to be involved in the generation of a certain phenotype is eliminated (and if this elimination changes the phenotype), the obligatory role of this protein is proven. Therefore, many attempts have been taken to disrupt specifically the components of AP-1 or AP-1-dependent genes. Elimination or functional inhibition was accomplished by the injection of antibodies directed against the components of AP-1 or of the binding sequence of AP-1 (Riabowol et al., 1988, 1992); by introducing into cells single-stranded oligonucleotides complementary to the ribosome binding sites on c-fos and c-jun mRNAs (van den Berg et al., 1991); by stably transfecting cells with constitutive as well as inducible gene constructs coding for c-jun or c-fos antisense RNA (Holt et al., 1986; Nishikura and Murray, 1987; Schönthal et al., 1988a, b); by stably transfecting gene constructs that code for dominant-negative Fos or Jun mutant proteins, which are still able to associate with endogenous Fos or Jun, but which inhibit their activity by preventing their binding to their enhancers (Deng and Karin, 1993); and, finally, by knocking out the genes coding for Fos and Jun by homologous recombination of mutant genes into the wild-type genes in embryonal stem cells. These cells, if injected into blastocysts and involved in germ cell formation, give rise to heterozygous animals (with one disrupted allele of the c-fos or c-jun gene). Once breeding occurs one can examine whether animals homozygous for the disrupted gene are viable. Only with this latter approach is it possible to eliminate one gene product at a given time. With the other strategies one cannot be sure which members of the AP-1 family will be affected and whether the elimination is complete.

The knockout of the c-fos and c-jun genes yielded surprises. The most spectacular was the finding that embryonal stem cells in which either c-fos or c-jun had been disrupted are viable, showing no abnormalities in cell culture (Wang et al., 1992; Hilberg and Wagner, 1992). In addition, *c-fos* null mice develop fairly normally within the first 10 days after birth, although, subsequently, they show a wide range of phenotypic deficits (Johnson et al., 1992; Wang et al., 1992). Future experiments must determine, whether other members of the Fos gene family can take over the role of c-Fos during early development and in the first days after birth.

c-Jun null mice are nonviable. Intrauterine death occurs around 12.5 days postcoitus. The surprise, however, is that the mice reach the midgestational stage in the absence of c-Jun (Johnson et al., 1993). The standard explanation, yet unproven, is that there is a tremendous degree of redundancy between the members of the jun family.

CONCLUSIONS

Environmental noxes, such as carcinogens, activate AP-1 by inducing the posttranslational modification of the constituent proteins Fos and Jun and by inducing their synthesis. As a consequence, the transcription of many genes may be activated and other genes may be repressed. Among these may be genes that drive cells through the cell cycle; thus, cell cycle check points may be disregarded, and mutations and cellular transformation may result. This may be counteracted through carcinogen-induced processes that establish a G_1-to-S block. Carcinogen-induced stabilization of the tumor suppressor protein p53 has been implicated in this process. For cells that do not seem to respond to certain environmental adversities by stabilization of this protein, such as cells derived from *Ataxia telangiectasia* (Kastan et al., 1992), or which do not express the protein because of deletion of the gene (Livingstone et al., 1992), AP-1 activation by environmental carcinogens may lead to an appreciable increase in the risk of carcinogenic transformation.

NOTE ADDED DURING PROOF

Since the submission of this manuscript, we have shown in cooperation with Alfred Nordheim that UV-C induces the rapid phosphorylation of p62TCF/elk-1 and that elk-1 is a necessary intermediate in UV-C-induced *c-fos* transcription. Further, in cooperation with Axel Ullrich, that cotransfection of a truncated EGF-receptor mutant with a UV-C-responsive promoter inhibits part of the UV-C response. We also found that UV-C, similar to EGF, induces the phosphorylation of the EGF receptor at tyrosine residues (Sachsenmaier et al., submitted). M. Karin's laboratory presented evidence that cytoplasts react to UV-C with the activation of NFκB and of a Jun-kinase, suggesting that this part of the UV-C response is not initiated in the cell nucleus (Devary et al., 1993). Moreover a Jun-kinase was isolated from UV-C treated cells whose binding to c-Jun is required for UV-C and Ras induced activation of this transcription factor (Hibi et al., 1993). The sequence of a molecular clone encoding this kinase demonstrates that it is related to MAP-kinases (Dérijard et al., 1994).

REFERENCES

Abate, C., Patel, L., Rauscher, F. J., III, and Curran, T., Redox regulation of Fos and Jun DNA-binding activity in vitro, *Science,* 249, 1157, 1990.

Amstad, P. A., Krupitza, G., and Cerutti, P. A., Mechanism of c-fos induction by active oxygen, *Cancer Res.,* 52, 3952, 1992.

Andrews, G. K., Harding, M. A., Calvet, J. P., and Adamson, E. D., The heat shock response in HeLa cells is accompanied by elevated expression of the c-fos proto-oncogene, *Mol. Cell. Biol.,* 7, 3452, 1987.

Angel, P. and Karin, M., The role of Jun, Fos and the AP-1 complex in cell-proliferation and transformation, *Biochim. Biophys. Acta*, 1072, 129, 1991.

Angel, P., Rahmsdorf, H. J., Pöting, A., and Herrlich, P., c-fos mRNA levels in primary human fibroblasts after arrest in various stages of the cell cycle, *Cancer Cells*, 3, 315, 1985.

Angel, P., Imagawa, M., Chiu, R., Stein, B., Imbra, R. J., Rahmsdorf, H. J., Jonat, C., Herrlich, P., and Karin, M., Phorbol ester-inducible genes contain a common cis element recognized by a TPA-modulated trans-acting factor, *Cell*, 49, 729, 1987.

Angel, P., Hattori, K., Smeal, T., and Karin, M., The jun proto-oncogene is positively autoregulated by its product, Jun/AP-1, *Cell*, 55, 875, 1988.

Axelrod, J. H., Reich, R., and Miskin, R., Expression of human recombinant plasminogen activators enhances invasion and experimental metastasis of H-*ras*-transformed NIH 3T3 cells, *Mol. Cell. Biol.*, 9, 2133, 1989.

Baker, S. J., Kerppola, T. K., Luk, D., Vandenberg, M. T., Marshak, D. R., Curran, T., and Abate, C., Jun is phosphorylated by several protein kinases at the same sites that are modified in serum-stimulated fibroblasts, *Mol. Cell. Biol.*, 12, 4694, 1992.

Barber, J. R. and Verma, I. M., Modification of fos proteins: Phosphorylation of c-fos, but not v-fos, is stimulated by 12-tetradecanoyl-phorbol-13-acetate and serum, *Mol. Cell. Biol.*, 7, 2201, 1987.

Betsholtz, C., Johnsson, A., Heldin, C.-H., and Westermark, B., Efficient reversion of simian sarcoma virus-transformation and inhibition of growth factor-induced mitogenesis by suramin, *Proc. Natl. Acad. Sci. U.S.A.*, 83, 6440, 1986.

Bialojan, C. and Takai, A., Inhibitory effect of a marine-sponge toxin, okadaic acid, on protein phosphatases, *Biochem. J.*, 256, 283, 1988.

Binétruy, B., Smeal, T., and Karin, M., Ha-Ras augments c-Jun activity and stimulates phosphorylation of its activation domain, *Nature*, 351, 122, 1991.

Boyle, W. J., Smeal, T., Defize, L. H. K., Angel, P., Woodgett, J. R., Karin, M., and Hunter, T., Activation of protein kinase C decreases phosphorylation of c-Jun at sites that negatively regulate its DNA-binding activity, *Cell*, 64, 573, 1991.

Brunet, S. and Giacomoni, P. U., Specific mRNAs accumulate in long-wavelength UV-irradiated mouse epidermis, *J. Photochem. Photobiol.*, B6, 431, 1990.

Bukh, A., Martinez-Valdez, H., Freedman, S. J., Freedman, M. H., and Cohen, A., The expression of c-fos, c-jun and c-myc genes is regulated by heat shock in human lymphoid cells, *J. Immunol.*, 144, 4835, 1990.

Büscher, M., Rahmsdorf, H. J., Litfin, M., Karin, M., and Herrlich, P., Activation of the c-fos gene by UV and phorbol ester: Different signal transduction pathways converge to the same enhancer element, *Oncogene*, 3, 301, 1988.

Cajot, J.-F., Schleuning, W.-D., Medcalf, R. L., Bamat, J., Testuz, J., Liebermann, L., and Sordat, B., Mouse L cells expressing human prourokinase-type plasminogen activator: Effects on extracellular matrix degradation and invasion, *J. Cell Biol.*, 109, 915, 1989.

Cochran, B. H., Reffel, A. C., and Stiles, C. D., Molecular cloning of gene sequences regulated by platelet-derived growth factor, *Cell*, 33, 939, 1983.

Colotta, F., Polentarutti, N., Staffico, M., Fincato, G., and Mantovani, A., Heat shock induces the transcriptional activation of c-fos protooncogene, *Biochem. Biophys. Res. Commun.*, 168, 1013, 1990.

Crawford, D., Zbinden, I., Amstad, P., and Cerutti, P., Oxidant stress induces the proto-oncogenes c-fos and c-myc in mouse epidermal cells, *Oncogene*, 3, 27, 1988.

Curran, T. and Morgan, J. I., Barium modulates c-fos expression and post-translational modification, *Proc. Natl. Acad. Sci. U.S.A.*, 83, 8521, 1986.

Curran, T., Miller, A. D., Zokas, L., and Verma, I. M., Viral and cellular fos proteins: A comparative analysis, *Cell*, 36, 259, 1984.

de Groot, R. P., Rijken, P. J., den Hertog, J., Boonstra, J., Verkleij, A. J., de Laat, S. W., and Kruijer, W., Microgravity decreases c-fos induction and SRE activity, *J. Cell Sci.*, 97, 33, 1990.

de Groot, R. P., Rijken, P. J., Boonstra, J., Verkleij, A. J., de Laat, S. W., and Kruijer, W., Epidermal growth factor induced expression of c-fos is influenced by altered gravity conditions, *Aviat. Space Environ. Med.*, 62, 37, 1991.

Demple, B., Regulation of bacterial oxidative stress genes, *Annu. Rev. Genet.*, 25, 315, 1991.

Deng, T. and Karin, M., JunB differs from c-Jun in its DNA-binding and dimerization domains, and represses c-Jun by formation of inactive heterodimers, *Genes Dev.*, 7, 479, 1993.

Dérijard, B., Hibi, M., Wu, I.-H., Barrett, T., Su, B., Deng, T., Karin, M., and Davies, R. J., JNK1: A protein kinase stimulated by UV light and Ha-Ras that binds and phosphorylates the c-Jun activation domain, *Cell*, 76, 1025, 1994.

Devary, Y., Gottlieb, R. A., Lau, L. F., and Karin, M., Rapid and preferential activation of the c-jun gene during the mammalian UV response, *Mol. Cell. Biol.*, 11, 2804, 1991.

Devary, Y., Gottlieb, R. A., Smeal, T., and Karin, M., The mammalian ultraviolet response is triggered by activation of Src tyrosine kinases, *Cell*, 71, 1081, 1992.

Devary, Y., Rosette, C., DiDonato, J. A., and Karin, M., NF-κB activation by ultraviolet light not dependent on a nuclear signal, *Science*, 261, 1442, 1993.

Fisch, T. M., Prywes, R., and Roeder, R. G., c-fos sequences necessary for basal expression and induction by epidermal growth factor, 12-O-tetradecanoyl phorbol-13-acetate, and the calcium ionophore, *Mol. Cell. Biol.*, 7, 3490, 1987.

Frisch, S. M., Reich, R., Collier, I. E., Genrich, L. T., Martin, G., and Goldberg, G. I., Adenovirus E1A represses protease gene expression and inhibits metastasis of human tumor cells, *Oncogene*, 5, 75, 1990.

Futscher, B. W. and Erickson, L. C., Changes in c-myc and c-fos expression in a human tumor cell line following exposure to bifunctional alkylating agents, *Cancer Res.*, 50, 62, 1990.

Gebel, S., Modifikationen an den Transkriptionsfaktoren c-Fos und c-Jun: ein Schlüssel zum Verständnis der schnellen Induktion von Genen, Dissertation, Universität Karlsruhe, 1992.

Gille, H., Sharrocks, A. D., and Shaw, P. E., Phosphorylation of transcription factor p62^{TCF} by MAP kinase stimulates ternary complex formation at c-fos promoter, *Nature*, 358, 414, 1992.

Greenberg, M. E. and Ziff, E. B., Stimulation of 3T3 cells induces transcription of the c-fos proto-oncogene, *Nature*, 311, 433, 1984.

Günthert, U., Hofmann, M., Rudy, W., Reber, S., Zöller, M., Haußmann, I., Matzku, S., Wenzel, A., Ponta, H. and Herrlich, P., A new variant of glycoprotein CD44 confers metastatic potential to rat carcinoma cells, *Cell*, 65, 13, 1991.

Hallahan, D. E., Sukhatme, V. P., Sherman, M. L., Virudachalam, S., Kufe, D., and Weichselbaum, R. R., Protein kinase C mediates x-ray inducibility of nuclear signal transducers EGR1 and JUN, *Proc. Natl. Acad. Sci. U.S.A.*, 88, 2156, 1991.

Herr, I., van Dam, H., and Angel, P., Binding of promoter associated AP-1 is not altered during induction and subsequent repression of the c-jun promoter by TPA and UV irradiation, *Carcinogenesis*, 15, 1105, 1994.

Herrera, R. E., Shaw, P. E., and Nordheim, A., Occupation of the c-fos serum response element in vivo by a multi-protein complex is unaltered by growth factor induction, *Nature*, 340, 68, 1989.

Herrlich, P., Ponta, H., and Rahmsdorf, H. J., DNA damage-induced gene expression: Signal transduction and relation to growth factor signaling, *Rev. Physiol. Biochem. Pharmacol.*, 119, 187, 1992.

Hescheler, J., Mieskes, G., Rüegg, J. C., Takai, A., and Trautwein, W., Effects of a protein phosphatase inhibitor, okadaic acid, on membrane currents of isolated guinea-pig cardiac myocytes, *Eur. J. Physiol.*, 412, 248, 1988.

Hibi, M., Lin, A., Smeal, T., Minden, A., and Karin, M., Identification of an oncoprotein- and UV-responsive protein kinase that binds and potentiates the c-Jun activation domain, *Genes & Dev.*, 7, 2135, 1993.

Hilberg, F. and Wagner, E. F., Embryonic stem (ES) cells lacking functional c-*jun*: Consequences for growth and differentiation, AP-1 activity and tumorigenicity, *Oncogene*, 7, 2371, 1992.

Hiraoka, M., Miyakoshi, J., Li, Y. P., Shung, B., Takebe, H., and Abe, M., Induction of c-fos gene expression by exposure to a static magnetic field in HeLaS3 cells, *Cancer Res.*, 52, 6522, 1992.

Hofmann, M., Rudy, W., Günthert, U., Zimmer, S. G., Zawadski, V., Zöller, M., Lichtner, R. B., Herrlich, P., and Ponta, H., A link between *ras* and metastatic behavior of tumor cells: *ras* induces CD44 promoter activity and leads to low-level expression of metastasis-specific variants of CD44 in CREF cells, *Cancer Res.*, 53, 1516, 1993.

Hollander, M. C. and Fornace, A. J., Jr., Induction of fos RNA by DNA-damaging agents, *Cancer Res.*, 49, 1687, 1989.

Holt, J. T., Venkat Gopal, T., Moulton, A. D., and Nienhuis, A. W., Inducible production of c-fos antisene RNA inhibits 3T3 cell proliferation, *Proc. Natl. Acad. Sci. U.S.A.*, 83, 4794, 1986.

Israël, N., Gougerot-Pocidalo, M.-A., Aillet, F., and Virelizier, J.-L., Redox status of cells influences constitutive or induced NF-κB translocation and HIV long terminal repeat activity in human T and monocytic cell lines, *J. Immunol.*, 149, 3386, 1992.

Johnson, R. S., Spiegelman, B. M., and Papaioannou, V., Pleiotropic effects of a null mutation in the c-fos proto-oncogene, *Cell*, 71, 577, 1992.

Johnson, R. S., van Lingen, B., Papaioannou, V. E., and Spiegelman, B. M., A null mutation at the c-*jun* locus causes embryonic lethality and retarded cell growth in culture, *Genes Dev.*, 7, 1309, 1993.

Kaina, B., Lohrer, H., Karin, M., and Herrlich, P., Overexpressed human metallothionein IIA gene protects Chinese hamster ovary cells from killing by alkylating agents, *Proc. Natl. Acad. Sci. U.S.A.*, 87, 2710, 1990.

Kastan, M. B., Zhan, Q., El-Deiry, W. S., Carrier, F., Jacks, T., Walsh, W. V., Plunkett, B. S., Vogelstein, B., and Fornace, A. J., Jr., A mammalian cell cycle checkpoint pathway utilizing p53 and GADD45 is defective in Ataxia-telangiectasia, *Cell*, 71, 587, 1992.

Kelley, S. L., Basu, A., Teicher, B. A., Hacker, M. P., Hamer, D. H., and Lazo, J. S., Overexpression of metallothionein confers resistance to anticancer drugs, *Science*, 241, 1813, 1988.

Khokha, R., Waterhouse, P., Yagel, S., Lala, P. K., Overall, C. M., Norton, G., and Denhardt, D. T., Antisense RNA-induced reduction in murine TIMP levels confers oncogenicity on swiss 3T3 cells, *Science*, 243, 947, 1989.

König, H., Ponta, H., Rahmsdorf, U., Büscher, M., Schönthal, A., Rahmsdorf, H. J., and Herrlich, P., Autoregulation of fos: The dyad symmetry element as the major target of repression, *EMBO J.*, 8, 2559, 1989.

König, H., Cell-type specific multiprotein complex formation over the c-fos serum response element in vivo: Ternary complex formation is not required for the induction of c-fos, *Nucleic Acids Res.*, 19, 3607, 1991.

Krämer, M., Stein, B., Mai, S., Kunz, E., König, H., Loferer, H., Grunicke, H. H., Ponta, H., Herrlich, P., and Rahmsdorf, H. J., Radiation-induced activation of transcription factors in mammalian cells, *Radiat. Environ. Biophys.*, 29, 303, 1990.

Krämer, M., Sachsenmaier, C., Herrlich, P., and Rahmsdorf, H. J., UV-irradiation induced interleukin-1 and basic fibroblast growth factor synthesis and release mediate part of the UV response, *J. Biol. Chem.*, 268, 6734, 1993.

Kruijer, W., Cooper, J. A., Hunter, T., and Verma, I. M., Platelet-derived growth factor induces rapid but transient expression of the c-fos gene and protein, *Nature*, 312, 711, 1984.

Lamph, W. W., Wamsley, P., Sassone-Corsi, P., and Verma, I. M., Induction of proto-oncogene Jun/AP-1 by serum and TPA, *Nature*, 334, 629, 1988.

Lautier, C., Luscher, P., and Tyrrell, R. M., Endogenous glutathione levels modulate both constitutive and UVA radiation/hydrogen peroxide inducible expression of the human heme oxygenase gene, *Carcinogenesis*, 13, 227, 1992.

Lee, W., Mitchell, P., and Tjian, R., Purified transcription factor AP-1 interacts with TPA-inducible enhancer elements, *Cell*, 49, 741, 1987.

Lee, W. M. F., Lin, C., and Curran, T., Activation of the transforming potential of the human fos proto-oncogene requires message stabilization and results in increased amounts of partially modified fos protein, *Mol. Cell. Biol.*, 8, 5521, 1988.

Livingstone, L. R., White, A., Sprouse, J., Livanos, E., Jacks, R., and Tlsty, T. D., Altered cell cycle arrest and gene amplification potential accompany loss of wild-type p53. *Cell*, 70, 923, 1992.

Lucibello, F. C., Lowag, C., Neuberg, M., and Müller, R., Trans-repression of the mouse c-fos promoter: A novel mechanism of Fos-mediated trans-regulation, *Cell*, 59, 999, 1989.

Marais, R., Wynne, J., and Treisman, R., The SRF accessory protein Elk-1 contains a growth factor-regulated transcriptional activation domain, *Cell*, 73, 381, 1993.

Meyer, M., Schreck, R., and Baeuerle, P. A., H_2O_2 and antioxidants have opposite effects on activation of NF-κB and AP-1 in intact cells: AP-1 as secondary antioxidant-responsive factor, *EMBO J.*, 12, 2005, 1993.

Mordacq, J. C. and Linzer, D. I. H., Co-localization of elements required for phorbol ester stimulation and glucocorticoid repression of proliferin gene expression, *Genes Dev.*, 3, 760, 1989.

Morgan, J. I. and Curran, T., Role of ion flux in the control of c-fos expression, *Nature*, 322, 552, 1986.

Mügge, K., Vila, M., Gusella, G. I., Musso, T., Herrlich, P., Stein, B., and Durum, S. K., Interleukin 1 induction of the c-jun promoter, *Proc. Natl. Acad. Sci. U.S.A.*, 90, 7054, 1993.

Müller, R., Bravo, R., Burckhardt, J., and Curran, T., Induction of c-fos gene and protein by growth factors precedes activation of c-myc, *Nature*, 312, 716, 1984.

Müller, R., Bravo, R., Müller, D., Kurz, C., and Renz, M., Different types of modification in c-fos and its associated protein p39: Modulation of DNA binding by phosphorylation, *Oncogene Res.*, 2, 19, 1987.

Munson, G. P. and Woloschak, G. E., Differential effect of ionizing radiation on transcription in repair-deficient and repair-proficient mice, *Cancer Res.*, 50, 5045, 1990.

Nerlov, C., De Cesare, D., Pergola, F., Caracciolo, A., Blasi, F., Johnsen, M., and Verde, P., A regulatory element that mediates co-operation between a PEA3-AP-1 element and an AP-1 site is required for phorbol ester induction of urokinase enhancer activity in HepG2 hepatoma cells, *EMBO J.*, 11, 4573, 1992.

Nishikura, K. and Murray, J. M., Antisense RNA of proto-oncogene c-fos blocks renewed growth of quiescent 3T3 cells, *Mol. Cell. Biol.*, 7, 639, 1987.

Nishizuka, Y., Studies and perspectives of protein kinase C, *Science*, 233, 305, 1986.

Norman, C., Runswick, M., Pollock, R., and Treisman, R., Isolation and properties of cDNA clones encoding SRF, a transcription factor that binds to the c-fos serum response element, *Cell*, 55, 989, 1988.

Nose, K., Shibanuma, M., Kikuchi, K., Kageyama, H., Sakiyama, S., and Kuroki, T., Transcriptional activation of early-response genes by hydrogen peroxide in a mouse osteoblastic cell line, *Eur. J. Biochem.*, 201, 99, 1991.

Oehler, T., Pintzas, A., Stumm, S., Darling, A., Gillespie, D., and Angel, P., Mutation of a phosphorylation site in the DNA binding domain is required for redox-independent transactivation of AP1-dependent genes by vJun, *Oncogene*, 8, 1141, 1993.

Offringa, R., Gebel, S., van Dam, H., Timmers, M., Smits, A., Zwart, R., Stein, B., Bos, J. L., van der Eb, A., and Herrlich, P., A novel function of the transforming domain of E1a: Repression of AP-1 activity, *Cell*, 62, 527, 1990.

Prywes, R. and Roeder, R. G., Inducible binding of a factor to the c-fos enhancer, *Cell*, 47, 777, 1986.

Prywes, R. and Roeder, R. G., Purification of the c-fos enhancer-binding protein, *Mol. Cell. Biol.*, 7, 3482, 1987.

Pulverer, B. J., Kyriakis, J. M., Avruch, J., Nikolakaki, E., and Woodgett, J. R., Phosphorylation of c-jun mediated by MAP kinases, *Nature*, 353, 670, 1991.

Radler-Pohl, A., Sachsenmaier, C., Gebel, S., Auer, H.-P., Bruder, J. T., Rapp, U., Angel, P., Rahmsdorf, H. J., and Herrlich, P., UV-induced activation of AP-1 involves obligatory extranuclear steps including Raf-1 kinase, *EMBO J.*, 12, 1005, 1993.

Rahmsdorf, H. J., Schönthal, A., Angel, P., Litfin, M., Rüther, U., and Herrlich, P., Posttranscriptional regulation of c-fos mRNA expression, *Nucleic Acids Res.*, 15, 1643, 1987.

Rahmsdorf, H. J., Gebel, S., Krämer, M., König, H., Lücke-Huhle, C., Radler-Pohl, A., Sachsenmaier, C., Stein, B., Auer, H.-P., Vanetti, M., and Herrlich, P., Ultraviolet irradiation and phorbol esters induce gene transcription by different mechanisms, in *Induced Effects of Genotoxic Agents in Eukaryotic Cells*, Rossman, T., Ed., Hemisphere, Washington, D.C., pp. 141-161, 1992.

Ran, W., Dean, M., Levine, R. A., Henkle, C., and Campisi, J., Induction of c-fos and c-myc mRNA by epidermal growth factor or calcium ionophore is cAMP dependent, *Proc. Natl. Acad. Sci. U.S.A.*, 83, 8216, 1986.

Reich, R., Thompson, E. W., Iwamoto, Y., Martin, G. R., Deason, J. R., Fuller, G. C., and Miskin, R., Effects of inhibitors of plasminogen activator, serine proteinases, and collagenase IV on the invasion of basement membranes by metastatic cells, *Cancer Res.*, 48, 3307, 1988.

Riabowol, K., Vosatka, R. J., Ziff, E. B., Lamb, N. J., and Feramisco, J. R., Microinjection of fos-specific antibodies blocks DNA synthesis in fibroblast cells, *Mol. Cell. Biol.*, 8, 1670, 1988.

Riabowol, K. T., Schiff, J., and Gilman, M. Z., Transcription factor AP-1 activity is required for initiation of DNA synthesis and is lost during cellular aging, *Proc. Natl. Acad. Sci. U.S.A.*, 89, 157, 1992.

Ronai, Z. A., Okin, E., and Weinstein, I. B., Ultraviolet light induces the expression of oncogenes in rat fibroblast and human keratinocyte cells, *Oncogene*, 2, 201, 1988.

Rotem, N., Axelrod, J. H., and Miskin, R., Induction of urokinase-type plasminogen activator by UV light in human fetal fibroblasts is mediated through a UV-induced secreted protein, *Mol. Cell. Biol.*, 7, 622, 1987.

Rubin, E., Kharbanda, S., Gunji, H., Weichselbaum, R., and Kufe, D., cis-diamminedichloroplatinum(II) induces c-jun expression in human myeloid leukemia cells: Potential involvement of a protein kinase C-dependent signaling pathway, *Cancer Res.*, 52, 878, 1992.

Sachsenmaier, C., Radler-Pohl, A., Müller, A., Herrlich, P., and Rahmsdorf, H. J., Damage to DNA by UV light and activation of transcription factors, *Biochem. Pharmacol.*, 47, 129, 1994.

Sassone-Corsi, P., Sisson, J. C., and Verma, I. M., Transcriptional autoregulation of the protooncogene fos, *Nature*, 334, 314, 1988.

Schönthal, A., Gebel, S., Stein, B., Ponta, H., Rahmsdorf, H. J., and Herrlich, P., Nuclear oncoproteins determine the genetic program in response to external stimuli, *Cold Spring Harbor Symp. Quant. Biol.*, 53, 779, 1988a.

Schönthal. A., Herrlich. P., Rahmsdorf. H. J., and Ponta. H.. Requirement for fos gene expression in the transcriptional activation of collagenase by other oncogenes and phorbol esters, *Cell,* 54, 325. 1988b.

Schönthal, A., Büscher. M., Angel. P., Rahmsdorf, H. J., Ponta, H., Hattori. K.. Chiu. R.. Karin. M., and Herrlich. P.. The Fos and Jun/AP-1 proteins are involved in the downregulation of Fos transcription. *Oncogene,* 4, 629, 1989.

Schönthal, A.. Alberts, A. S., Frost. J. A.. and Feramisco. J. R.. Differential regulation of jun family gene expression by the tumor promoter okadaic acid. *New Biol.,* 3, 977, 1991a.

Schönthal, A., Sugarman, J.. Brown, J. H., Hanley. M. R.. and Feramisco. J. R.. Regulation of c-fos and c-jun protooncogene expression by the Ca2+-ATPase inhibitor thapsigargin. *Proc. Natl. Acad. Sci. U.S.A.,* 88, 7096. 1991b.

Schönthal. A., Tsukitani. Y.. and Feramisco. J. R.. Transcriptional and post-transcriptional regulation of c-fos expression by the tumor promoter okadaic acid, *Oncogene,* 6, 423. 1991c.

Schorpp. M.. Mallick, U.. Rahmsdorf. H. J.. and Herrlich. P.. UV-induced extracellular factor from human fibroblasts communicates the UV response to nonirradiated cells. *Cell,* 37. 861, 1984.

Schröter. H., Shaw, P. E., and Nordheim, A.. Purification of intercalator-released p67, a polypeptide that interacts specifically with the c-fos serum response element. *Nucleic Acids Res.,* 15. 10145. 1987.

Schröter. H.. Mueller. C. G. F.. Meese. K.. and Nordheim. A., Synergism in ternary complex formation between the dimeric glycoprotein p67SRF, polypeptide p62TCF and the c-fos serum response element. *EMBO J.,* 9, 1123, 1990.

Shah. G.. Ghosh. R.. Amstad. P. A., and Cerutti. P. A.. Mechanism of induction of c-fos by ultraviolet B (290–320 nm) in mouse JB6 epidermal cells, *Cancer Res.,* 53. 38. 1993.

Shaw. P. E.. Frasch. S.. and Nordheim. A.. Repression of c-fos transcription is mediated through p67SRF bound to the SRE. *EMBO J.,* 8, 2567, 1989a.

Shaw. P. E.. Schröter. H.. and Nordheim. A.. The ability of a ternary complex to form over the serum response element correlates with serum inducibility of the human c-fos promoter. *Cell,* 56. 563, 1989b.

Sherman. M. L.. Datta. R.. Hallahan. D. E.. Weichselbaum. R. R., and Kufe, D. W.. Ionizing radiation regulates expression of the c-jun protooncogene. *Proc. Natl. Acad. Sci. U.S.A.,* 87. 5663. 1990.

Shibanuma. M.. Kuroki, T.. and Nose. K.. Induction of DNA replication and expression of proto-oncogene c-myc and c-fos in quiescent Balb/3T3 cells by xanthine/xanthine oxidase. *Oncogene,* 3. 17. 1988.

Shibanuma. M.. Kuroki, T., and Nose. K.. Stimulation by hydrogen peroxide of DNA synthesis, competence family gene expression and phosphorylation of a specific protein in quiescent Balb/3T3 cells. *Oncogene,* 5, 1025. 1990.

Sirum-Conolly. K. and Brinckerhoff. C. E.. Interleukin-1 or phorbol induction of the stromelysin promoter requires an element that cooperates with AP-1. *Nucleic Acids Res.,* 19. 335. 1991.

Smeal. T.. Binétruy. B.. Mercola. D. A.. Birrer. M.. and Karin. M.. Oncogenic and transcriptional cooperation with Ha-ras requires phosphorylation of c-Jun on serines 63 and 73. *Nature,* 354, 494. 1991.

Sreenath. T.. Matrisian. L. M.. Stetler-Stevenson. W.. Gattoni-Celli, S., and Pozzatti. R. O.. Expression of matrix metalloproteinase genes in transformed rat cell lines of high and low metastatic potential, *Cancer Res.,* 52. 4942. 1992.

Stein. B.. Krämer. M.. Rahmsdorf. H. J.. Ponta. H.. and Herrlich. P.. UV-induced transcription from the human immunodeficiency virus type 1 (HIV-1) long terminal repeat and UV-induced secretion of an extracellular factor that induces HIV-1 transcription in nonirradiated cells, *J. Virol.,* 63. 4540. 1989a.

Stein, B., Rahmsdorf, H. J., Steffen, A., Litfin, M., and Herrlich, P., UV-induced DNA damage is an intermediate step in UV-induced expression of human immunodeficiency virus type 1, collagenase, c-fos, and metallothionein, *Mol. Cell. Biol.*, 9, 5169, 1989b.

Stein, B., Angel, P., van Dam, H., Ponta, H., Herrlich, P., van der Eb, A., and Rahmsdorf, H. J., Ultraviolet-radiation induced c-jun gene transcription: Two AP-1 like binding sites mediate the response, *Photochem. Photobiol.*, 55, 409, 1992.

Thastrup, O., Cullen, P., Drøbak, B. K., Hanley, M. R., and Dawson, A. P., Thapsigargin, a tumor promoter, discharges intracellular Ca^{2+} stores by specific inhibition of the endoplasmic reticulum Ca^{2+}-ATPase, *Proc. Natl. Acad. Sci. U.S.A.*, 87, 2466, 1990.

Thornalley, P. and Vasak, M., Possible role for MT in protection against radiation-induced oxidation stress. Kinetics and mechanism of its reaction with superoxide and hydroxyl radicals, *BBA*, 827, 36, 1985.

Treisman, R., Transient accumulation of c-fos RNA following serum stimulation requires a conserved 5' element and c-fos 3' sequences, *Cell*, 42, 889, 1985.

Treisman, R., Identification and purification of a polypeptide that binds to the c-fos serum response element, *EMBO J.*, 6, 2711, 1987.

van Dam, H., Duyndam, M., Rottier, R., Bosch, A., de Vries-Smits, L., Herrlich, P., Zantema, A., Angel, P., and van der Eb, A. J., Heterodimer formation of cJun and ATF-2 is responsible for induction of c-jun by the 243 amino acid adenovirus E1A protein, *EMBO J.*, 12, 479, 1993.

van den Berg, S., Kaina, B., Rahmsdorf, H. J., Ponta, H., and Herrlich, P., Involvement of Fos in spontaneous and ultraviolet light induced genetic changes, *Molec. Carcinogenesis*, 4, 460, 1991.

van den Berg, S., Rahmsdorf, H. J., Herrlich, P., and Kaina, B., Overexpression of c-fos increases recombination frequency in human osteosarcoma cells, *Carcinogenesis*, 14, 925, 1993.

Vaughan, P., Sedgwick, B., Hall, J., Gannon, J., and Lindahl, T., Environmental mutagens that induce the adaptive response to alkylating agents in Escherichia coli, *Carcinogenesis*, 12, 263, 1991.

Walker, G. C., Inducible DNA repair systems, *Annu. Rev. Biochem.*, 54, 425, 1985.

Wang, Z-Q., Ovitt, C., Grigoriadis, A.E., Möhle-Steinlein, U., Rüther, U., and Wagner, E. F., Bone and haematopoietic defects in mice lacking c-fos, *Nature*, 360, 741, 1992.

Wasylyk, C., Imler, J. L., Perez-Mutul, J., and Wasylyk, B., The c-Ha-ras oncogene and a tumor promoter activate the polyoma virus enhancer, *Cell*, 48, 525, 1987.

Xanthoudakis, S. and Curran, T., Identification and characterization of Ref-1, a nuclear protein that facilitates AP-1 DNA-binding activity, *EMBO J.*, 11, 653, 1992.

Zinck, R., Hipskind, R. A., Pingoud, V., and Nordheim, A., c-fos transcriptional activation and repression correlate temporally with the phosphorylation status of TCF, *EMBO J.*, 12, 2377, 1993.

Section IV

FUNCTIONS IN PATHOLOGY

The components of the AP-1 family are regulatory proteins. Their abundance and activity are tightly controlled. If this control is overruled, either by prolongation of the lifetime of the proteins, by excessive expression, or by enhanced signal flow to the AP-1 proteins, the borderline to pathology is crossed. Target genes will be turned on or off in non-physiological dose or duration. The cellular fate may range from apoptosis to transient arrest to transient hyperproliferation and immortalization/transformation. Presumably the outcome depends on the exact member of the AP-1 family that escaped control, and on other regulatory proteins. Although there is now evidence of a role for Fos in apoptosis (Smeyne et al., *Nature,* 363, 166, 1993), the subsequent chapters concentrate on cell proliferation and transformation, simply because more relevant data are available. A change in subunit abundance or activity may result in a number of other pathologies that could be deduced from the physiological roles addressed in Section III.

Chapter 14

Role of Fos Proteins in the Induction of Transformation and DNA Synthesis

Rolf Müller

CONTENTS

THE ONCOGENIC POTENTIAL OF FOS PROTEINS *IN VIVO* AND *IN VITRO*

The *fos* oncogene was discovered in two murine osteosarcoma viruses, the Finkel-Biskis-Jinkins mouse osteosarcoma virus (FBJ-MSV), isolated from a spontaneous osteosarcoma-like tumor in a CF-1 mouse, and the Finkel-Biskis-Reilly mouse osteosarcoma virus (FBR-MSV), recovered from a radiation (^{90}Sr)-induced osteosarcoma in an X/Gf mouse (Finkel et al., 1966, 1975; Ward and Young, 1976). Both viruses cause chondro-osseous sarcomas in newborn mice (Finkel et al., 1966, 1975) and induce morphological transformation in nonestablished and immortalized murine fibroblast cell lines (Levy et al., 1973; Curran et al., 1982; Jenuwein et al., 1985). The FBR-MuSV *gag-fos-fox* fusion gene is, however, a more potent oncogene than FBJ-MSV *in vitro*, in that it induces higher numbers of foci after a shorter latency period (Jenuwein et al., 1985). In addition, the FBR-MuSV gene product can trigger the establishment of low-passage fetal mouse fibroblasts, a property that is totally lacking in the FBJ-MSV counterpart (Jenuwein et al., 1985). A detailed structure–function analysis showed that the immortalizing potential of the FBR-MSV *fos* oncogene is due to a point mutation close to the basic amino acid stretch encompassing the DNA contact site (Jenuwein and Müller, 1987). In contrast, the molecular basis for the enhanced transforming potential of FBR-MSV could not be as clearly delineated and is presumably due to multiple alterations (Jenuwein and Müller, 1987; Lucibello et al., 1991).

fos oncogenes transform not only fibroblasts but also other connective tissue cells, including cartilage, bone, and muscle cells. The transformation of muscle cells by Fos leads to a block of myogenesis, apparently by an inhibition of *myo*D expression (Lassar et al., 1989). The oncogenic potential of *fos* genes seems, however, to be restricted to connective tissue cells, since the malignant transformation of other cell types *in vitro* has not been achieved to date. In agreement with this hypothesis, transgenic mice expressing an ectopic c-*fos* gene have been shown to develop specifically osteogenic and chondrogenic sarcomas (Rüther et al., 1989; Wang et al., 1991). It has, however, been shown that inducible Fos–estrogen receptor (Fos–ER) fusion proteins can induce morphological alterations and other transformation parameters in human carcinoma cells in culture (Schuermann et al., unpublished observations), and can trigger an epithelial-fibroblastoid cell conversion (Reichmann et al., 1992). In addition, *fos* genes have been reported to stimulate the proliferation of avian neuroretina cells in culture (Iba et al., 1988).

The oncogenic potential of other *fos*-related genes that have been identified on the basis of their structural relatedness and have been termed *fos*B (Zerial et al., 1989) and *fos*-related antigens-1 and -2 (*fra*-1, *fra*-2) has not been studied extensively (Cohen and Curran, 1988; Matsui et al., 1990; Nishina et al., 1990). Oncogenic properties have been shown for the *fos*B gene (Schuermann et al., 1991; Mumberg et al., 1991; Yen et al., 1991; Kovary et al., 1991). The *fos*B proto-oncogene is of particular interest because of the alternative splicing of its transcript, which results in the synthesis of two proteins of different length (Nakabeppu and Nathans, 1991; Mumberg et al., 1991; Yen et al., 1991; Kovary et al., 1991; Dobrzanski et al., 1991), and apparently antagonistic properties with respect to transregulation and transformation (Nakabeppu and Nathans, 1991; Mumberg et al., 1991; Yen et al., 1991). The two forms of FosB are consecutively expressed after serum stimulation of fibroblasts, the oncogenic longer form preceding expression of the C-terminally truncated anti-oncogenic protein, indicating that a specific mechanism controls the activity of FosB (Mumberg et al., 1991).

ACTIVATION OF THE ONCOGENIC POTENTIAL OF CELLULAR *fos* GENES

v-*fos* genes have the capacity to transform murine fibroblasts in culture efficiently, but c-*fos* is nontransforming (Miller et al., 1984). The molecular mechanisms leading to the activation of the c-*fos* proto-oncogene have been extensively analyzed using v-*fos*/c-*fos* chimeric molecules (Miller et al., 1984; Meijlink et al., 1985). A chimeric gene containing all 5' and coding sequences from c-*fos* and the 3' noncoding region from FBJ-MSV efficiently induces transformation *in vitro* and causes sarcomas in newborn mice. Therefore, no structural alterations in the coding region of c-*fos* are required to activate its transforming potential. The presence of a long terminal repeat is not sufficient, however, for the oncogenic activation of c-*fos* (Miller et al., 1984; Meijlink et al., 1985). In addition, a 67-bp element in the 3' noncoding region (123–189 bp upstream from the poly A addition site), which results in a destabilization of the mRNA, must be deleted (Miller et al., 1984; Meijlink et al., 1985; Rahmsdorf et al., 1987; Lee et al., 1988). Deletion of a 130-bp region of the C-terminal coding sequence has the same effect as the removal of the 67-bp element (Meijlink et al., 1985). It has been proposed that the mechanism by which the latter region inhibits *fos* mRNA and protein expression may involve the formation of a mRNA hairpin structure (Meijlink et al., 1985). It should, however, be noted that the C-terminus of c-Fos proteins has been reported to contain inhibitory serine phosphorylation sites that have a dramatic effect on the transforming potential of the protein (Tratner et al., 1992). This observation is in agreement with the fact that the transforming potential of c-Fos is low compared to its viral homologues, which both lack the C-terminal inhibitory sequences (van Beveren et al., 1983, 1984; Miller et al., 1984; Lee et al., 1988). It can, therefore, be concluded that the full oncogenic activation of c-*fos* requires both the deregulation of its expression and structural changes to its encoded product.

TRANSCRIPTIONAL TRANSREGULATION AND TRANSFORMATION

The most crucial region of Fos for the induction of transformation lies between amino acids 111 and 220 (Jenuwein and Müller, 1987). Comprehensive mutagenesis experiments showed that this region harbors the leucine zipper, the DNA contact site, and sequences involved in transactivation (Kouzarides and Ziff, 1988; Gentz et al., 1989; Turner and Tjian, 1989; Schuermann et al., 1989; Neuberg et al., 1989a,b; Abate et al.,

1991; Neuberg et al., 1991; Lucibello et al., 1991; Sutherland et al., 1992) and confirmed that both a functional leucine zipper and an intact DNA contact site are indispensable for transformation (Schuermann et al., 1989; Neuberg et al., 1989b; Neuberg et al., 1991). These correlations suggest that transactivation of AP-1-regulated genes is a (or the) key mechanism in Fos-induced transformation. This hypothesis is supported by the observation that transfected *fos* and *jun* genes cooperate in the induction of transformation *in vitro* (Neuberg et al., 1991; Schütte et al., 1989). Furthermore, it was shown that the fusion of heterologous transactivation domains from the transcription factors VP16, SP1, CTF, or the estrogen receptor (ER) rescues the transforming potential of FosB-S and enhances the transforming properties of c-Fos (Wisdom et al., 1992; Superti-Furga et al., 1991; Braselmann et al., 1992; Schuermann et al., 1993). In addition, the region between amino acid positions 40 and 111 in c-Fos/v-Fos is crucial for the induction of transformation (Jenuwein and Müller, 1987), and contains an autonomous transactivation domain (Jooss et al., 1994). On the other hand, there is some evidence that there is no direct correlation between transformation and the activation of AP-1-dependent transcription. Thus, a number of Fos proteins with mutations in the DNA contact site or other domains showed a lack of correlation between transformation and transactivation (Neuberg et al., 1991; Lucibello et al., 1991). Likewise, the temperature-dependent transforming properties of a wild-type v-Fos protein and of a temperature-sensitive derivative did not match temperature behavior in transactivation (Jooss and Müller, 1992). It must, however, be pointed out that in all published studies addressing this issue, transformation and transactivation assays were not performed in the same cell line. Thus, if the transactivating properties of Fos are cell line specific, this might explain the discrepancies, at least in part. In addition, the 5×TRE reporter construct used in most studies may not reflect the natural situation closely enough to allow for the establishment of clear conclusions. In this respect the identification of Fos-regulated transformation-relevant genes is of paramount importance.

A lack of correlation between transformation and AP-1-dependent transactivation is also suggested by other studies. First, a number of cell lines stably expressing Fos–ER fusion proteins have been isolated and analyzed for hormone-dependent transformation and transregulation (Lucibello et al., unpublished observations). In many cases, transformation was hormone-inducible, but transactivation of a 5×TRE reporter construct was not. In addition, there was no correlation between the basal or induced levels of AP-1 activity and the induction of transformation. Finally, there was no consistent induction in the hormone-treated, transformed cells of any of the known genes thought to be regulated by AP-1, such as collagenase or c-*jun*. The only exception in this context was the *fra*-1 gene, which in many cases showed elevated levels of expression after hormone treatment (Braselmann et al., 1992; Schuermann et al., 1993). Another study used a different approach but realized a similar conclusion (Oliviero et al., 1991). In this study, the level of cellular AP-1 activity was elevated by the ectopic expression of GCN4 in rat embryo fibroblasts. This increase in AP-1 activity, again monitored by the transactivation of a 5×TRE reporter construct, was, however, insufficient to induce transformation in cooperation with a *ras* oncogene, in contrast to combinations of *ras* plus *fos* or *ras* plus *jun*. From the results of both these studies one might conclude that TREs of the type present in the collagenase gene may not be the most critical targets in *fos*-induced transformation. In this context it is noteworthy that several transcription factors have recently been shown to interact with AP-1 family members, including Fos proteins. Examples of this kind are NF-κB (Stein et al., 1993) and NF-AT (Boise et al., 1993; Northrop et al., 1994) whose activity can be stimulated by, or is dependent on, the interaction with Fos/Jun.

The data discussed above suggest that transactivation is crucial in the induction of transformation, even though the target(s) are not known. On the other hand, one might also speculate that transregulatory mechanisms other than transactivation are involved in the induction of transformation, such as the transrepression of the glucocorticoid receptor (Jonat et al., 1990; Lucibello et al., 1990; Touray et al., 1991) or serum response elements (SREs) (Sassone-Corsi et al., 1988; König et al., 1989; Lucibello et al., 1989; Gius et al., 1990). Interestingly, inhibition of the glucocorticoid receptor requires the stretch between amino acids 40 and 111 (Lucibello et al., 1990), which is also needed for transformation (Jenuwein and Müller, 1987; Lucibello et al., 1991). It must, however, be pointed out that DNA binding, which is an absolute prerequisite of transformation, is not required for this type of transrepression (Lucibello et al., 1990). In addition, there does not seem to be any correlation between transformation and transrepression of SREs, in that on the one hand viral Fos proteins are unable to repress, and on the other hand DNA binding is not required for SRE repression (Lucibello et al., 1989; Gius et al., 1990). It is possible, however, that transformation by Fos may involve more than one molecular mechanism, e.g., the transactivation and transrepression of specific sets of genes by mechanisms that may be different, at least in part, from those described to date.

Some insight into the transformation-relevant targets of Fos may be derived from the analysis of revertants of *fos*-transformed cells (Zarbl et al., 1987). Such cells have been isolated and found to be resistant to the transforming potential of other oncogenes such as H-*ras*, *abl*, and *mos*, suggesting a defect in a transformation effector gene that is under the control of the Ras → Raf → MAP kinase → AP-1 signal transduction pathway (Leevers and Marshall, 1992). One of the genes defective in the revertants and responsible for the block in transformation has recently been cloned and termed *fte*-1 (derived from *fos* *t*ransformation *e*ffector) (Kho and Zarbl, 1992). Its encoded product has been shown to possess a remarkable homology to a yeast protein involved in mitochondrial protein import (Kho and Zarbl, 1992). This finding is intriguing, since it points to a potential role of mitochondria in the induction of transformation, a speculation that is substantiated by the proposed mitochondrial localization of the *bcl*-2 proto-oncogene product (Grivell and Jacobs, 1991) and the mitogen-induced expression of mitochondrial chaperonine *hsp*-60 (Wick et al., 1994).

REQUIREMENT FOR FOS PROTEINS IN CELL CYCLE PROGRESSION

All the results discussed above concern the ability of fos genes to induce morphological transformation. Another and perhaps even more important hallmark of malignant transformation is the deregulation of the molecular mechanisms controlling DNA replication and mitogenesis. The following section will, therefore, address the role of Fos proteins and AP-1 complexes in cell proliferation and, more specifically, in G0/G1-to-S progression.

During prenatal development the c-*fos* gene is expressed at high levels specifically in the cells of the cartilaginous growth zones of the bones (Dony and Gruss, 1987), suggesting a role in cell proliferation. This hypothesis is supported by the fact that many genes of the immediate early response encode transcription factors belonging to the AP-1 family (Hershman, 1991), and the functional significance of their induction has been shown in several studies using loss-of-function approaches, i.e., the expression of antisense RNA or the microinjection of antibodies. In the first two studies of this kind, dexamethasone-inducible antisense vectors were introduced into 3T3 cells (Holt et al., 1986; Nishikura and Murray, 1987). It was shown that in the presence of the inducer both the reentry of quiescent cells into the cell cycle after serum stimulation and the proliferation of asynchronous cells were inhibited, although the latter effect was seen only in one of the studies (Holt et al., 1986). These findings are in agreement with the results obtained by microinjecting antibodies directed against Fos protein, which also led to an inhibition of G1-to-S progression (Riabowol et al., 1988). In addition, it could be shown that the expression of antisense *fos* RNA or transdominant negative mutant Fos proteins inhibited and even reversed Ras-induced transformation, which also suggested a crucial role for c-Fos in the control of cell proliferation (Ledwith et al., 1990; Wick et al., 1992). Such a key role for c-Fos is, however, unlikely in view of more recent results. Bravo and co-workers produced antibodies that discriminate between the different known members of the Fos and Jun families and used these for microinjection into serum-stimulated NIH3T3 cells (Kovary and Bravo, 1991). These results showed that the inhibition of c-Fos alone has no significant effect on G0-to-S progression, while S-phase entry was clearly inhibited after injection of Jun-specific antibodies. In addition, DNA synthesis could be efficiently blocked by coinjecting antibodies against different members of the Fos family. This suggests that Fos family members may be functionally redundant with respect to the induction of S-phase entry after mitogen stimulation of quiescent cells, while Jun proteins seem to play more crucial roles. This may be related to the fact that, in contrast to Fos proteins, Jun family members can form transcriptionally active AP-1 homodimers (Nakabeppu et al., 1988; Halazonetis et al., 1988; Rauscher et al., 1988). An important role for AP-1 in this context is indeed suggested by the fact that the microinjection of TRE containing competitor oligonucleotides leads to the inhibition of DNA synthesis (Riabowol et al., 1992). It is not clear what the functional significance of the diversity of Fos and Jun proteins is, but for at least two proteins of this group, JunB (Schütte et al., 1989) and the truncated splice variant of FosB (FosB, FosB-S) described above (Nakabeppu and Nathans, 1991; Mumberg et al., 1991; Yen et al., 1991), antagonistic properties with respect to transregulation and transformation have been reported. It is, therefore, possible that one function of some of the Fos and Jun proteins is the establishments of finely tuned regulatory loops.

The results of the antibody microinjection and antisense RNA studies described above are partly contradictory, which is most likely due to some undesired side effects of the experimental manipulations used. Such artefacts may stem from (1) the formation of double-stranded RNA and the ensuing induction of the interferon system, (2) the intra-cellular decay of oligonucleotides resulting in the release of thymidine that could lead to an inhibition of DNA synthesis and compete for labeled DNA precursors used for quantitation, (3) the cross-hybridization of antisense RNA to transcripts from related or even unknown genes, or (4) the cross-reaction of antibodies with immunologically and, thus, perhaps, functionally related proteins. This also shows that the results of studies using antisense RNA, antisense oligonucleotides, or antibodies for the inhibition of protein expression or function are not easy to interpret, especially if the measured parameter is DNA synthesis. Considering these shortcomings, one probably has to conclude that the most promising strategy to obtain clear results regarding the function of *fos* gene products in G0-to-S and G1-to-S progression is the elimination of the respective genes by homologous recombination. First results along these lines have been obtained with embryonal stem (ES) cells and transgenic mice lacking c-Fos protein (Johnson et al., 1992; Wang et al., 1992). These ES cells showed normal proliferation *in vitro*, and the transgenic mice showed defects or alterations only in specific organs, most notably in bone hematopoietic tissue. These observations underscore the conclusion that, with regard to its function in cell proliferation, c-Fos is not as crucial as suggested by the earlier studies mentioned above.

ACKNOWLEDGMENTS

I am grateful to Dr. F. C. Lucibello for critically reading the manuscript. The work in the author's laboratory is supported by grants from the Deutsche Forschungs-gemeinschaft (SFB215/D8, Mu601/5-2, Mu601/5-3, and Mu601/7-1) and the Dr. Mildred Scheel-Stiftung für Krebsforschung.

REFERENCES

Abate. C.. Luk. D.. and Curran. T.. Transcriptional regulation by Fos and Jun in vitro: Interaction among multiple activator and regulatory domains, *Mol. Cell. Biol.*, 11. 3624, 1991.

Boise. L. H., Petryniak. X. M.. June. C. H.. Wang, C.-Y.. Lindsten, R. B.. Kovary, K.. Leiden. J. M., and Thompson. C. B.. The NFAT-1 DNA binding complex in activated T cells contains Fra-1 and JunB, *Mol. Cell. Biol.*, 13. 1911, 1993.

Braselmann. S.. Bergers. G.. Wrighton. C.. Graninger. P.. Superti-Furga. G.. and Busslinger. M.. Identification of Fos target genes by the use of selective induction systems. *J. Cell Sci.*, Suppl. 16. 97. 1992.

Cohen. D. R. and Curran. T.. fra-1: A serum-inducible. cellular immediate-early gene that encodes a fos-related antigen, *Mol. Cell. Biol.*, 8. 2063, 1988.

Curran. T.. Peters. G.. Van Beveren. C.. Teich. N. M.. and Verma. I. M.. FBJ murine osteosarcoma virus: Identification and molecular cloning of biologically active proviral DNA. *J. Virol.*, 44, 674, 1982.

Dobrzanski. P.. Noguchi, T.. Kovary. K.. Rizzo. C. A., Lazo. P. S., and Bravo, R.. Both products of the fosB gene, FosB and its short form. FosB/SF, are transcriptional activators in fibroblasts, *Mol. Cell. Biol.*, 11. 5470, 1991.

Dony. C. and Gruss. P.. Proto-oncogene c-fos expression in growth regions of fetal bone and mesodermal web tissue, *Nature*, 328, 711, 1987.

Finkel. M. P.. Biskis. B. O.. and Jinkins. P. B., Virus induction of osteosarcomes in mice. *Science*, 151. 698, 1966.

Finkel. M. P.. et al.. Viral ethology of bone cancer. *Front. Radiat. Ther. Oncol.*, 10. 28. 1975.

Gentz. R.. Rauscher. F. J.. III. Abate. C.. and Curran. T., Parallel association of Fos and Jun leucine zippers juxtaposes DNA binding domains, *Science*, 243. 1695, 1989.

Gius. D.. Cao. X., Rauscher, F. J., III. Cohen. D. R.. Curran. T.. and Sukhatme. V. P.. Transcriptional activation and repression by Fos are independent functions: The C terminus represses immediate-early gene expression via CArG elements, *Mol. Cell. Biol.*, 10. 4243, 1990.

Grivell. L. A. and Jacobs. H. T.. Oncogenes. mitochondria and immortality, *Curr. Biol.*, 1, 94, 1991.

Halazonetis. T. D.. Georgopoulos. K.. Greenberg. M. E.. and Leder, P.. c-Jun dimerizes with itself and with c-Fos, forming complexes of different DNA binding affinities. *Cell*, 55. 917, 1988.

Hershman. H. R.. Primary response genes induced by growth factors and tumor promoters. *Annu. Rev. Biochem.*, 60. 281. 1991.

Holt. J. T.. Venkat Gopal. T.. Moulton. A. D.. and Nienhus. A. W.. Inducible production of c-fos antisense RNA inhibits 3T3 cell proliferation. *Proc. Natl. Acad. Sci. U.S.A.*, 83. 4794, 1986.

Iba. H.. Shindo. Y., Nishina. H.. and Yoshida. T.. Transforming potential and growth stimulating activity of the v-fos and c-fos genes carried by avian retrovirus vectors. *Oncogene Res.*, 2. 121, 1988.

Jenuwein. T. and Müller. R.. Structure-function analysis of fos protein: A single amino acid change activates the immortalizing potential of v-fos, *Cell*, 48. 647, 1987.

Jenuwein. T.. Müller. D.. Curran. T.. and Müller. R.. Extended life span and tumorigenicity of nonestablished mouse connective tissue cells transformed by the *fos* oncogene of FBR-MuSV, *Cell*, 41. 629. 1985.

Johnson. R. S., Spiegelman. B. M.. and Papaionnaou. V., Pleiotropic effects of a null mutation in the c-fos proto-oncogenes. *Cell*, 71, 577. 1992.

Jonat. C.. Rahmsdorf. H. J.. Park. K.-K.. Cato. A. C. B.. Gebel, S., Ponta, H., and Herrlich. P., Anti-tumor promotion and antiinflammation: Down-modulation of AP-1 (Fos/Jun) activity by glucocorticoid hormone. *Cell*, 62. 1189. 1990.

Jooss, K. and Müller, R., Analysis of temperature-sensitive functions of Fos: Lack of a correlation between transformation and TRE-dependent trans-activation, *Oncogene*, 7, 1933, 1992.

Kho, C.-J. and Zarbl, H., Fte-1, a v-fos transformation effector gene, encodes the mammalian homologue of a yeast gene involved in protein import into mitochondria, *Proc. Natl. Acad. Sci. U.S.A.*, 89, 2200, 1992.

König, H., Ponta, H., Rahmsdorf, U., Büscher, M., Schönthal, A., Rahmsdorf, H. J., and Herrlich, P., Autoregulation of fos: The dyad symmetry element as the major target of repression, *EMBO J.*, 8, 2559, 1989.

Kouzarides, T. and Ziff, E., The role of the leucine zipper in the fos-jun interaction, *Nature*, 336, 646, 1988.

Kovary, K. and Bravo, R., The Jun and Fos protein families are both required for cell cycle progression in fibroblasts, *Mol. Cell. Biol.*, 11, 4466, 1991.

Kovary, K., Rizzo, C. A., Ryseck, R.-P., Noguchi, T., Raynoschek, C., Pelosin, J.-M., and Bravo, R., Constitutive expression of FosB and its short form, FosB/SF, induces malignant cell transformation in rat-1A cells, *New Biol.*, 3, 870, 1991.

Lassar, A. B., Thayer, M. J., Overell, R. W., and Weintraub, H., Transformation by activated ras or fos prevents myogenesis by inhibiting expression of MyoD1, *Cell*, 58, 659, 1989.

Ledwith, B. J., Manam, S., Kraynak, A. R., Nichols, W. W., and Bradley, M. O., Antisense-fos RNA causes partial reversion of the transformed phenotypes induced by the c-Ha-ras oncogenes, *Mol. Cell. Biol.*, 10, 1545, 1990.

Lee, W. M. F., Lin, C., and Curran, T., Activation of the transforming potential of the human fos proto-oncogene requires message stabilization and results in increased amounts of partially modified fos protein, *Mol. Cell. Biol.*, 8, 5521, 1988.

Leevers, S. J. and Marshall, C. J., MAP kinase regulation — the oncogene connection, *Trends Cell Biol.*, 2, 283, 1992.

Levy, J. A., Hartley, J. W., Rowe, W. P., and Huebner, R. J., Studies of FBJ osteosarcoma virus in tissue culture. I. Biological characteristics of the "C"-type virus, *J. Natl. Cancer Inst.*, 51, 525, 1973.

Lucibello, F. C., Lowag, C., Neuberg, M., and Müller, R., Trans-repression of the mouse c-fos promoter: A novel mechanism of Fos-mediated trans-regulation, *Cell*, 59, 999, 1989.

Lucibello, F. C., Slater, E. P., Jooss, K. U., Beato, M., and Müller, R., Mutual transrepression of Fos and the glucocorticoid receptor: Involvement of a functional domain in Fos which is absent in FosB, *EMBO J.*, 9, 2827, 1990.

Lucibello, F. C., Neuberg, M., Jenuwein, T., and Müller, R., Multiple regions of v-Fos protein involved in the activation of AP1-dependent transcription: Is trans-activation crucial for transformation? *New Biol.*, 3, 671, 1991.

Matsui, M., Tokuhara, M., Konuma, Y., Nomura, N., and Ishizaki, R., Isolation of human fos-related genes and their expression during monocyte-macrophage differentiation, *Oncogene*, 5, 249, 1990.

Meijlink, F., Curran, T., Miller, A. D., and Verma, I. M., Removal of a 67 base pair sequence in the non-coding region of proto-oncogene, *Proc. Natl. Acad. Sci. U.S.A.*, 82, 4987, 1985.

Miller, A. D., Curran, T., and Verma, I. M., c-fos protein can induce cellular transformation: A novel mechanism of activation of a cellular oncogene, *Cell*, 36, 51, 1984.

Mumberg, D., Lucibello, F. C., Schuermann, M., and Müller, R., Alternative splicing of fosB transcripts results in differentially expressed mRNAs encoding functionally antagonistic proteins, *Genes Dev.*, 5, 1212, 1991.

Nakabeppu, Y. and Nathans, D., A naturally occurring truncated form of FosB that inhibits Fos/Jun transcriptional activity, *Cell*, 64, 751, 1991.

Nakabeppu, Y., Ryder, K., and Nathans, D., DNA binding activities of three murine jun proteins: Stimulation by fos, *Cell*, 55, 907, 1988.

Neuberg, M., Adamkiewicz, J., Hunter, J. B., and Müller, R. A, Fos protein containing the Jun leucine zipper forms a homodimer which binds to the AP1 binding site, *Nature*, 341, 243, 1989a.

Neuberg, M., Schuermann, M., Hunter, J. B., and Müller, R., Two functionally different regions in Fos are required for the sequence-specific DNA interaction of the Fos/Jun protein complex, *Nature*, 338, 589, 1989b.

Neuberg, M., Schuermann, M., and Müller, R., Mutagenesis of the DNA contact site in Fos protein: Compatibility with the scissors grip model and requirement for transformation. *Oncogene*, 6, 1325, 1991.

Nishikura, K. and Murray, J. M., Antisense RNA of proto-oncogene c-fos blocks renewed growth of quiescent 3T3 cells. *Mol. Cell. Biol.*, 7, 639, 1987.

Nishina, H., Sato, H., Suzuki, T., Sato, M., and Iba, H., Isolation and characterization of fra-2, an additional member of the fos gene family. *Proc. Natl. Acad. Sci. U.S.A.*, 87, 3619, 1990.

Northrop, J. P., Ho, S. H., Chen, L., Thomas, D. J., Timmerman, L. A., Nolan, G. P., Admon, A., and Crabtree, G. R., NF-AT components define a family of transcription factors targeted in T-cell activation, *Nature*, 369, 497, 1994.

Oliviero, S., Robinson, G. S., Struhl, K., and Speigelman, B. M., Yeast GCN4 as a probe for oncogenesis by AP-1 transcription factors: Transcriptional activation through AP-1 sites is not sufficient for cellular transformation. *Genes Dev.*, 6, 1799, 1991.

Rahmsdorf, H. J., Schönthal, A., Angel, P., Litfin, M., Rüther, U., and Herrlich, P., Posttranscriptional regulation of c-fos mRNA expression, *Nucleic Acids Res.*, 15, 1643, 1987.

Rauscher, F. J., III, Voulalas, P. J., Franza, R. J., and Curran, T., Fos and Jun bind cooperatively to the AP-1 site: Reconstitution in vitro. *Genes Dev.*, 2, 1687, 1988.

Reichmann, E., Schwarz, H., Deiner, E. M., Leitner, I., Eilers, M., Berger, J., Busslinger, M., and Beug, H., Activation of an inducible c-FosER fusion protein causes loss of epithelial polarity and triggers epithelial-fibroblastoid cell conversion. *Cell*, 71, 1103, 1992.

Riabowol, K. T., Vosatka, R. J., Ziff, E. B., Lamb, N. J., and Feramisco, J. R., Microinjection of fos-specific antibodies blocks DNA synthesis in fibroblast cells. *Mol. Cell. Biol.*, 8, 1670, 1988.

Riabowol, K. T., Schiff, J., and Gilman, M. Z., Transcription factor AP-1 activity is required for initiation of DNA synthesis and is lost during cellular aging. *Proc. Natl. Acad. Sci. U.S.A.*, 89, 157, 1992.

Rüther, U., Komitowski, D., Schubert, F. R., and Wagner, E. F., c-fos expression induces bone tumors in transgenic mice. *Oncogene*, 4, 861, 1989.

Sassone-Corsi, P., Sisson, J. C., and Verma, I. M., Transcriptional autoregulation of the proto-oncogene fos. *Nature*, 334, 314, 1988.

Schuermann, M., Neuberg, M., Hunter, J. B., Jenuwein, T., Ryseck, R.-P., Bravo, R., and Müller, R., The leucine repeat motif in fos protein mediates complex formation with Jun/AP-1 and is required for transformation. *Cell*, 56, 507, 1989.

Schuermann, M., Jooss, K., and Müller, R., fosB is a transforming gene encoding a transcriptional activator. *Oncogene*, 6, 567, 1991.

Schuermann, M., Hennig, G., and Müller, R., Transcriptional activation and transformation by chimeric Fos–estrogen receptor proteins: Altered properties as a result of gene fusion. *Oncogene*, 8, 2781, 1993.

Schütte, J., Viallet, J., Nau, M., Segal, S., Fedorko, J., and Minna, J., jun-B inhibits and c-fos stimulates the transforming and trans-activating activities of c-jun. *Cell*, 59, 987, 1989.

Stein, B., Baldwin, A. S., Ballard, D. W., Greene, W. C., Angel, P., and Herrlich, P., Cross-coupling of the NF-κB p65 and Fos/Jun transcription factors produces potentiated biological function, *EMBO J.*, 12, 3879, 1993.

Superti-Furga, G., Berger, G., Picard, D., and Busslinger, M., Hormone-dependent transcriptional regulation and cellular transformation by Fos-steroid receptor fusion proteins, *Proc. Natl. Acad. Sci. U.S.A.*, 88, 5114, 1991.

Sutherland, J. A., Cook, A., Bannister, A. J., and Kouzarides, T., Conserved motifs in Fos and Jun define a new class of activation domain, *Genes Dev.*, 6, 1810, 1992.

Touray, M., Ryan, F., Jaggi, R., and Martin, F., Characterisation of functional inhibition of glucocorticoid receptor by Fos/Jun, *Oncogene*, 6, 1227, 1991.

Tratner, I., Ofir, R., and Verma, I. M., Alteration of a cyclic AMP-dependent protein kinase phosphorylation site in c-Fos protein augments its transforming potential, *Mol. Cell. Biol.*, 12, 998, 1992.

Turner, R. and Tjian, R., Leucine repeats and an adjacent DNA binding domain mediate the formation of functional cFos-cJun heterodimers, *Science*, 243, 1689, 1989.

van Beveren, C., van Straaten, F., Curran, T., Müller, R., and Verma, I. M., Analysis of FBJ-MuSV provirus and c-fos (mouse) gene reveals that viral and cellular fos gene products have different carboxy termini, *Cell*, 32, 1241, 1983.

van Beveren, C., Enami, S., Curran, T., and Verma, I. M., FBR murine osteosarcoma virus. II. Nucleotide sequence of the provirus reveals that the genome contains sequences acquired from two cellular genes, *Virology*, 135, 229, 1984.

Wang, Z.-Q., Grigoriadis, A. E., Möhle-Steinlein, U., and Wagner, E. F., A novel target cell for c-fos-induced oncogenesis: Development of chondrogenic tumours in embryonic stem cell chimeras, *EMBO J.*, 10, 2437, 1991.

Wang, Z.-Q., Ovitt, C., Grigoriadis, A. E., Möhle-Steinlein, U., Rüther, U., and Wagner, E. F., Bone and haematopoietic disorders in mice lacking c-fos, *Nature*, 360, 741, 1992.

Ward, J. M. and Young, D. M., Histogenesis and morphology of periosteal sarcomas induced by FBJ virus in NIH-Swiss mice, *Cancer Res.*, 36, 3985, 1976.

Wick, M., Lucibello, F. C., and Müller, R., Inhibition of Fos- and Ras-induced transformation by mutant Fos proteins with structural alterations in functionally different domains, *Oncogene*, 7, 859, 1992.

Wick, M., Bürger, C., Brüsselbach, S., Lucibello, F. C., and Müller, R., Identification of serum-inducible genes: different patterns of gene regulation during G_0->S and G_1->S progression, *J. Cell Sci.*, 107, 227, 1994.

Wisdom, R., Yen, J., Rashid, D., and Verma, I. M., Transformation by FosB requires a trans-activation domain missing in FosB2 that can be substituted by heterologous activation domains, *Genes Dev.*, 6, 667, 1992.

Yen, J., Wisdom, R. M., Tratner, I., and Verma, I. M., An alternative spliced form of FosB is a negative regulator of transcriptional activation and transformation by Fos proteins, *Proc. Natl. Acad. Sci. U.S.A.*, 88, 5077, 1991.

Zarbl, H., Latreille, J., and Jolicoeur, P., Revertants of v-fos-transformed fibroblasts have mutations in cellular genes essential for transformation by other oncogenes, *Cell*, 51, 357, 1987.

Zerial, M., Toschi, L., Ryseck, R.-P., Schuermann, M., Müller, R., and Bravo, R., The product of a novel growth factor activated gene, fos B, interacts with Jun proteins enhancing their DNA binding activity, *EMBO J.*, 8, 805, 1989.

Chapter 15

Oncogenic Transformation by Jun

Peter K. Vogt

CONTENTS

INTRODUCTION

Genes coding for transcription factors make unique oncogenes because of their position in cellular signal transduction. Transcription factors are the ultimate recipients of afferent growth signals and convert these signals into specific programmed responses of the cell consisting of patterns of gene expression (Herrlich and Ponta, 1989). In the process of this conversion they must interact directly with the genome and the cellular transcriptional machinery. Because of this direct effect on gene expression, transcription factors hold much promise for an understanding of oncogenesis and of the oncogenic cellular phenotype at the level of gene regulation (Forrest and Curran, 1992).

The *jun* gene belongs to this category of transcription factor oncogenes (Vogt and Tjian, 1988; Curran and Vogt, 1991). *jun* is a widely conserved eucaryotic gene coding for a transcriptional regulator of strikingly simple structure and organization that dimerizes with related proteins, binds to a short DNA sequence (TGACTCA) known as TRE, and affects the transcription of genes containing that sequence in their promoters (Vogt and Bos, 1989, 1990). The mechanism by which *jun* induces cancer also appeared deceptively simple at first; it was widely assumed that *jun* would upregulate the transcription of growth-promoting genes and in this way transform the cell (Lewin, 1991).

Yet the functions of the Jun protein are not unidirectional nor on a single track; they are multiple and surprisingly complex, defining a nodal point in transcriptional regulation. Jun interacts with related and unrelated transcription factors, gaining influence on different and seemingly unrelated signal pathways and controlling diverse targets (Lamph, 1991; Miner et al., 1991; Schüle and Evans, 1991; Cato et al., 1992; Sassone-Corsi, 1992). This protean ability of Jun to participate in various modes of gene regulation and to act as a positive or negative element is still not completely understood. Similarly, the mechanism of Jun-induced oncogenic transformation emerges as a more complex process than was initially expected (Angel and Karin, 1991). Defining the precise nature and role of the presumed transcriptional interference has become the central problem for understanding Jun-induced transformation. This review will seek to integrate and reconcile the sometimes contradictory data that relate regulatory functions of Jun to oncogenesis.

0-8493-4573-1/94/$0.00+$.50
© 1994 by CRC Press, Inc.

jun INITIATES TRANSFORMATION OF
AVIAN CELLS AS A SINGLE ONCOGENE

The original *jun*-carrying retrovirus, avian sarcoma virus 17 (ASV 17), was isolated from a spontaneous sarcoma in an adult chicken (Cavalieri et al., 1985; Maki et al., 1987). ASV 17 transforms chicken embryo fibroblasts (CEF) in culture and induces fibrosarcomas in young chickens. Focal areas of transformed CEF, appearing in dishes that have been infected with appropriate dilutions of ASV 17, consist of elongated, needle-like cells growing in a characteristic whorl-like, parallel arrangement. In contrast to surrounding normal CEF, the transformed cells pile up in several layers. ASV 17 also confers on infected CEF the ability to grow and form colonies in nutrient agar suspension. The virus is defective in replication because it lacks the entire *pol* gene and contains only partial *gag* and *env* sequences. It replicates in conjunction with an avian leukosis helper virus that supplies the missing viral functions and proteins in *trans*. Solitary infection of a cell by ASV 17 without helper results in transformation in the absence of infectious progeny virus synthesis. Such ASV 17 nonproducers can be cultured over several weeks of active growth but they are not immortal, and eventually they cease to divide. As is common with nonproducer cells transformed by a replication defective retrovirus, synthesis of infectious viral progeny can be activated by superinfection with a related replication competent retrovirus. Focus formation by ASV 17 in CEF follows one hit kinetics indicating that a single viral particle is sufficient to produce a focus. In avian cells, *jun* does not require the cooperation of a second exogenous oncogene to induce transformation.

The transforming gene in the ASV 17 genome is the chicken cell-derived insert *jun;* the *jun* sequences excised from ASV 17 and inserted into the expression vector RCAS make this vector oncogenic for CEF (Maki et al., 1987; Bos et al., 1990). RCAS is a vector developed from the genome of the Prague strain of Rous sarcoma virus, essentially by replacing the *src* gene of that virus with a cloning site (Hughes and Kosik, 1984; Hughes et al., 1987). Since RCAS contains all necessary genetic information for retrovirus replication, CEF transfected with this vector produce infectious viral progeny carrying the inserted gene. RCAS and similar expression vectors have been widely used to probe the oncogenicity of various mutant *jun* genes. ASV 17 and RCAS constructs expressing the viral (v-)*jun* gene transform, in addition to CEF, fibroblasts of the Japanese quail, *Coturnix coturnix japonica* (Vogt, 1985; Castellazzi et al., 1990), chicken neuroretinal cells (Tsuchie and Vogt, 1989), and chicken skeletal muscle myoblasts (Grossi et al., 1991; Su et al., 1991). They do not transform chicken yolk sac cultures that contain macrophages as well as primitive cells of the myeloid and erythroid hematopoietic lineages (Wong and Vogt, 1989). The chicken cellular (c-)*jun* gene expressed in RCAS is quantitatively less transforming than v-*jun* for CEF cultures, but it too induces foci of cells that show altered morphology and increased growth resembling v-*jun* foci (Wong et al., 1992). c-*jun* genes of chicken, mouse, and Japanese quail also stimulate anchorage-independent growth of CEF (Castellazzi et al, 1990, 1991; Wong et al., 1992).

ASV 17 and the RCAS–v-*jun* constructs are oncogenic in chickens, inducing fibrosarcomas at the site of injection (Wong et al., 1992). The latent period for tumor induction by ASV 17 is about three weeks — relatively long compared to Rous sarcoma virus which can produce a tumor within one week. RCAS–v-*jun* constructs induce sarcomas somewhat more rapidly than ASV 17 possibly because the vector replicates more efficiently or expresses *jun* to higher levels than ASV 17. The long latent period for ASV 17-induced sarcomas suggested the possibility that these tumors may arise from a rare transforming event and, therefore, might be clonal in origin. Tests for clonality of sarcoma tissue with respect to *jun* integration sites have, however, given negative results (Nagata and Vogt, 1989). These observations show that *jun* is an efficient oncogene for certain avian cell types, being sufficient on its own to cause transformation. Nevertheless, cofactors can enhance this oncogenicity. If chickens are injected in one wing with ASV 17, and a small cut penetrating the skin is made on the other wing, a *jun*-induced sarcoma arises not only at the site of injection but also at the site of the wound that did not receive the virus (Marshall et al., 1992). Similar wound-related tumors were first observed in chickens injected with Rous sarcoma virus, where this phenomenon has been analyzed in detail (Sieweke et al., 1989, 1990). Wounding as a cofactor for *jun*-dependent oncogenesis is particularly apparent in v-*jun* transgenic mice and will be discussed below.

Tests for tumorigenicity in chickens also reveal an important qualitative difference between v-*jun* and c-*jun*. Although both v-*jun* and c-*jun* transform CEF in culture, albeit with different efficiencies, only v-*jun* induces tumors in the chicken; c-*jun* is nontumorigenic (Wong et al., 1992). This is true not only of chicken but also of mouse c-*jun*. Rare tumors that arise in chickens inoculated with a c-*jun* construct are probably due to mutations of the *jun* gene that occur in the course of RCAS virus replication and that confer oncogenicity *in vivo*. This interpretation is suggested by the fact that the change to tumorigenicity is heritable in the virus recovered from such rare tumors. Sequence analyses of such presumptive mutants could yield information about the changes in *jun* that lead to oncogenicity in the animals.

TRANSFORMATION OF MAMMALIAN CELLS BY *jun* REQUIRES THE OVERT COOPERATION OF EXOGENOUS OR ENDOGENOUS GROWTH SIGNALS

Human c-*jun* expression vectors can transform primary rat embryo fibroblasts if they are cotransfected with a *ras* oncogene that carries an activating mutation (Schütte et al., 1988, 1989; Oliviero et al., 1992). *ras* alone is an inefficient transformer in this cell system; *jun* by itself is inactive. The two oncogenes together complement each other, leading to a substantial increase in the number of foci produced per unit DNA transfected. The cotransformed cells are oncogenic in nude mice. Interestingly, in this system of mammalian cell transformation it is a c-*jun* version that is more active than v-*jun*, and in conjunction with *ras* participates in tumor formation (M. Birrer, 1992, personal communication). The reason for the reversal of the oncogenic difference between viral and cellular *jun* as compared to the avian system is not known. The cotransformation test of human *jun* and *ras* in primary rat fibroblasts has become a standard for the analysis of *jun* functions that are important in mammalian cell transformation. However, the respective roles of *jun* and *ras* in this system need more precise definition. The transformation related events triggered by *jun* in cotransformation with *ras* may not be the same as those in *jun*-induced oncogenesis in avian cells.

Questions about the cotransformation assay are also raised by the observation that mouse c-*jun*, in contrast to human c-*jun*, strongly inhibits transformation by *ras* in primary rat fibroblasts (Ginsberg et al., 1991). Deletion analysis suggests that the inhibitory domain is located in the *carboxy*-terminal region of mouse c-*jun* but does not include the leucine zipper.

Unlike primary rat embryo cells, the continuous rat cell line Rat 1 can be transformed by human c-*jun* without the aid of another oncogene (Schütte et al., 1988). The transformants are able to grow in agar and are tumor producing in nude mice. Again, c-*jun* is more active in this transformation assay than v-*jun* (M. Birrer, 1992, personal communication). As a continuous cell line, Rat 1 probably contains or produces cofactors that are required by *jun* in the transformation of mammalian cells and that must be induced by *ras* in primary cells.

A dramatic demonstration of the role of cofactors in *jun*-induced oncogenesis is seen in mice that are transgenic for the v-*jun* gene. The transgene in these animals is under the control of the major histocompatibility H2K promoter, with elevated v-*jun* expression evident in thymus, spleen, testes, and skin (Schuh et al., 1990). These mice do not show an increased incidence of spontaneous tumors. However, if they suffer a wound that penetrates the dermis, the wound heals with scar tissue that is highly hyperplastic compared to wounds in nontransgenic mice, and a high percentage of these animals develop sarcomas at the site of wounding. Expression of the *jun* transgene is increased in the hyperplastic scar tissue compared to normal skin, and it is further elevated in the tumor. In this situation, *jun* is necessary but not sufficient for tumor formation. Cofactors that are activated during wounding and wound healing appear to interact with the transgene to cause oncogenic transformation. In cell culture, v-*jun* transgenic mouse embryo fibroblasts and revertant tumor cells from wound-related tumors of v-*jun* transgenic animals show increased potential for anchorage-independent growth when exposed to one of two wound-related growth factors, TNFα or IL-1. The effect is not seen with several other cytokines, including PDGF and EGF, and is specific for *jun* transgenic cells. It is inhibited by retinoic acid, dexamethasone, and forskolin (Vanhamme et al., 1993). These observations suggest that TNFα or IL-1 may play a part in *jun* tumorigenesis in the mammal. However, initially at least, the effect of these cytokines is phenotypic and reversible. In tumor formation it must be followed by a genotypic change that remains to be defined.

JUN MUTANTS AND CHIMERAS IDENTIFY DOMAINS THAT ARE IMPORTANT IN TRANSFORMATION

The Jun protein can be roughly divided into three domains: the leucine zipper, which mediates dimerization, the amino terminally adjacent basic region, which forms the DNA contact surface, and three acidic regions in the amino terminal half of the molecule, which act as the major transactivator domains (Vogt and Bos, 1990; Vogt and Morgan, 1990; Angel and Karin, 1991; Curran and Vogt, 1991; Ransone and Verma, 1990). When Jun functions as a component of the AP-1 transcription factor complex, these three regions operate in a hierarchical fashion: dimerization is required for DNA binding, and DNA binding, in turn, is needed for transcriptional activation from the TRE. Oncogenic transformation seems positioned at the end of this hierarchy, dependent on dimerization, DNA binding, and some form of transcriptional control.

Amino acid substitutions in the leucine zipper region that decrease dimerization of Jun also reduce transforming potential for CEF. Deletion of the leucine zipper results in a nontransforming Jun protein (Morgan et al., 1992). Similarly, amino acid substitutions in the basic region that weaken DNA binding also diminish focus-forming potential for CEF. Deletion of this region abolishes transformation altogether and can turn the truncated protein into a transdominant negative, inhibiting transformation (Gaynor et al., 1991; Morgan et al., 1992). A deletion of the 124 amino terminal amino acids of Jun affecting the transactivation domains also leaves the protein without ability to transform CEF (Morgan et al., 1992).

With some Jun proteins carrying amino acid substitutions in the basic region, wild-type levels of DNA binding can be restored by coexpressed *fos*. The restoration of DNA binding also reactivates the transforming potential of the mutants (Morgan et al., 1992). The Fos protein can thus affect Jun transformation. However, overexpression of the *fos* gene does not enhance transformation by highly oncogenic *jun* mutants, nor are there increased levels of *fos* expressed in *jun*-transformed cells. Indeed, mutant Jun proteins have been constructed that can only form homodimers, and these mutants still retain oncogenic potential, though at a reduced level compared to v-Jun (Hartl and Vogt, 1992b; Hughes et al., 1992; Oliviero et al., 1992). Therefore, oncogenic transformation by Jun does not depend on heterodimerization of Jun with Fos, with Fos-related proteins, or with other bZIP proteins.

v-*jun* transforms CEF more efficiently than overexpressed chicken c-*jun*, the direct ancestral form of the viral gene. This enhanced transformation potential of v-*jun* must reflect changes in protein structure and function. A comparison between viral and cellular versions of *jun* reveals the following differences (Nishimura and Vogt, 1988; Bos et al., 1990). First, the viral gene is fused at its amino terminus to 222 amino acids derived from the v-*gag* gene. Second, within the *jun* coding region the viral version has suffered a 27 amino acid deletion in the amino terminal transactivation domain this deletion defines the delta region. Third, in the *C*-terminal half of Jun there are two amino acid substitutions, an S-to-F at position 222 of the chicken c-Jun (position 246 of rodent and 243 of human Jun proteins) and a C-to-S substitution at position 248 of the chicken c-Jun protein (272 of rodent and 269 of human c-Jun). Some clones of v-Jun also have an amino acid substitution at position 181 (G-to-R), but since this mutation is not present in all clones of v-Jun, it is probably of no consequence and has not been studied. Fourth, v-*jun* also lacks the long 3' untranslated region present in the cellular *jun* mRNA. This region is AU rich and carries instability signals, e.g., a copy of the AUUUA sequence that probably marks the cellular *jun* message for rapid degradation.

The mutations and alterations identified in v-*jun* have been introduced alone and in combination onto a chicken c-*jun* background, and the effect on transforming potential has been studied by expressing the constructs from the RCAS retroviral vector (Bos et al., 1990; Morgan et al., 1993). These experiments have identified the delta region as a major factor determining efficient focus formation in CEF. Introducing this deletion on a chicken c-Jun background yields a Jun protein that produces foci in CEF with an efficiency comparable to the viral version. The amino terminal Gag tail of the v-Jun protein, has no measurable effect on transforming efficiency of a v-Jun protein but it may mediate elevated levels of protein expression and, therefore, could enhance transformation by Jun mutants that are poorly expressed. The point mutations in the *C*-terminal half of v-Jun affect transformation in a more subtle way and will be discussed below. Removal of the 3' untranslated region of c-Jun is a requirement for obtaining significant focus-forming activity by the cellular protein. In the presence of this region, expression remains below the threshold level needed for transformation (Cavalieri et al., 1985; Wong et al., 1992).

A possible link between transformation and transactivation was first suggested by observations on the delta deletion. The delta region has a controlling effect on transactivation. In an *in vitro* transcription system using HeLa cell nuclear extracts, deletion of delta was found to increase AP-1 activity, suggesting that c-Jun may be negatively regulated through the delta domain (Bohmann and Tjian, 1989; Baichwal and Tjian, 1990). Delta could either bind a cellular inhibitor or it could be affected by some posttranslational modification, e.g., phosphorylation. In support of this suggestion, v-Jun has been found to be underphosphorylated in CEF as compared to c-Jun (Black et al., 1991). Delta also suppresses the phorbol ester-stimulated phosphorylation of Jun (Adler et al, 1992). Additional analyses of the delta domain have, however, not fully supported a correlation of enhanced transactivation by Jun with transformation. Chicken c-Jun with delta deleted fails to transactivate the collagenase promoter (containing a TRE) in CEF (Adler et al., 1992), and various partial deletions and amino acid substitutions in the delta region show no or an even inverse correlation between transactivation of the collagenase promoter in CEF and transformation potential for the same cell (Håvarstein et al., 1992). The delta domain is clearly important in the regulation of Jun but its effect depends on additional factors that are cell type-specific and have yet to be defined.

The C-terminal amino acid substitutions of Jun also affect various Jun functions (Morgan et al., 1993; Oehler et al., 1993). The serine at position 243 in human Jun (position 222 in the chicken protein) is a substrate for a still unidentified kinase, and in the phosphorylated state, reduces DNA binding and transactivation (Boyle et al., 1991; Lin et al., 1992). Similar effects on Jun function are caused by the phosphorylation of nearby threonine 231 and serine 249 by casein kinase II. In v-Jun the homologous serine 222 is mutated to phenylalanine and is no longer subject to this downregulation. The same S-to-F mutation introduced on a c-Jun background constitutively increases DNA binding and transactivation from a collagenase promoter in F9 embryonal carcinoma cells (Boyle et al., 1991; Morgan et al., 1993). Interestingly, the S-to-F mutation in mammalian c-Jun also reduces the phosphorylation of serine 249 and threonine 231 by casein kinase II (Lin et al., 1992). In CEF the S-to-F change at the homologous 222 position has very different consequences. There is a slight reduction in the efficiency of focus formation compared to wild-type c-Jun, and the transactivation potential of the mutant is diminished. However, the originally nontumorigenic chicken c-Jun protein becomes weakly tumorigenic with the S-to-F mutation in position 222 (Morgan and Vogt, unpublished observations).

The C-to-S mutation at position 269 of human Jun corresponding to position 272 of rodent or 248 of chicken Jun also has multiple consequences. This cysteine is the target of a redox control mechanism (Abate et al., 1990). For efficient DNA binding, it needs to be reduced, a task carried out by a cellular protein called REF (Xanthoudakis and Curran, 1992; Xanthoudakis et al., 1992). The C-to-S mutation relieves DNA binding of redox control and elevates it constitutively. This mutation also makes the nuclear translocation of the Jun protein cell cycle dependent (Chida and Vogt, 1992). Whereas Jun proteins containing the wild-type cysteine in this position are translocated to the nucleus throughout the cell cycle, v-Jun, with its C-to-S mutation, enters the nucleus most rapidly during the S/M phase. The C-to-S mutation generates a potential new phosphorylation site in a PKC consensus sequence. A cell-cycle dependent phosphorylation event may, therefore, determine nuclear translocation of the mutant Jun protein; this suggestion needs to be examined experimentally. The C-to-S mutation also affects transformation. This single amino acid exchange makes the c-Jun protein tumorigenic for chickens. It increases focus-forming efficiency in CEF about fourfold. Transactivation of the collagenase promoter is reduced by this mutation both in CEF and in F9 mouse embryonal carcinoma cells (Morgan et al., 1993; Oehler et al., 1993; Morgan and Vogt, unpublished observations).

The transforming potential of Jun has also been investigated by constructing chimeras between Jun and other transcription factors. Functional domains of transcription factors have been defined by mutational analysis and by excision and exchange for similar domains derived from other proteins. The results of such "domain switching" experiments suggest a modular organization for many transcription factors: distinct domains for dimerization, DNA binding, and transactivation remain functional when transplanted to a different protein. While this exchangeability is the general rule, there are exceptions (Golemis and Brent, 1992). In the folded protein, regions that are distant on the linear map may be close together and may cooperate or interfere with each other. Properties of protein chimeras must, therefore, be interpreted with caution, but they do offer some insights into Jun functions that are important in transformation.

An example is the replacement of the Jun transactivation domain with that from other transcription factors. Transactivation domains can be roughly divided into three categories: proline-rich, glutamine-rich, and acidic. There is some evidence that domains belonging to the same category can to some extent substitute for each other, interacting with the same or similar auxiliary factors, whereas those from different categories cannot be exchanged (Tanese et al., 1991; Oehler and Angel, 1992). VP16 is a highly active transcription factor coded for by the herpes simplex virus. Although Jun and VP16 show no sequence relationship, their transactivation domains belong to the acidic category. Chimeras in which the transactivation domain of Jun is replaced by that of VP16 still transform, although they do not transactivate the collagenase promoter in CEF (Schuur et al., 1993). Of these chimeric constructs, the one containing the C-terminus of c-Jun is far more oncogenic than that carrying the corresponding region of v-Jun, which carries mutations that on a purely Jun background enhance oncogenicity. This result shows that domain switching can alter the functions of domains in the chimeric construct. Similar chimeras have been constructed with the yeast transcription factor GCN4, a close relative of v-Jun and c-Jun (Oliviero et al., 1992). GCN4 by itself activates transcription from a TRE-containing promoter in mammalian cells but fails to transform such cells. It also does not transform CEF (Monteclaro and Vogt, 1991). Chimeras in which the transactivation domain is derived from c-Jun, and dimerization and DNA binding domains come from GCN4, are highly transforming and transactivating; however, the reciprocal constructs are poor transformers but good transactivators (Yang-Yen et al., 1990).

The JunD protein, a relative of Jun, does not transform CEF and fails to transactivate the collagenase promoter in these cells. A series of chimeras between the chicken JunD and chicken c-Jun and v-Jun yielded some transforming constructs but all except one of these failed to acquire transactivation potential for CEF (Castelazzi et al., 1991; Hartl and Vogt, 1992). Although wild-type *jun*D does not transform CEF, spontaneous mutants of this gene arising during replication of the RCAS *jun*D viral vector gain transforming ability (Hartl and Vogt, 1992). All these mutants show a duplication of an amino terminal region that is homologous to a transactivation domain of Jun. These mutants are also transactivating in CEF.

Little is known about the transforming activity of *jun*B, the other cousin of *jun*. A human *jun*B construct transfected into rat cells, together with human c-*jun* plus activated *ras*, inhibits the co-transformation induced by the latter two genes (Schütte et al., 1989). It also reduces transactivation in F9 cells and is, therefore, considered a *jun* antagonist (Lisitsyn et al., 1993). Murine *jun*B, on the other hand, can stimulate anchorage-independent growth of CEF (Castelazzi et al., 1991). Clearly, more work on *jun*B and *jun*D is needed before their basic functions are known (Schütte et al., 1989).

JUN IS A COMPONENT OF POSITIVE AND OF NEGATIVE GROWTH SIGNALS

Jun can be thought of as a converter of signals. It receives regulatory input originating outside the cell, traversing plasma membrane, cytoplasm, and nuclear envelope in a cascade of biochemical reactions (Herrlich and Ponta, 1989; Ransone and Verma, 1990; Gutman et al., 1991; Karin and Smeal, 1992). These signals can modify the transcription of the *jun* gene or affect the activity of the Jun protein posttranslationally. Altered AP-1 activity then induces changes in a multiplicity of gene expression programs, the selection of which is determined by ancillary factors. In keeping with this role as a signal converter, *jun* belongs to the category of immediate early genes. Expression and activity of these genes is transiently stimulated by mitogenic signals. Expression and activity of the Jun protein are exquisitely responsive to a long list of growth factors including, PDGF, FGF, EGF, and TNFα (Lamph et al., 1988; Quantin and Breathnach, 1988; Ryder and Nathans, 1988; Brenner et al., 1989; Oemar et al., 1991). Rapidly dividing cells show increased levels of Jun/AP-1 transcriptional activation, some of which may be due to the stimulation of *jun*-related genes rather than *jun* itself (Angel et al., 1988; Sakai et al., 1989; Alcorn et al., 1990; Hsu et al., 1992). The Jun protein is required for the initiation of DNA synthesis during S phase, which may be the reason why Jun is essential for the cell cycle.

Both microinjected antibody and antisense Jun RNA stop cell growth and division (Carter et al., 1991; Kovary and Bravo, 1991a, 1991b; Riabowol et al., 1992; Smith and Prochownik, 1992). Transforming oncogenes including *src, ras, raf* and E1A increase AP-1/Jun activity (Wasylyk et al., 1988; Müller et al., 1989; Binétruy et al., 1991; de Groot et al., 1991; van Dam et al., 1993). Mutationally activated Ras protein induces phosphorylation of Jun on two serines at position 63 and 73, and this phosphorylation is required for Ras-induced transformation (Smeal et al., 1991, 1992). E1A also causes the phosphorylations at serines 63 and 73 (Hagmeyer et al., 1993). The viral Src protein has been proposed to counteract a cellular inhibitor that works through the amino terminal delta domain of Jun (Baichwal et al., 1991).

But Jun is not connected exclusively to positive growth signals (Sassone-Corsi, 1992). It is induced by TGFβ in cells that are growth inhibited by the cytokine, and Jun, in turn, can induce transcription of TGFβ (Pertovaara et al., 1988; Birchenall-Roberts, 1990; Kim et al., 1990) Jun also does not invariably function as a transcriptional activator. It can act as a repressor, especially in conjunction with its ability to influence other transcriptional control signals like those involving steroid hormone and retinoic acid receptors (Lamph, 1991; Miner et al., 1991; Schüle and Evans, 1991, Cato et al., 1992; Sassone-Corsi, 1992).

The potential of Jun to induce seemingly opposite growth regulatory effects is best illustrated in systems of cell differentiation. The differentiation of skeletal myoblasts into myotubes and muscle is inhibited by Jun, notably v-Jun (Vogt, 1985; Tsuchie and Vogt, 1989; Endo, 1992). Jun-expressing myoblasts do not fuse into multinucleated myotubes, they continue to divide and do not initiate transcription of several muscle-specific genes. Jun inhibits expression of the muscle-specific transcriptional regulator MyoD and interferes with activation of muscle-specific gene transcription by MyoD and myogenin. The interference with MyoD appears to result from an interaction of the Jun and MyoD proteins, which targets the helix-loop-helix region of MyoD. Evidence regarding the Jun effector domains in this interaction is contradictory; it is either the amino terminal domain or the leucine zipper (Bengal et al., 1992; Li et al., 1992).

In a number of other systems increased c-Jun expression and AP-1 activity are correlated with cessation of cell growth and differentiation. In the rat pheochromocytoma cell line PC12, NGF induces neuronal differentiation accompanied by increased Jun expression (Bartel et al., 1989). Similar observations were made with differentiating neuroblastoma cells (de Groot and Kruijer, 1991). P19 mouse embryonal carcinoma cells are induced to differentiate by overexpression of c-Jun (de Groot et al., 1990). In F9 mouse embryonal carcinoma cells differentiation is also accompanied by increased Jun expression (Yang-Yen et al., 1990). TPA, okadaic acid, diacylglycerol, and GMCSF all stimulate c-Jun mRNA induction and AP-1 activity in U937 human leukemia cells and at the same time induce differentiation to macrophages (William et al., 1990; Adunyah et al., 1991, 1992; Kharbanda et al., 1992). Differentiation induced into macrophages in HL60 cells by TPA or vitamin D is also accompanied by elevated levels of Jun mRNA (Gaynor et al., 1991). Growth hormone induces the differentiation of preadipocytes (3T3-F442A) into adipocytes and at the same time induces c-Jun (Sumantran et al., 1992). However, c-Jun expression is induced by growth hormone even in nondifferentiating cell lines, suggesting that changes in the expression of the proto-oncogene are correlated with the response to the hormone rather than with the adipocyte differentiation program.

The picture that emerges from these studies is that of Jun as a general and, perhaps, neutral transcriptional regulator that can function as a growth stimulator or a growth attenuator. It can activate transcription or repress it. The direction of these activities is determined by additional, largely unknown factors.

JUN TRANSFORMS BY INTERFERING
WITH TRANSCRIPTIONAL REGULATION

Deletion of the major transactivation domain of Jun abolishes the transforming and oncogenic potential of the molecule in mammalian and avian cells (Morgan et al., 1992; Lloyd et al., 1991; Pulverer et al., 1991; Granger-Schnarr et al., 1992). In mammalian cells, such a deletion mutant even acts as a transdominant negative for transformation, inducing reversion of the neoplastic phenotype of *ras*-transformed cells. It is, therefore, likely that the regulation of transcription plays an essential role in the transformation process. Attempts have been made with avian and mammalian cell systems to correlate changes in transforming potential with changes in transactivation. The results with these two cell systems are not in agreement, possibly because the mechanism of *jun* transformation in mammalian and avian cells is not the same. In mammalian cells, *jun*-induced transformation requires cofactors that may be endogenous to a permanent cell line or, in primary cells, must be supplied by an exogenous oncogene such as *ras*. Cotransfection of primary rat embryo cells with *jun* and activated *ras* is widely used as a test for transformation (Schütte et al., 1988, 1989; Oliviero et al., 1992). In this test, the Ras protein stimulates the transactivation potential of the cotransfected Jun through phosphorylation at two amino terminal sites, serines 63 and 73 (Binétruy et al., 1991; Smeal et al., 1991). Mutational analysis suggests that phosphorylation of serines 63 and 73 is a required step in *ras*-dependent transformation. These serines are substrates of MAP kinase (Pulverer et al., 1991). This stimulation of Jun-dependent transactivation has been proposed as an essential step in the transformation process. In accord with the observation linking Ras/Jun cotransformation to a phosphorylation-induced increase of transactivation, there is a perfect correlation between transformation and transactivation of the collagenase promoter in this system (Alani et al., 1991). Yet, although transactivation by AP-1 is required, it is not sufficient to induce transformation in mammalian cells. The yeast transcriptional activator GCN4 stimulates transcription from TREs in rat cells but does not transform alone or in conjunction with transfected activated *ras*. Replacement of the GCN4 transactivation domain with that of Jun generates a construct that can cotransform rat embryo cells (Oliviero et al., 1992). The Jun transactivation domains must, therefore, fulfill more specific functions in addition to transcriptional activation, and these functions are essential for transformation. It is not known whether GCN4 can be a substrate for *ras*-induced phosphorylation, however, the particular serines 63 and 73 are not conserved in GCN4. GCN4 may lack transforming potential because it cannot receive the upstream signals or because it cannot interact with crucial downstream targets. Of interest would be tests for Jun phosphorylation at serines 63 and 73 in Rat-1A cells in which no exogenous cofactor is required for transformation, presumably because these continuously growing cells already harbor the requisite transformation promoting activity.

An alternative model for the activation of Jun by oncogenes such as *src* or *ras* proposes a cellular inhibitor that interacts with negative regulatory domains in the amino terminal half of the Jun molecule (Baichwal et al., 1991). *src* or *ras* would initiate a signal that leads to a disruption of the interaction between Jun and its cellular inhibitor. This model is based on the observation that, in some test systems, v-Jun is a better transactivator than c-Jun and that the higher activity is correlated with the absence of the delta domain that is deleted in v-Jun (Bohmann and Tjian, 1989; Baichwal and Tjian, 1990). There is evidence that delta and an adjacent region termed epsilon are important in the negative control of Jun transactivation (Baichwal and Tjian, 1990; Baichwal et al., 1992). This control is cell type specific, e.g., it is seen in HeLa cells but not in F9 mouse embryonal carcinoma cells. The activated Ras protein has been found to increase transactivation potential of Jun in those cell types that show delta- and epsilon-dependent negative control.

Although these two models are not completely incompatible, they differ in important assumptions and predictions that need to be verified or refuted by future work. As mentioned above, v-Jun is the more effective transforming gene in the avian system, compared to c-Jun, but shows lower oncogenic activity in the mammalian cotransformation system. An explanation for this difference would discriminate between the models.

In contrast to mammalian cells, avian primary cells can be transformed by c-Jun and v-Jun, and tumors can be effectively induced in birds by v-Jun (Maki et al., 1987; Castellazzi et al., 1990, 1991; Lamph, 1991; Wong et al., 1992). A need for additional tumor promoters is not evident in this system, although it could be argued that this role is covertly played by serum stimulation in cell culture or wounding in the animal. Transformation of avian cells is not linked to increased transactivation from TRE sites (Adler et al., 1992; Hartl and Vogt, 1992a; Håvarstein et al., 1992; Schuur et al., 1993) In fact, Jun mutants functioning as effective transactivators in CEF are often poorly oncogenic, and some of the most potently transforming Jun constructs show no or very low transactivation potential. For instance, deletion of the delta domain eliminates transactivation potential for the collagenase TRE in CEF while greatly increasing trans-forming efficiency (Morgan and Vogt, unpublished observations). This same delta dele-tion upregulates TRE transactivation in the HeLa cell transcriptional system (Bohmann and Tjian, 1989; Baichwal and Tjian, 1990). In numerous chimeras linking partial Jun sequences to domains derived from other transcription factors, there is no correlation between transactivation of the collagenase TRE and transformation (Hartl and Vogt, 1992a; Schuur et al., 1993). Mutants of serines 63 and 73 do not abolish the transforming activity of murine c-*jun* for chicken embryo fibroblasts (Metivier et al., 1993). Only occasionally is transformation linked to increased transactivation in avian cells, as in mutants of the JunD protein, which acquire oncogenicity through a genetic rearrangement that also endows the molecule with a well-functioning transactivation domain (Hartl and Vogt, 1992a).

To understand the data on Jun transformation and transactivation and design experi-ments that could resolve the apparent contradictions, a consideration of three aspects of Jun function appears helpful: cell type specificity, modulation by partner molecules, and promoter context. Transcriptional regulation by Jun is highly cell type dependent (Imler et al., 1988; Bohmann and Tjian, 1989; Baichwal and Tjian, 1990), because of different upstream activators, partner molecules, or downstream targets. Different populations of partner bZIP and other proteins that interact with Jun are known to modulate Jun function. ATF family members associate with Jun and alter its binding specificity for DNA (Benbrook and Jones, 1990; Ivashkiv et al., 1990; Macgregor et al., 1990; Hai and Curran, 1991). In regenerating liver, a new bZIP protein, LRF, forms dimers with Jun and Fos, and these heterodimers target sites that are not recognized by Jun–Fos heterodimers (Alcorn et al., 1990). Similarly, the increased transcription of *jun* caused by the adenovi-rus oncoprotein E1A is due to heterodimers between Jun and an ATF2-like protein that transactivates a site in the Jun promoter not affected by Jun–Fos heterodimers. Jun–ATF2 heterodimers in E1A-treated cells also provide examples of context dependence of Jun-induced transactivation (Müller et al., 1990; Hagmeyer et al., 1993; van Dam et al., 1993). They would affect genes that are not touched by the conventional Jun–Fos partnership. Directly related to the question of Jun transformation is the observation that v-Jun is a more effective transactivator than c-Jun in CEF if the reporter contains the 21-bp repeat of the HTLV 1 promoter (a CRE-like site), while the reverse is true if transactivation is measured with a collagenase promoter (a true TRE) (Seiki et al., 1992).

A possible interpretation of the data on transformation and transactivation would be that avian and mammalian cells differ with respect to the prevalent or effective Jun heterodimer populations. This may result in different spectra of target genes that are regulated, including differences in the targets that are crucial for transformation. For instance, in mammalian cells cotransformation with *ras* may involve genes with promoters marked by classical TRE sites; in avian cells a different set of genes with CRE containing promoters may be targeted by *jun* acting alone. Transformation in the mammalian cells would then be correlated with AP-1 activity as measured on the TRE containing collagenase promoter, while the same promoter may not be an appropriate indicator for the transactivation potential of Jun heterodimers that are oncogenic in avian cells.

Available experimental evidence is in accord with the broad hypothesis that *jun* transforms by altering normal transcriptional controls. Jun-induced stimulation of DNA synthesis remains an alternative mechanism, but this possibility has not gathered strong support (Murakami et al., 1991; Wasylyk et al., 1990; Morgan et al., 1993). Regarding transcriptional control, it is not clear whether the changes that make a *jun* mutant more transforming correspond to a gain or a loss of normal function. In theory, transformation could be caused by increased expression of growth-promoting genes or a decrease in the expression of growth-attenuating genes, possibly also by a combination of both. A gain of function is generally favored, especially in view of the correlation between transformation and increased AP-1 activity in mammalian cells, but loss of function as the oncogenic change in *jun* cannot be definitely ruled out. Such a loss would be in accord with the observed downregulation of c-Jun in v-Jun-transformed cells and with the recently observed attenuation of c-*jun* induction in cells that are transformed by tyrosine kinase oncogenes (Hartl and Vogt, 1992; Yu et al., 1993; Kretzschmar et al., 1993). To determine whether loss or gain of function is important requires the identification of those target genes whose altered expression induces the oncogenic phenotype. So far, no dominant oncogenes that act downstream of transforming transcription factors have been identified, probably because at this level complex programs of gene expression maintain the neoplastic phenotype, while current methods are designed to reveal only single, dominant transforming genes that function closer to the start of the mitogenic signal. Rapid progress in this area may now be at hand, prompted by recent technical advances in the detection of small differences between otherwise identical genomes (Liang and Pardee, 1992; Lisitsyn et al., 1993).

ACKNOWLEDGMENTS

Work of the author is supported by U.S. Public Health Service Research Grant CA 42564 and Grant 1951 from the Council for Tobacco Research. The help of Sarah Olivo, Esther Olivo, Arianne Helenkamp, and Martha TerMaat in producing the manuscript is gratefully acknowledged. Iain Morgan and Azel Schönthal provided useful critical comments on the manuscript.

REFERENCES

Abate, C., Patel, L., Rauscher, F. J., and Curran, T., Redox regulation of Fos and Jun DNA-binding activity in vitro, *Science*, 249, 1157, 1990.

Adler, V., Franklin, C. C., and Kraft, A. S., Phorbol esters stimulate the phosphorylation of c-Jun but not v-Jun — regulation by the N terminal δ domain, *Proc. Natl. Acad. Sci. U.S.A.*, 89, 5341, 1992.

Adunyah, S. E., Unlap, T. M., Wagner, F., and Kraft, A. S., Regulation of c-*jun* expression and AP-1 enhancer activity by granulocyte-macrophage colony-stimulating factor, *J. Biol. Chem.*, 266, 5670, 1991.

Adunyah, S. E., Unlap, T. M., Franklin, C. C., and Kraft, A. S., Induction of differentiation and c-*jun* expression in human leukemic cells by okadaic acid, an inhibitor of protein phosphatases, *J. Cell. Phys.*, 151, 415, 1992.

Alani, R., Brown, P., Binétruy, B., Dosaka, H., Rosenberg, R. K., Angel, P., Karin, M., and Birrer, M. J., The transactivating domain of the c-Jun proto-oncoprotein is required for cotransformation of rat embryo cells, *Mol. Cell. Biol.*, 11, 6286, 1991.

Alcorn, J. A., Feitelberg, S. P., and Brenner, D. A., Transient induction of c-Jun during hepatic regeneration, *Hepatology*, 11, 909, 1990.

Angel, P. and Karin, M., The role of Jun, Fos and the AP-1 complex in cell-proliferation and transformation, *Biochim. Biophys. Acta*, 1072, 129, 1991.

Angel, P., Hattori, K., Smeal, T., and Karin, M., The *jun* proto-oncogene is positively autoregulated by its product, Jun/AP-1, *Cell*, 55, 875, 1988.

Baichwal, V. R. and Tjian, R., Control of c-Jun activity by interaction of a cell-specific inhibitor with regulatory domain d: Differences between v- and c-Jun, *Cell*, 63, 815, 1990.

Baichwal, V. R., Park, A., and Tjian, R., v-Src and EJ Ras alleviate repression of c-Jun by a cell-specific inhibitor, *Nature*, 352, 165, 1991.

Baichwal, V. R., Park, A., and Tjian, R., The cell-type-specific activator region of c-Jun juxtaposes constitutive and negatively regulated domains, *Genes Dev.*, 6, 1493, 1992.

Bartel, D. P., Sheng, M., Lau, L. F., and Greenberg, M. E., Growth factors and membrane depolarization activate distinct programs of early response gene expression: Dissociation of *fos* and *jun* induction, *Genes Dev.*, 3, 301, 1989.

Benbrook, D. M. and Jones, N. C., Heterodimer formation between CREB and JUN proteins, *Oncogene*, 5, 295, 1990.

Bengal, E., Ransone, L., Scharfmann, R., Dwarki, V. J., Tapscott, S. J., Weintraub, H., and Verma, I. M., Functional antagonism between c-Jun and MyoD proteins — a direct physical association, *Cell*, 68, 507, 1992.

Binétruy, B., Smeal, T., and Karin, M., Ha-Ras augments c-Jun activity and stimulates phosphorylation of its activation domain, *Nature*, 351, 122, 1991.

Birchenall-Roberts, Ruscetti, F. A., Kasper, J., Lee, H.-D., Friedman, R., Geiser, A., Sporn, M. B., Roberts, A. B., and Kim, S.-J., Transcriptional regulation of the transforming growth factor b1 promoter by v-*src* gene products is mediated through the AP-1 complex, *Mol. Cell. Biol.*, 10, 4978, 1990.

Black, E. J., Street, A. J., and Gillespie, D. A. F., Protein phosphatase 2A reverses phosphorylation of c-Jun specified by the delta domain *in vitro*: Correlation with oncogenic activation and deregulated transactivation activity of v-Jun, *Oncogene*, 6, 1949, 1991.

Bohmann, D. and Tjian, R., Biochemical analysis of transcriptional activation by Jun: Differential activity of c- and v-Jun, *Cell*, 59, 709, 1989.

Bos, T. J., Monteclaro, F. S., Mitsunobu, F., Ball, A. R., Jr., Chang, C. H. W., Nishimura, T., and Vogt, P. K., Efficient transformation of chicken embryo fibroblasts by c-Jun requires structural modification in coding and noncoding sequences, *Genes Dev.*, 4, 1677, 1990.

Boyle, W. J., Smeal, T., Defize, L. H. K., Angel, P., Woodgett, J. R., Karin, M., and Hunter, T., Activation of protein kinase C decreases phosphorylation of c-Jun at sites that negatively regulate its DNA-binding activity, *Cell*, 64, 573, 1991.

Brenner, D. A., O'Hara, M., Angel, P., Chojkier, M., and Karin, M., Prolonged activation of *jun* and collagenase genes by tumour necrosis factor-α, *Nature*, 337, 661, 1989.

Carter, R., Cosenza, S. C., Pena, A., Lipson, K., Soprano, D. R., and Soprano, K. J., A potential role for c-jun in cell cycle progression through late G_1 and S, *Oncogene*, 6, 229, 1991.

Castellazzi, M., Dangy, J.-P., Mechta, F., Hirai, S.-I., Yaniv, M., Samarut, J., Lassailly, A., and Brun, G., Overexpression of avian or mouse c-*jun* in primary chick embryo fibroblasts confers a partially transformed phenotype, *Oncogene*, 5, 1541, 1990.

Castellazzi, M., Spyrou, G., La Vista, N., Dangy, J.-P., Piu, F., Yaniv, M., and Brun, G., Overexpression of c-*jun*, *junB*, or *junD* affects cell growth differently, *Proc. Natl. Acad. Sci. U.S.A.*, 88, 8890, 1991.

Cato, A. C. B., König, H., Ponta, H., and Herrlich, P., Steroids and growth promoting factors in the regulation of expression of genes and gene networks, *J. Steroid Biochem. Mol. Biol.*, 43, 63, 1992.

Cavalieri, F., Ruscio, T., Tinoco, R., Benedict, S., Davis, C., and Vogt, P. K., Isolation of three new avian sarcoma viruses: ASV9, ASV17 and ASV25, *Virology*, 143, 680, 1985.

Chida, K. and Vogt, P. K., Nuclear translocation of viral Jun but not of cellular Jun is cell cycle dependent, *Proc. Natl. Acad. Sci. U.S.A.*, 89, 4290, 1992.

Chiu, R., Angel, P., and Karin, M., Jun-B differs in its biological properties from, and is a negative regulator of, c-Jun, *Cell*, 59, 979, 1989.

Curran, T. and Vogt, P. K., Dangerous liasons: Fos and Jun — oncogenic transcription factors, in *Transcriptional Regulation*, Vol. 2, McKnight, S. L. and Yamamoto, K., Eds., Cold Spring Harbor Laboratory Press, Cold Spring Harbor, NY, 1992, 797.

de Groot, R. P. and Kruijer, W., Up-regulation of Jun/AP-1 during differentiation of N1E-115 neuroblastoma cells, *Cell Growth Differ.*, 2, 631, 1991.

de Groot, R. P., Kruyt, F. A. E., van der Saag, P. T., and Kruijer, W., Ectopic expression of c-*jun* leads to differentiation of P19 embryonal carcinoma cells, *EMBO J.*, 9, 1831, 1990.

de Groot, R., Foulkes, N., Mulder, M., Kruijer, W., and Sassone-Corsi, P., Positive Regulation of *jun*/AP-1 by E1A, *Mol. Cell. Biol.*, 11, 192, 1991.

Endo, T., SV40 Large-T inhibits myogenic differentiation partially through inducing c-*jun*. *J. Biochem.*, 112, 321, 1992.

Forrest, D. and Curran, T., Crossed signals: Oncogenic transcription factors, *Curr. Biol.*, 2, 19, 1992.

Gaynor, R., Simon, K., and Koeffler, P., Expression of c-*jun* during macrophage differentiation of HL-60 cells, *Blood*, 77, 2618, 1991.

Ginsberg, D., Hirai, S.-I., Pinhasi-Kimhi, O., Yaniv, M., and Oren, M., Transfected mouse c-*jun* can inhibit transformation of primary rat embryo fibroblasts, *Oncogene*, 6, 669, 1991.

Golemis, E. A. and Brent, R., Fused protein domains inhibit DNA binding by LexA, *Mol. Cell. Biol.*, 12, 3006, 1992.

Granger-Schnarr, M., Benusiglio, E., Schnarr, M., and Sassone-Corsi, P., Transformation and transactivation suppressor activity of the c-Jun leucine zipper fused to a bacterial repressor, *Proc. Natl. Acad. Sci. U.S.A.*, 89, 4236, 1992.

Grossi, M., Calconi, A., and Tatò, F., v-*jun* oncogene prevents terminal differentiation and suppresses muscle-specific gene expression in ASV-17-infected muscle cells, *Oncogene*, 6, 1767, 1991.

Gutman, A. and Wasylyk, B., Nuclear targets for transcription regulation by oncogenes, *TIG*, 7, 49, 1991.

Hagmeyer, B., König, H., Herr, I., Offringa, R., Zantema, A., van der Eb, A. J., Herrlich, P., and Angel, P., Adenovirus E1A negatively and positively modulates transcription of AP-1 dependent genes by dimer-specific regulation of the DNA binding and transactivation activities of Jun, *EMBO J.*, 12, 3559, 1993.

Hai, T. and Curran, T., Cross-family dimerization of transcription factors Fos/Jun and ATF/CREB alters DNA binding specificity, *Proc. Natl. Acad. Sci. U.S.A.*, 88, 3720, 1991.

Hartl, M. and Vogt, P. K., A rearranged JunD transforms chicken embryo fibroblasts, *Cell Growth Differ.*, 3, 909, 1992a.

Hartl, M. and Vogt, P. K., Oncogenic transformation by Jun: Role of transactivation and homodimerization, *Cell Growth Differ.*, 3, 899, 1992b.

Håvarstein, L. S., Morgan, I. M., Wong, W.-Y., and Vogt, P. K., Mutations in the Jun-Delta region suggest an inverse correlation between transformation and transcriptional activation, *Proc. Natl. Acad. Sci. U.S.A.*, 89, 618, 1992.

Herrlich, P. and Ponta, H., 'Nuclear' oncogenes convert extracellular stimuli into changes in the genetic program. *TIG*, 5, 112, 1989.

Hsu, J. C., Bravo, R., and Taub, R., Interactions among LRF-1, JunB, c-Jun, and c-Fos define a regulatory program in the G_1 phase of liver regeneration, *Mol. Cell. Biol.*, 12, 4654, 1992.

Hughes, M., Sehgal, A., Hadman M., and Bos, T., Heterodimerization with cFos is not required for cell transformation of chicken embryo fibroblasts by Jun, *Cell Growth Differ.*, 3, 889, 1992.

Hughes, S. and Kosik, E., Mutagenesis of the region between env and src of the SR-A strain of Rous sarcoma virus for the purpose of constructing helper-independent vectors, *Virology*, 136, 89, 1984.

Hughes, S., Greenhouse, J. J., Petropoulos, C. J., and Sutrave, P., Adaptor plasmids simplify the insertion of foreign DNA into helper-independent retroviral vectors, *J. Virol.*, 61, 3004, 1987.

Imler, J. L., Ugarte, E. Wasylyk, C., and Wasylyk, B., v-jun is a transcriptional activator, but not in all cell-lines, *Nucleic Acids Res.*, 16, 3005, 1988.

Ivashkiv, L. B., Liou, H.-C., Kara, C. J., Lamph, W. W., Verma, I. M., and Glimcher, L. H., mXBP/CRE-BP2 and c-Jun form a complex which binds to the cyclic AMP, but not to the 12-*O*-tetradecanoylphorbol-13-acetate, response element, *Mol. Cell. Biol.*, 10, 1609, 1990.

Karin, M. and Smeal, T., Control of transcription factors by signal transduction pathways: The beginning of the end, *TIBS*, 17, 418, 1992.

Kharbanda, S., Datta, R., Rubin, E., Nakamura, T., Hass, R., and Kufe, D., Regulation of c-*jun* expression during induction of monocytic differentiation by okadaic acid, *Cell Growth Differ.*, 3, 391, 1992.

Kim, S.-J., Angel, P., Lafyatis, R., Hattori, K., Kim, K. Y., Sporn, M. B., Karin M., and, Roberts, A. B., Autoinduction of transforming growth factor b1 is mediated by the AP-1 complex, *Mol. Cell. Biol.*, 10, 1492, 1990.

Kovary, K. and Bravo, R., Expression of different Jun and Fos proteins during the G_0-to-G_1 transition in mouse fibroblasts — in vitro and in vivo associations, *Mol. Cell. Biol.*, 11, 2451, 1991a.

Kovary, K. and Bravo, R., The Jun and Fos protein families are both required for cell cycle progression in fibroblasts, *Mol. Cell. Biol.*, 11, 4466, 1991b.

Kretszschmar, D. E., Morgan, I. M., and Vogt, P. K., unpublished observations, 1993.

Lamph, W. W., Cross-coupling of AP-1 and intracellular hormone receptors, *Cancer Cells*, 3, 183, 1991.

Lamph, W. W., Wamsley, P., Sassone-Corsi, P., and Verma, I. M., Induction of proto-oncogene *JUN*/AP-1 by serum and TPA, *Nature*, 334, 629, 1988.

Lewin, B., Oncogenic conversion by regulatory changes in transcription factors, *Cell*, 64, 303, 1991.

Li, L., Chambard, J.-C., Karin, M., and Olson, E. N., Fos and Jun repress transcriptional activation by myogenin and MyoD: The amino terminus of Jun can mediate repression, *Genes Dev.*, 6, 676, 1992.

Liang, P. and Pardee, A. B., Differential display of eukaryotic messenger RNA by means of the polymerase chain reaction, *Science*, 257, 967, 1992.

Lin, A. N., Frost, J., Deng, T. L., Smeal, T., Al-Alawi, N., Kikkawa, U., Hunter, T., Brenner, D., and Karin, M., Casein kinase-II is a negative regulator of c-Jun DNA binding and AP-1 activity, *Cell*, 70, 777, 1992.

Lisitsyn, N., Lisitsyn, N., and Wigler, M. Cloning the differences between two complex genomes, *Science*, 259, 946, 1993.

Lloyd, A., Yancheva, N., and Wasylyk, B., Transformation suppressor activity of a Jun transcription factor lacking its activation domain, *Nature*, 352, 632, 1991.

Macgregor, P. F., Abate, C., and Curran, T., Direct cloning of leucine zipper proteins: Jun binds cooperatively to the CRE with CRE-BP-1, *Oncogene*, 5, 451, 1990.

Maki, Y., Bos, T. J., Davis, C., Starbuck, M., and Vogt, P. K., Avian sarcoma virus 17 carries the *jun* oncogene, *Proc. Natl. Acad. Sci. U.S.A.*, 84, 2848, 1987.

Marshall, G. M., Vanhamme, L., Wong, W.-Y., Su, H., and Vogt, P. K., Wounding acts as a tumor promoter in chickens inoculated with avian sarcoma virus 17, *Virology*, 188, 373, 1992.

Metivier, C., Piu, F., Pfarr, C. M., Yaniv, M., Loiseau, L., and Castellazzi, M., *In vitro* ttransforming capacities of mouse c-*jun: jun*D chimeric genes, *Oncogene*, 8, 2311, 1993.

Miner, J. N., Diamond, M. I., and Yamamoto, K. R., Joints in the regulatory lattice: Composite regulation by steroid receptor-AP1 complexes, *Cell Growth Differ.*, 2, 525, 1991.

Monteclaro, F. and Vogt, P. K., Unpublished observations, 1991.

Morgan, I. M. and Vogt, P. K., Unpublished observations.

Morgan, I. M., Ransone, L. J., Bos, T. J., Verma, I. M., and Vogt, P. K., Transformation by Jun — requirement for leucine zipper, basic region and transactivation domain and enhancement by Fos, *Oncogene*, 7, 1119, 1992.

Morgan, I. M., Asano, M., Ishikawa, H., Håvarstein, L. S., Hiiragi, T., Ito, Y., and Vogt, P. K., Amino acid substitutions modulate the effect of Jun on transformation, transcriptional activation and on DNA replication, *Oncogene*, 8, 1134, 1993.

Müller, U., Roberts, M. P., Engel, D. A., Doerfler, W., and Shenk, T., Induction of transcription factor AP-1 by adenovirus E1A protein and cAMP, *Genes Dev.*, 3, 1991, 1989.

Müller, U., Roberts, M. P., and Shenk, T., Adenovirus E1A protein and cAMP act in concert to activate transcription of AP-1 and ATF/CREB binding-site containing promoters, in *Structure and Function of Nucleic Acids and Proteins*, Wu, F. Y.-H. and Wu, C.-W., Eds., Raven Press, New York, p. 249, 1990.

Murakami, Y., Satake, M., Yamaguchi-Iwai, Y., Sakai, M., Muramatsu, M., and Ito, Y., The nuclear oncogenes c-*jun* and c-*fos* as regulators of DNA replication, *Proc. Natl. Acad. Sci. U.S.A.*, 88, 3947, 1991.

Nagata, L. and Vogt, P. K., Unpublished observations, 1989.

Nishimura, T. and Vogt, P. K., The avian cellular homolog of the oncogene *jun*, *Oncogene*, 3, 659, 1988.

Oehler, T. and Angel, P., A common intermediary factor (p52/54). Recognizing "acidic blob"-type domains is required for transcriptional activation by the Jun proteins, *Mol. Cell. Biol.*, 12, 5508, 1992.

Oehler, T., Pintzas, A., Stumm, S., Darling, A., Gillespie, D., and Angel, P., Mutation of a phosphorylation site in the DNA-binding domain is required for redox-independent transactivation of AP1-dependent genes by v-Jun, *Oncogene*, 4, 1141, 1993.

Oemar, B. S., Law, N. M., and Rosenzweig, S. A., Insulin-like growth factor I induces tyrosyl phosphorylation of nuclear proteins, *J. Biol. Chem.*, 266, 2424, 1991.

Oliviero, S., Robinson, G. S., Struhl, K., and Spiegelman, B. M., Yeast GCN4 as a probe for oncogenesis by AP-1 transcription factors: Transcriptional activation through AP-1 sites is not sufficient for cellular transformation, *Genes Dev.*, 6, 1799, 1992.

Pertovaara, L., Sistonen, L., Bos, T., Vogt, P., Keski-Oja, J., and Alitalo, K., Enhanced *jun* gene expression is an early genomic response to transforming growth factor-ß stimulation, *Mol. Cell. Biol.*, 9, 1255, 1988.

Pulverer, B. J., Kyriakis, J. M., Avruch, J., Nikolakaki, E., and Woodgett, J. R., Phosphorylation of c-*jun* mediated by MAP kinases, *Nature*, 353, 670, 1991.

Quantin, B. and Breathnach, R., Epidermal growth factor stimulates transcription of the c-*jun* proto-oncogene in rat fibroblasts, *Nature*, 334, 538, 1988.

Ransone, L. J. and Verma, I. M., Nuclear proto-oncogenes *Fos* and *Jun*, *Annu. Rev. Cell Biol.*, 6, 539, 1990.

Riabowol, K., Schiff, J., and Gilman, M. Z., Transcription factor AP-1 activity is required for initiation of DNA synthesis and is lost during cellular aging, *Proc. Natl. Acad. Sci. U.S.A.*, 89, 157, 1992.

Ryder, K. and Nathans, D., Induction of protooncogene c-*jun* by serum growth factors, *Proc. Natl. Acad. Sci. U.S.A.*, 85, 8464, 1988.

Sakai, M., Okuda, A., Hatayama, I., Sato, K., Nishi, S., and Muramatsu, M., Structure and expression of the rat c-*jun* messenger RNA: Tissue distribution and increase during chemical hepatocarcinogenesis, *Cancer Res.*, 49, 5633, 1989.

Sassone-Corsi, P., Fatal attraction: The *fos-jun* affair with nuclear receptors, *Intl. J. Oncol.*, 1, 5, 1992.

Schuh, A. C., Keating, S. J., Monteclaro, F. S., Vogt, P. K., and Breitman, M. L., Obligatory wounding requirement for tumorigenesis in v-*jun* transgenic mice, *Nature*, 346, 756, 1990.

Schüle, R. and Evans, R. M., Cross-coupling of signal transduction pathways: Zinc finger meets leucine zipper, *TIG*, 7, 377, 1991.

Schütte, J., Minna, J. D., and Birrer, M. J., Deregulated expression of human transcription factor c-*jun* transforms primary rat embryo cells in cooperation with an activated c-Ha-*ras* gene and Rat 1a cells as a single gene, *Proc. Natl. Acad. Sci. U.S.A.*, 86, 2257, 1988.

Schütte, J., Viallet, J., Nau, M., Segal, S., Fedorko, J., and Minna, J., *jun-B* inhibits and c-*fos* stimulates the transforming and *trans*-activating activities of c-*jun*, *Cell*, 59, 987, 1989.

Schuur, E. R., Parker, E. J., and Vogt, P. K., Chimerus of herpes simplex viral VP16 and Jun are oncogenic, *Cell Growth Differ.*, 4, 761, 1993.

Seiki, M., Morgan, I. M., and Vogt, P. K., Unpublished observations, 1992.

Sieweke, M. H., Stoker, A. W., and Bissell, M. J., Evaluation of the carcinogenic effect of wounding in rous sarcoma virus tumorigenesis, *Cancer Res.*, 49, 6419, 1989.

Sieweke, M. H., Thompson, N. L., Sporn, M. B., and Bissell, M. J., Mediation of Wound-Related Rous Sarcoma Virus Tumorigenesis by TGFb, *Science*, 248, 1656, 1990.

Smeal, T., Binétruy, B., Mercola, D. A., Birrer, M., and Karin, M., Oncogenic and transcriptional cooperation with Ha-Ras requires phosphorylation of c-Jun on serines 63 and 73, *Nature*, 354, 494, 1991.

Smeal, T., Binétruy, B., Mercola, D., Grover-Bardwick, A., Heidecker, G., Rapp, U. R., and Karin, M., Oncoprotein-mediated signalling cascade stimulates c-Jun activity by phosphorylation of serines 63 and 73, *Mol. Cell. Biol.*, 12, 3507, 1992.

Smith, M. J. and Prochownik, E. V., Inhibition of c-*jun* causes reversible proliferative arrest and withdrawal from the cell cycle, *Blood*, 79, 2107, 1992.

Su, H., Bos, T. J., Monteclaro, F. S., and Vogt, P. K., *Jun* inhibits myogenic differentiation, *Oncogene*, 6, 1759, 1991.

Sumatran, V. N., Tsai, M., and Schwartz, J., Growth hormone induces c-jun and c-fos expression in cells with varying requirements for differentiation, *Endocrinology*, 130, 2016, 1992.

Tanese, N., Pugh, F. P., and Tjian, R., Coactivators for a proline-rich activator purified from the multi-subunit human TF2D complex, *Genes Dev.*, 5, 2212, 1991.

Tsuchie, H. and Vogt, P. K., Unpublished observations, 1989.

van Dam, H., Duyndam, M., Rottier, R., Bosch, A., de Vries-Smits, L., Herrlich, P., Zantema, A., Angel, P., and van der Eb, A. J., Heterodimer formation of cJun and ATF-2 is responsible for induction of c-*jun* by the 243 amino acid adenovirus E1A protein, *EMBO J.*, 12, 479, 1993.

Vanhamme, L., Marshall, G. M., Schuh, A. C., Breitman, M. L., and Vogt, P. K., TNFα and IL-1α induce anchorage-independent growth and tumorigenicity in v-Jun transgenic murine cells, *Cancer Res.*, 53, 615, 1993.

Vogt, P. K., Unpublished observations, 1985.

Vogt, P. K. and Bos, T. J. The oncogene *jun* and nuclear signalling, *Trends Biochem. Sci.*, 14, 172, 1989.

Vogt, P. K. and Bos, T. J., Jun: Oncogene and transcription factor, *Adv. Cancer Res.*, 55, 1, 1990.

Vogt, P. K. and Morgan, I., The genetics of *jun*, *Cancer Biol.*, 1, 27, 1990.

Vogt, P. K. and Tjian, R., *Jun:* A transcriptional regulator turned oncogenic, *Oncogene*, 3, 3, 1988.

Wasylyk, C., Imler, J. L., and Wasylyk, B., Transforming but not immortalizing oncogenes activate the transcription factor PEA1, *EMBO J.*, 7, 2475, 1988.

Wasylyk, C., Schneickert, J., and Wasylyk, B., The oncogene v-*jun* modulates DNA replication, *Oncogene*, 5, 1055, 1990.

William, F., Wagner, F., Karin, M., and Kraft, A. S., Multiple doses of diacylglycerol and calcium ionophore are necessary to activate AP-1 enhancer activity and induce markers of macrophage differentiation, *J. Biol. Chem.*, 265, 18166, 1990.

Wong, W. Y. and Vogt, P. K., Unpublished observations, 1989.

Wong, W. Y., Håvarstein, L. S., Morgan, I. M., and Vogt, P. K., c-Jun causes focus formation and anchorage-independent growth in culture but is non-tumorigenic, *Oncogene*, 7, 2077, 1992.

Xanthoudakis, S. and Curran, T., Identification and characterization of Ref-1, a nuclear protein that facilitates AP-1 DNA-binding activity, *EMBO J.*, 11, 653, 1992.

Xanthoudakis, S., Miao, G., Wang, F., Pan, Y. C. E., and Curran, T., Redox activation of Fos-Jun DNA binding activity is mediated by a DNA repair enzyme, *EMBO J.*, 11, 3323, 1992.

Yang-Yen, H.-F., Chui, R., and Karin, M., Elevation of AP-1 activity during F9 cell differentiation is due to increased c-*jun* transcription, *New Biol.*, 2, 351, 1990.

Yu, C., Prochownik, E., Imperiale, M. J., and Jove, R., Attenuation of serum inducibility of immediate early genes by oncoproteins in tyrosine kinase signaling pathways, *Mol. Cell. Biol.*, 13, 2011, 1993.

Chapter 16

Role of the Raf Signal Transduction Pathway in Fos/Jun Regulation and Determination of Cell Fates

Ulf R. Rapp, Joseph T. Bruder, and Jakob Troppmair

CONTENTS

INTRODUCTION

Fos and Jun and most of their regulators were originally identified as oncogenes (Curran and Franza, 1988; Vogt and Bos, 1990). In fact, a major accomplishment of the last 8 years of molecular oncology was the identification of functional connections between different classes of oncogenes as well as their negative control elements, the tumor suppressor genes. As a result, we now know that the activities of probably hundreds of these genes can be reduced to a small number of signal transduction pathways onto which they converge. The evidence in support of convergence of function has been recently reviewed (Rapp, 1991). Briefly, these findings include: first, the apparent functional equivalence for growth induction by transforming versions of protein tyrosine kinases (PTKs) (Cleveland et al., 1989), which make up the largest contingent of the oncogenes; second, the observation that Ras proteins control the coupling of growth-regulating PTKs to phosphorylation of cytosolic Raf-1 protein serine/threonine kinase (PSK) (Troppmair et al., 1992a; Wood et al., 1992; Grugel et al., unpublished), which parallels activation of kinase activity for all three Raf isozymes (Troppmair et al., 1992a; Grugel et al., unpublished). Raf kinases are essential for induced growth (Carroll et al., 1991; Kolch et al., 1991) and several developmental pathways (Ambrosio et al., 1989; Dickson et al., 1992). These enzymes function by activating a protein kinase cascade (Kyriakis et al., 1992) that culminates in the phosphorylation of oncogene class transcription factors (Pulverer et al., 1991; Gille et al., 1992; Smeal et al., 1992) while simultaneously integrating pathways of energy metabolism (Cohen, 1992). Finally, the altered transcription factor activity leads to induction of a cascade of response genes (early, intermediate, and late) some of which are known to promote cell cycle progression (Eilers et al., 1991) and are able to substitute, at least in part, for upstream growth inducers (Armelin et al., 1984; Rapp et al., 1985c; Dean et al., 1987; Cleveland et al., 1993). Our observations on Raf/ Myc synergism in transformation and growth factor abrogation, in addition to the abrogation with ectopic PTKs, initially prompted us to propose a dual pathway kinase cascade model for mitogenic signal transduction from the membrane to the nucleus: the oncogene cascade described above (Raf pathway) and a second one that regulates expression of the oncogene class transcription factor Myc (Rapp et al., 1986). We now have data indicating that the balance of signal flow through these two (Raf and Myc) pathways is not limited to regulation of growth but apparently also determines cell fates such as apoptosis (Rapp et al., 1987; Askew et al., 1991; Evan et al., 1992; Troppmair et al., 1992b; Cleveland et al., 1993) and differentiation (Klinken et al., 1988, 1989; Principato et al., 1990; Troppmair et al., 1992a; Rapp et al., in preparation). Of these two signaling cascades, the Raf pathway is the best understood.

There are three raf family PSK genes, A-raf, B-raf, and c-raf-1, which encode cytosolic phosphoproteins of 70–72, 90–95, and 72–74 kDa, respectively (Grugel et al., unpublished; Oshima et al., 1991; Heidecker et al., 1992). All three genes are proto-oncogenes, i.e., specific mutational changes render them oncogenic (Ikawa et al., 1988; Rapp et al., 1988a; Heidecker et al., 1990; Sithanandam et al., 1990). Activating mutations include 5′ deletions that remove the N-terminal negative autoregulatory domain, conserved region one and two (CR1 and CR2), gene fusions that extend the intact N-terminus as well as point mutations in CR2 (Heidecker et al., 1990) and CR3 (Storm and Rapp, 1993; Storm and Rapp, submitted), the kinase domain. Deletion activation has been observed for the oncogenic viral form, v-raf (Rapp et al., 1983; Bonner et al., 1985, 1986), but rarely in human tumors (Rapp et al., 1987). Activation by specific point mutations occurs regularly in a mouse model for induction of lung adenocarcinoma as well as T cell lymphoma (Storm and Rapp, 1993; Storm and Rapp, submitted), and similar mutations are apparently also present in certain human lung carcinomas (Lyons et al., unpublished data). However, the most frequent alteration in human tumors is overexpression of Raf-1 kinase, which was found in many types of human tumors, including lung carcinoma (Rapp et al., 1987, 1988c), and which can contribute to the transformed phenotype by lowering the threshold for transformation by Ras and, by implication, tyrosine kinase oncogenes (Cuadrado et al., 1993). Normally, all three Raf kinases are activated by growth factors or the protein kinase C (PKC) ligand TPA (Rapp, 1991; Oshima et al., 1991; Stephens et al., 1992; Grugel et al., unpublished) through a process that requires activated receptor kinase, Ras-GTP (Troppmair et al., 1992a; Grugel et al., 1993) and presumably a receptor kinase-induced ligand for the CR1 domain of Raf (Bruder et al., 1992). These regulators conspire to activate Raf kinases, which is concurrent with hyperphosphorylation on specific sites (Heidecker et al., 1992). There appear to be several Raf kinase kinases that are growth factor regulated and can mediate Raf kinase activation, including classic forms of PKC (Lee et al., 1991; Sozeri et al., 1992; Kolch et al., 1993a). Figure 16-1 shows a summary of growth- and differentiation-inducing receptors that regulate Raf kinase activity and others that do not. The Raf-negative receptors such as IL-4R did not support full growth in the cells examined (Turner et al., 1991). Work on the muscarinic receptors and the endothelin receptor is ongoing (Gutkind et al.; Force et al., 1994).

As there are many growth factors that regulate Raf kinases, activated growth factor receptors such as the PDGF receptor make contact with a large number of cellular candidate second messenger proteins (Figure 16-2). This is a result of receptor autophosphorylation on a series of tyrosine residues that become docking sites for specific SH2-containing proteins (Heldin, 1991; Pawson and Gish, 1992), which in turn can bring in additional proteins with which they interact. The complexity of receptor-regulated functions has reinforced a widely held belief that multiple parallel pathways are being activated through which growth is induced. The striking findings of the last two years, however, argue against extravagant levels of redundancy and instead are consistent

Raf-1 Coupled Receptors

PTK RECEPTORS

INTRINSIC: PDGFR InsR EGFR FGFR NGFR
CSF-1R
SCFR

ASSOCIATED: IL-1R TCR CD4 THY-1 BCR
IL-2R
IL-3R
EPOR

G-PROTEIN RECEPTORS

ENDOTHELIN-R   M1,M3 MUSCARINIC-R

PHORBOL ESTER RECEPTORSS

PKCα

Raf-1 negative Receptors

IL-4R IL-6R M2, M4 MUSCARINIC-R

Figure 16-1   Receptors regulating Raf kinase activity.

with the hourglass model for mitogenic signal transduction (Rapp, 1991). In fact, it appears that most receptor complex proteins do not function independently but form a small number of functional linkage groups. For example, PLC-gamma and PKC are connected through the formation of the PKC-activating ligand diacylglycerol. PKC has multiple substrates including receptor PTKs (Hunter et al., 1984; Davis et al., 1985), which it inhibits as part of a negative feedback loop, and Raf-1, which it activates and on which it depends for growth stimulation. However, PKC is not required for receptor activation of Raf, indicating the existence of redundant pathways through which Raf activation is achieved. A second functional linkage group consists of four or more proteins (Trahey et al., 1988; Vogel et al., 1988; Clark et al., 1992; Lowenstein et al., 1992; Rozakis-Adcock et al., 1992; Shou et al., 1992) that converge on the regulation of Ras activity. Ras function, in turn, is integrated into a third chain that connects the receptor PTK via Src and Ras with Raf (Kypta et al., 1990; Kolch et al., 1991; Kremer et al., 1992; Troppmair et al., 1992a). Thus, three of the oncoproteins in the complex, which were prime candidates for activating redundant pathways in fact turned out to function in hierarchical order. Recently, we have identified MAP kinase kinase (MAPKK, MEK) as a substrate for Raf kinase (Kyriakis et al., 1992), and consistent findings have been reported by others (Dent et al., 1992; Howe et al., 1992). Thus, activation of Raf kinase eventually leads to activation of MAPK and phosphorylation of oncogene class transcription factors p62^{TCF}(elk-1, sap-1) and c-Jun (Karin and Smeal, 1992; Treisman, 1992; see also Chapters 5 and 6). Certain natural promoters that contain binding sites for these factors are Raf responsive in transient assays (Kaibuchi et al., 1989; Wasylyk et al., 1989a, 1989b; Jamal and Ziff, 1990; Qureshi et al., 1991a; Karin and Smeal, 1992), and the corresponding genes show altered expression in v-raf transformed cells (Siegfried and Ziff, 1990; Kolch et al., 1993b). Genes with Raf-responsive promoters include early and

PDGF-R

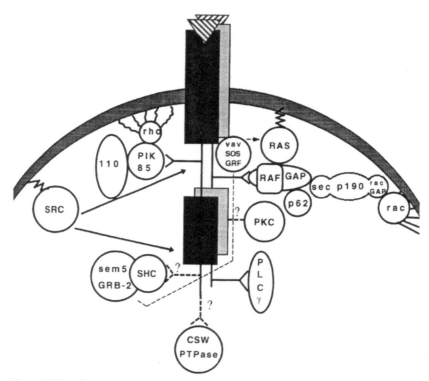

Figure 16-2 Components of activated PTK receptor complexes. Most of the proteins have been shown to interact with the PDGF receptor. Exceptions include: corkscrew180 (CSW), shc (Rozakis-Adcock et al., 1992), sem5/GRB-2 (Clark et al., 1992; Lowenstein et al., 1992), and, perhaps, GRF (Shou et al., 1992), for which binding positions remain to be established. rho and rac (Hall, 1992) are shown to interact with the PI-3 kinase regulatory subunit p85 (Parker and Waterfield, 1992) and the GAP-associated protein p190 (Settleman et al., 1992) to indicate potential binding to their respective homology regions with small G protein GAP. p62 is a second GAP-associated protein (Wong et al., 1992).

late growth response genes such as *c-fos* (early) (Kaibuchi et al., 1989; Jamal and Ziff, 1990) and *CAD, rep* (late) (Farnham and Kollmar, 1990; Farnham and Means, 1990; P. G. Farnham, personal communication).

Of the oncogene class of transcription factors that are regulated through the Raf pathway, p62[TCF] (elk-1, sap-1) (Treisman, 1992) belongs to the Ets family of transcription factors (Macleod et al., 1992), while Fos and Jun family member are components of the AP-1 complex (Karin and Smeal, 1992). In some instances, Ets and AP-1 proteins have been found to act synergistically to promote transcription (Wasylyk et al., 1990). The AP-1 factors are part of an important group of signal-regulated transcription factors, the bZip proteins (Karin and Smeal, 1992), which also includes myc (Lüscher and Eisenman, 1990) and which are described in detail in Chapters 1 and 2.

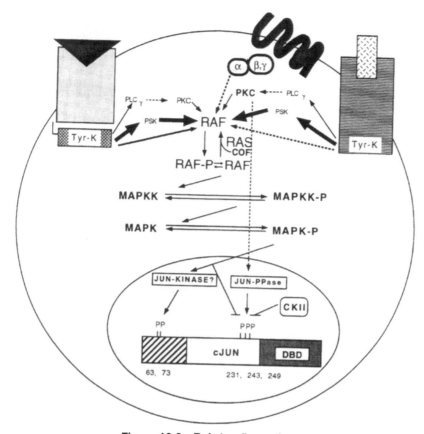

Figure 16-3 Raf signaling pathway.

ROLE OF THE RAF PATHWAY IN
MITOGEN REGULATION OF FOS AND JUN

The components of the Raf signaling cascade, which connects mitogen receptors at the membrane with activity regulation of oncogene class transcription factors in the nucleus, are shown in Figure 16-3. Important reagents that allowed the delineation of this chain include gain of function and dominant negative mutants of Ras (Barbacid, 1987; Feig and Cooper, 1988) and of Raf kinase (Heidecker et al., 1990; Kolch et al., 1991; Bruder et al., 1992). Substitution of Asn for Ser at position 17 of Ha-Ras confers a dominant inhibitory function on the mutant protein. In NIH3T3 cells, expression of 17N Ras[H] blocks cell growth and growth factor-induced DNA synthesis as well as expression from the *fos* promoter in transient assays (Cai et al., 1990). The use of this mutant to inhibit Ras function has indicated that Ras functions to activate Raf-1 following growth factor induction of NIH3T3 cells and NGF induction of PC12 cells (Troppmair et al., 1992a). In addition, expression of 17N Ras[H] also blocked NGF-induced neurite outgrowth (Szeberenyi et al., 1990) and MAPK activation in PC12 cells (Thomas et al., 1992; Wood et al., 1992). This mutant also blocked insulin-induced activation of Erk2 in NIH3T3 cells that overexpress the insulin receptor (de Vries-Smits et al., 1992). The dominant negative phenotype appears to result from improper coordination of Mg^{2+}, rendering Ras locked in an inactive conformation. The dominant negative nature of 17N Ras[H] may be explained

226

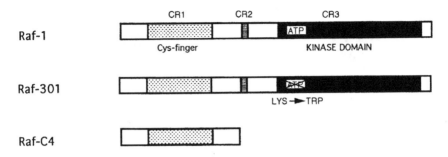

Figure 16-4 Schematic representation of Raf-1, Raf-301, and Raf-C4. The position of the three conserved regions, CR1, CR2, and CR3, are indicated. Raf-301 contains a single amino acid substitution (K-to-W) at position 375 in the ATP binding site of Raf-1 (Heidecker et al., 1990; Kolch et al., 1991). Raf-C4 is a *carboxy*-terminal deletion mutant of Raf-1 (Bruder et al., 1992). It contains amino acids 1–256 of Raf-1, which includes the cysteine finger motif in CR1. Both Raf-301 and Raf-C4 function as dominant inhibitory Raf mutants.

by competition for a guanine nucleotide-releasing factor (Farnsworth and Feig, 1991).

The Raf-1 mutant Raf-301 has a single amino acid substitution K-to-W in the ATP binding site (Heidecker et al., 1990; Kolch et al., 1991). Expression of this mutant blocks cell proliferation and transformation by activated Raf and v-Ha-Ras (Kolch et al., 1991). Deletion analysis has revealed that expression of the amino-terminal 257 amino acids of Raf-1 is sufficient for the dominant negative effect. This mutant, termed Raf-C4, inhibits Raf activation by serum growth factors, TPA, and the v-H-Ras oncogene product (Bruder et al., 1992). The amino-terminal 257 amino acids of Raf-1 include a cysteine finger motif in CR1 (Figure 16-4). Point mutations in the cysteine finger motif abolished the dominant negative phenotype observed with Raf-C4, suggesting that the cysteine finger is necessary for this effect. In addition, overexpression of Raf-1 and Ras resulted in a synergistic increase in transformation (Cuadrado et al., 1993) and in AP-1/Ets driven reporter gene expression (Bruder et al., 1992). Since Raf-1/Ras synergy depends on an intact cysteine finger domain of Raf-1, it appears that Ras mediates Raf-1 kinase activation through interactions with the cysteine finger domain. We have speculated that this domain constitutes the binding site for a lipid-derived cofactor (Rapp et al., 1988b; Bruder et al., 1992) and that this cofactor as well as Ras may be required for transient translocation of Raf from the cytosol to Raf kinase kinases in activated growth factor receptor complexes (Grugel et al., 1993). Experiments are underway to test this hypothesis.

In addition to activating Raf-1, oncogenic Ras activates MAPK in a variety of cell types (Leevers and Marshall, 1992; Nori et al., 1992), suggesting that this might be a branchpoint in receptor signaling. To address the question as to whether MAPK functions upstream, downstream, or in a pathway parallel to Raf, we have determined MAPK activity in Raf-transformed cells and found it to be elevated. This work was extended to include examination of MAPKK, through a collaboration with Avruch's laboratory, which led to the identification of MAPKK as an *in vitro* and probable *in vivo* substrate of Raf-1 kinase (Kyriakis et al., 1992). Although the reverse order, Ras–MAPK–Raf-1, had been proposed by others on the basis of experiments in PC12 pheochromocytoma cells (Wood et al., 1992), data consistent with our findings have since been reported by three other laboratories (Dent et al., 1992; Howe et al., 1992; Gallego et al., 1992). We are now working on the genetic analysis of this connection by determining Raf-specific phosphorylation sites on MAPKK and evaluating the activity of proteins mutated at these sites.

In order to determine whether oncogenes, TPA, or serum have Raf-independent access to MAPK, transient transfection assays were performed with an epitope-tagged version of ERK-1 and the Raf-C4 dominant negative mutant (Troppmair and Rapp, unpublished findings). As Raf-C4 blocked the ability of these upstream activators to stimulate ERK kinase activity it appears that Raf is an obligatory entry point for stimulation of the MAPK-inducing protein kinase cascade in 293 and NIH3T3 cells. Although no oncogenic versions of MAPK have thus far been described, we have recently observed cooperation between activated Raf and MAPK in NIH3T3 cell transformation assays and in transcriptional activation of AP-1-driven reporter gene expression. Moreover, a kinase negative MAPK mutant, Erk-2-B3, inhibited v-Raf-induced transformation (Troppmair et al., 1994). These results demonstrate that MAPK activation is the major efferent pathway for Raf transforming activity in NIH3T3 cells. Thus, in contrast to the wide range of upstream growth regulators and downstream substrates of this cascade, the central part of the pathway, Raf–MAPKK–MAPK, does not appear to have branchpoints. However, it was reported that Rat-1a cells transformed by v-Ras and v-Raf did not display elevated MAPK activity. In these cells it appears that another oncogene product GIP-2 activates MAP kinase (Gallego et al., 1992). These data are difficult to evaluate as they did not use an inducible vector system, but they do raise the possibility that alternate MAPK activation pathways exist. A study that examined regulation of IL-2 production by activated Raf led to a similar conclusion. Jurkat cells expressing activated Raf-1 did not display elevated MAPK activity. In this system activated Raf relieved the requirement for TPA in anti-CD3 or anti-CD28 induced IL-2 production (Owaki et al., 1993). These combined studies raise the possibility that Ras as well as Raf have downstream effectors other than MAPKK.

Candidates for alternative MAPKKs in mammalian cells include Mos and byr-2/STE11 homologues. *mos* is a PSK oncogene that activates MAPK in transient transfection assays, and Mos-transformed NIH3T3 cells contain activated MAPKK as well as MAPK (Troppmair et al., unpublished findings). In yeast, MAPK homologues have also been found to be part of a signaling cascade that connects a G protein coupled membrane receptor to transcription factor activation in the nucleus (Marsh et al., 1991). The recent cloning of MAPKK has revealed that it is homologous to byr-1 in *Schizosaccharomyces pombe* and STE7 in *Schizosaccharomyces cerevisiae* (Crues et al., 1992; Ashworth et al., 1992). These protein kinases appear to lie upstream of spk-1 and FUS3/KSS1, which are yeast MAPK homologues (Teague et al., 1986; Courchesne et al., 1989; Elion et al., 1990; Toda et al., 1991), and downstream of byr-2 (*S. pombe*) and STE11 (*S. cerevisiae*), which are yeast protein kinases that relay a signal for mating from the yeast GTPase, Ras-1, or STE4, respectively (Wang et al., 1991; Carin et al., 1992). Whether or not mammalian homologues of byr-2/STE11 exist and in what signaling context they function remains to be determined.

The importance of the pathway, PTK–Ras–Raf, in the regulation of ets- and AP-1-dependent transcription has come into focus by work in transient assay systems with two different endpoints: transcriptional activation of reporter gene expression or c-Jun phosphorylation. We have shown that expression of the dominant inhibitory Raf mutant, Raf-C4, blocks v-Ha-Ras- (Bruder et al., 1992) and Src- (Bruder and Rapp, unpublished results) induced expression of AP-1/Ets-driven promoters. However, the Raf-301 dominant negative mutant did not block v-Ha-Ras induced expression from the AP-1/Ets-driven reporter (Bruder and Rapp, unpublished results) or v-Src-induced expression from a reporter driven by five AP-1 binding sites (Qureshi et al., 1992). These and other findings (Bruder et al., 1992; Bruder and Rapp, unpublished results) suggest that the Raf-

C4 mutant is a much more potent dominant inhibitory mutant than is Raf-301. The Raf-301 mutant efficiently blocked Src-induced expression from the SRE binding sites in the Egr-1 promoter (Qureshi et al., 1991c, 1992). Thus, it appears that different levels of Raf activity are required for oncogene-induced transactivation of SRE versus AP-1 driven promoters.

Analysis of c-Jun phosphopeptide maps following transient expression of various oncogenes has revealed that v-Src, v-Ras, and activated Raf induced the phosphorylation of Jun on serines 63 and 73 (Smeal et al., 1992). These oncogenes also increase the transcriptional activity of a Jun/GHF1 fusion protein but fail to activate a Jun/GHF1 protein carrying alanine substitutions at Ser 63 and 73, indicating that oncogene activation of Jun occurs through phosphorylation of Ser 63 and 73 in the transactivation domain of Jun. Experiments by Smeal et al. (1992) have extended previous work (Cai et al., 1990; Binétruy et al., 1991; Kolch et al., 1991; Bruder et al., 1992) by showing that dominant negative mutants of Ras (17N Ras^H) and Raf-1 (Raf-301) efficiently blocked Src- and Ras-induced activation and phosphorylation of Jun suggesting that Jun is a nuclear target of the Raf-1 signaling pathway. Taken together with recent findings, showing that dominant negative Jun mutants block transformation by oncogenes that activate the Raf-1 pathway (Lloyd et al., 1992; Rapp et al., 1993), these results suggest that Raf-induced phosphorylation and activation of Jun is necessary for transformation of NIH3T3 cells.

The nature of the kinase that phosphorylates Jun at Ser 63 and 73 is controversial. Using Raf-1 protein overexpressed in the bacculovirus system, we have shown that activated Raf-1 did not phosphorylate c-Jun efficiently *in vitro*. Efficient c-Jun phosphorylation was observed, however, when purified MAPKK and ERK-1 were added to the *in vitro* kinase reaction (Bruder and Rapp, unpublished data). Pulverer et al. (1991) have demonstrated that pp42 and pp54 MAPK phosphorylate Jun at the two activating serine residues, Ser 63 and 73. In addition, Adler et al. (1992) have purified an activity from TPA-stimulated U937 cells that associates with and phosphorylates c-Jun at these *N*-terminal activating serine residues. This kinase phosphorylates Ser 63 and 73 on c-Jun but not v-Jun or a c-Jun mutant carrying a deletion in the δ domain, demonstrating a requirement for the δ domain (amino acids 34–60) in these phosphorylation events. It is not clear whether this kinase is related to MAPK; however, the requirement of the δ domain for Ser 63 and 73 phosphorylation was also observed when using partially purified pp42 and pp54 MAP kinases *in vitro* (Adler et al., 1992). Conflicting reports have demonstrated that immunopurified preparations of ERK kinase (pp42 and pp44 MAP kinases) and pp44 MAPK immunoprecipitations contain an activity that phosphorylates c-Jun on Ser 246 (in mouse c-Jun), a residue that lies adjacent to the c-Jun DNA binding domain (Alvarez et al., 1991; Baker et al., 1992). It is possible that different MAP/ERK kinases phosphorylate Jun at different sites and contribute to the regulation of Jun activity in cells. The use of cloned MAPK family members will help sort out this controversy. Finally, MAPK phosphorylates and activates pp90^{rsk} protein kinase (Sturgill et al., 1988; Chung et al., 1991). pp90^{rsk} has been shown to phosphorylate c-Fos *in vitro* (Chen, R. H., et al., 1992), although the physiological role of this phosphorylation event is unknown. Therefore, MAPKs may modulate AP-1 activity indirectly by activating pp90^{rsk} or other protein kinases.

Both c-Jun and c-Fos appear to be under the control of negative regulators. Under serum-starved conditions c-Fos is sequestered in the cytoplasm by a docking protein. Serum stimulation relieves this inhibition by an as yet unknown mechanism (Roux et al., 1990; Kolch et al., 1993b). It has been reported that c-Jun is negatively regulated by a cell type-specific inhibitor (Baichwal and Tjian, 1990). Interestingly, Src and Ras appear to increase the transactivator activity of c-Jun by modulating the activity of this inhibitor, which is thought to interact with the δ and A1 regions of c-Jun (Baichwal et al., 1991).

Additionally, an inhibitor of Jun/Fos termed IP-1 has been identified in cytoplasmic fractions, and phosphorylation of IP-1 blocks its inhibitory activity (Auwerx and Sassone-Corsi, 1991). Identification of the signal transduction pathways involved in relieving negative regulation of Fos and Jun family members awaits further investigation.

In addition to its role in the posttranslational modification of c-Jun, Raf-1 may modulate AP-1 activity by activating expression from the *c-fos* promoter (Kaibuchi et al., 1989; Jamal and Ziff, 1990). The serum response element (SRE) is sufficient for Raf-induced Fos expression, but other elements in the *c-fos* promoter appear to be Raf responsive as well (Kaibuchi et al., 1989; Jamal and Ziff, 1990). v-Src (Fujii et al., 1989; Qureshi et al., 1991b), v-Fps (Alexandropoulos et al., 1992), and v-Ras (Sassone-Corsi et al., 1989; Fukumoto et al., 1990; Gutman et al., 1991) have also been shown to activate expression from SREs. The *c-fos* SRE contains binding sites for serum response factor (SRF) and for $p62^{TCF}$ (ternary complex factor), which, together with SRF and the c-fos SRE, form a ternary complex that appears to be necessary for serum- and TPA-induced transactivation of the *c-fos* promoter (Shaw et al., 1989; Graham and Gilman, 1991; see also Chapter 8, this volume). *In vitro* analysis has demonstrated that MAPK phosphorylates $p62^{TCF}$ and that this phosphorylation event results in ternary complex formation at the *c-fos* SRE (Gille et al., 1992). Since MAPK activation by EGF (Ballou et al., 1991) correlates with ternary complex formation and transcriptional activation of the *c-fos* gene (Gille et al., 1992) this signal transduction pathway is thought to be physiologically significant.

$p62^{TCF}$ has been shown to be similar or identical to Elk-1 (Hipskind et al., 1991) and SAP-1 (Dalton and Treisman, 1992), two Ets family DNA binding proteins. All three proteins bind to the SRE in an SRF-dependent manner and Elk-1 is immunologically related to $p62^{TCF}$. Consistent with MAPK phosphorylation of $p62^{TCF}$, both Elk-1 and SAP-1 have consensus MAPK phosphorylation sites. We have shown that Raf activates transcription from Ets binding sites and that Raf kinase is required for Ets-dependent transcriptional activation (Bruder et al., 1992). Thus, it is likely that Raf-induced MAPK activation is a common mechanism for transactivation through AP-1 and Ets enhancer elements and that the *c-fos* promoter is a major point of cross-talk between the Ets family and AP-1 family transcription factors.

Like *c-fos* and *c-jun*, *egr-1* is a cellular immediate early gene (Sukhatme et al., 1987, 1988) whose gene product functions as a transcription factor (Lemaire et al., 1990; Patwardhan et al., 1991). v-Raf activates expression of Egr-1 through a cluster of SREs in the *egr-1* promoter (Rim et al., 1992). In collaboration with Foster and colleagues, we have shown that both Ras and Raf are required for v-Src- (Qureshi et al., 1991c, 1992) and v-Fps- (Alexandropoulos et al., 1992) induced expression, and Raf is necessary for Ras-induced expression, from the *egr-1* promoter (Alexandropoulos et al., 1992). Thus, signaling through the Raf pathway is essential for *egr-1* induction by upstream PTKs and Ras. Although no Ets family binding proteins have been shown to bind to sequences adjacent to the *egr-1* SREs, several potential Ets binding sites are present and may be required for Egr-1 expression induced via the Raf pathway.

Recently we observed that Raf activates expression through the NF-κB binding sites in the HIV-LTR (Bruder et al., 1993). Our preliminary results suggest that Raf may be functioning in this system through another Ets family member that binds to the Ets core binding site, which overlaps the HIV NF-κB binding sites. We are presently testing the ability of PSKs of the Raf-initiated kinase cascade to phosphorylate this Ets binding protein *in vitro* and are especially interested in the functional consequences of this phosphorylation event.

MAPK has a broad range of substrates (Sturgill et al., 1988; Chung et al., 1991; Pulverer et al., 1991; Baker et al., 1992; Gille et al., 1992; Stokoe et al., 1992), including other oncogene class transcription factors such as Myc, which it phosphorylates *in vitro* on Ser 62 within the transactivation domain of c-Myc (Alvarez et al., 1991). The use of GAL4/Myc fusion proteins in cotransfection studies with p41^{MAPK} has suggested that MAPK induced phosphorylation of Ser 62 results in an increase in the transactivation functions of c-Myc (Seth et al., 1992). Since Myc cooperates with activated Raf in transformation (Blasi et al., 1985; Rapp et al., 1985d; Morse and Rapp. 1988) and growth factor abrogation experiments (Blasi et al., 1985; Rapp et al., 1985c; Cleveland et al., 1986a, 1993; Morse and Rapp, 1988; Troppmair et al., 1992b) we initially speculated that the basis for Raf/Myc synergism, as well as synergism between myc and oncogenes that work through the Raf pathway, may be activity regulation of Myc or Myc-interacting factors by phosphorylation (Rapp et al., 1985a, 1986). Preliminary experiments did not support a role for Raf-1 in direct phosphorylation of Myc (Lüscher and Rapp, unpublished data). However, the possibility that Raf may activate Myc via MAPK deserves further attention. In addition, Raf may function in the modulation of *c-myc* transcription. We have observed elevated levels of *c-myc* expression in Raf-revertant cells (Kolch et al., 1993b), suggesting a possible role of Raf kinase in the negative regulation of *c-myc*. Work in other systems points to a potential positive role. Stimulation of the CSF-1 receptor results in the Ets-dependent induction of *c-myc* expression (Roussel et al., 1991; Langer et al., 1992). Since CSF-1 stimulates Raf-1 (Baccarini et al., 1990) and Raf-1 activates Ets family DNA binding proteins, the Raf pathway may play a role in *c-myc* induction in this system.

DIFFERENTIAL SENSITIVITY OF TRANSFORMED CELLS WITH A CONSTITUTIVELY ACTIVATED RAF PATHWAY TO GROWTH INHIBITION BY DOMINANT NEGATIVE JUN

Based on our observation of Raf responsiveness of an AP-1-dependent early response gene promoter in the polyoma enhancer (Wasylyk et al., 1989b; Bruder et al., 1992) and subsequent work on the *fos* (Kaibuchi et al., 1989; Jamal and Ziff, 1990), *egr-1* (Qureshi et al., 1991a, 1991b; Rim et al., 1992), and *egr-2* (Cortner et al., 1993) promoters by others, we have recently investigated whether v-*raf* oncogene-transformed cells are stably altered in the regulation of Fos and Jun expression. Since this turned out to be the case (Kolch et al., 1993b; Rapp et al., 1993) and similar findings have been reported for *fos* in *ras*-transformed cells (Nose et al., 1989), it appeared that Fos/Jun might mediate Raf transforming activity at least in part. We, therefore, determined whether v-*raf* transformation can be blocked by a dominant negative mutant of c-Jun (TAM67). These studies were extended to transformation by oncogenes known to facilitate Raf-1 kinase activation *in vivo*, such as *abl* and *ras*, as well as others that may function in interacting, parallel pathways or downstream of Raf, such as v-*mos*, v-*fos*, v-*myc*, and SV40 (Table 16-1) (Rapp et al., 1993). For growth and transformation of established rodent fibroblast cell lines, the data demonstrate that oncogene transformation increases levels of constitutive Jun expression and renders cells differentially sensitive to growth inhibition by TAM67 (Rapp et al., 1993) (see Table 16-1, first column of data). The dominant negative mutant of c-Jun strongly inhibited the establishment of transformation by representatives of all classes of oncogenes (tyrosine kinase group, intracellular signal transducer- and transcription-factor class), as judged from two types of focus reduction assays, cotransfection, and virus titration (Table 1, second and third columns of data). These data suggest that transformation enforces and/or requires an altered balance of expressed Fos/Jun family proteins and that this shift in balance allows differential growth inhibition by the AP-1 antagonist TAM67.

The finding of downregulated Fos and upregulated cJun protein expression in v-*raf*-transformed NIH3T3 cells seems surprising in light of *in vitro* studies that showed

Table 16-1   % Inhibition by TAM67 of Growth and
Transformation of Oncogene Expressing Cells

Oncogene	Yield of neoR Colonies	Cotransfection Assay	Titration on TAM67 Jun Cells
v-raf	87 ± 5	88 ± 4	100
v-abl	92 ± 2	N.D.	100
v-ras	57 ± 1	79 ± 2	90
v-mos	88 ± 1	77 ± 1	100
v-myc	N.D.[a]	90 ± 9	N.D.
v-raf/v-myc	N.D.	85 ± 2	N.D.
v-fos	N.D.	86 ± 2	100
SV40	N.D.	89 ± 1	N.D.

[a] N.D.. no data.

Note: Three types of assays were used to assay TAM67 inhibition of oncogene-induced growth and differentiation. The first column of values shows comparison of the yield of neoR colonies between normal and oncogene-transformed NIH3T3 cells cotransfected with the neoR marker, TAM67, or vector control. The second assay was a cotransfection/focus reduction assay on NIH3T3 cells. Oncogene DNAs were cotransfected with vector control or TAM67. The third assay used stable TAM67 expressing Rat cells together with Rat-1a control cells as target cells for titration of transforming virus.

induction of both c-fos (Kaibuchi et al., 1989; Jamal and Ziff, 1990) and c-jun (P. Herrlich, personal communication) promoter constructs by v-Raf. It is consistent, however, with other findings on early response gene expression in v-raf-transformed cells (Siegfried and Ziff, 1990; Kolch et al., 1993), although the mechanism of downregulation of fos and induction of jun family genes is poorly understood (Curran and Franza, 1988; Vogt and Bos, 1990). The involvement of Fos protein in negative autoregulation (Sassone-Corsi et al., 1988; Wilson and Treisman, 1988; König et al., 1989) might explain a contribution to the initiation of Fos downregulation but the basis for its maintenance remains to be established. The pattern of altered early response gene expression is cell type dependent and controlled by multiple regulators, as we did not detect constitutive c-jun expression in v-Raf expressing 32D myeloid cells in the absence of IL-3 (Cleveland et al., 1993). Irrespective of the mechanism involved in the reprogramming of early response genes in stable v-raf-expressing NIH3T3 cells, the lack of c-Fos presumably favors Jun–Jun homodimer or heterodimer formation with other bZip proteins, which may somehow contribute to the increased sensitivity to TAM67 inhibition.

As shown in Table 16-1, increased sensitivity to growth/transformation inhibition by TAM67 was not restricted to v-raf transformation but includes many oncogenes. For several, such as v-abl (Rapp et al., 1993) and v-ras (Lloyd et al., 1992), this is not surprising because they are components of the Raf pathway (Figure 16-3). Of the other oncogenes that were inhibited by TAM67, v-mos deserves special mention. Although response elements for transcriptional transactivation by v-raf and v-mos overlap and include AP-1-dependent elements (Bruder et al., 1992; Wasylyk et al., 1989a, 1989b), suggesting similar functions of the two oncogenes, the prevailing view in the field holds that mos transforms cells by a unique mechanism that involves premature induction of a mitotic phenotype in the absence of gene induction (Propst et al., 1993). However, the data summarized here suggest that like abl, ras, and raf, mos induces and depends on AP-1 activity for transformation. mos presumably induces AP-1 activity, at least in part, through activation of the MAPK cascade. Troppmair et al. (1993) have recently determined the activity of MAPKK and MAPK in mos-transformed cells and found them to be activated. Moreover, *v-mos* activates MAPK activity in a Raf-1-dependent fashion, as determined in transient cotransfection assays using human 293 embryonic kidney cells. Finally, the fact that *v-raf* revertant cells (Kolch et al., 1993b) are also *mos* resistant is consistent with a function of Mos in a pathway that converges with that of Raf.

More unexpected than Mos dependence on *jun* was the finding that *v-myc* transformation also was TAM67 sensitive. *v-myc* alone does not morphologically transform NIH3T3 cells (Rapp et al., 1985d; Troppmair et al., 1989). However, it allows continued growth in confluent monolayer cultures, leading to focal areas of dense growth (focal overgrowth) by blocking postconfluence inhibition of division without inducing plating efficiency for growth on monolayers (Troppmair et al., 1989). Cells derived from such foci tend to form only small colonies in soft agar that can be expanded by addition of a "progression" factor such as EGF (Stern et al., 1986), which is a Raf-activating ligand. Thus, *v-myc* appears to partially transform cells through a pathway different from but synergistic with the Raf pathway. However, the data on elevated *c-jun* expression in *v-myc* focus-derived cells as well as their growth dependence on Jun/AP-1 (Rapp et al., 1993) suggests that Myc also requires Jun/AP-1 activity for this partial transformation of NIH3T3 cells.

The finding in rodent fibroblasts (Lloyd et al., 1992; Rapp et al., 1993) of apparently universal Jun dependence of oncogene transformation raises the hope that regulation of AP-1 activity may be a common target for differential downregulation of tumor cell growth. One way to approach this is to downregulate upstream PSKs, such as Raf, on which AP-1 activity depends (Smeal et al., 1992; Bruder et al., 1992). Small molecular weight inhibitors that block either Raf activation or its active site may be easier to develop than Fos/Jun inhibitors.

ROLE OF THE RAF PATHWAY IN CELL FATE DETERMINATION: THE LEVELS OF RAF AND MYC SET THE BALANCE BETWEEN APOPTOSIS, GROWTH, AND DIFFERENTIATION

Clearly, the work reviewed so far has established that the Raf signaling pathway is essential for induction of growth by most strong mitogens in mammalian cells (Carroll et al., 1991; Kolch et al., 1991; Cleveland et al., 1993), as well as Drosophila (Ambrosio et al., 1989; Dickson et al., 1992) and *Caenorhabditis elegans* (Han et al., 1993). But is activated Raf sufficient to replace the growth factor? This is apparently the case in NIH3T3 cells, where microinjection of activated but not control Raf proteins induced DNA synthesis in the absence of serum (Smith et al., 1990). In the *Drosophila* system, activated Raf was also sufficient to overcome a block in signaling by sevenless receptor PTK (Dickson et al., 1992). However, in myeloid 32D cells, which specifically depend on IL-3 for growth, oncogenic Raf did not supply a complete mitogenic signal (Blasi et al., 1985; Rapp et al., 1985c, 1988b, 1990; Cleveland et al., 1993). The findings with this cell system can be summarized as follows. First, any of a number of ectopically expressed oncogenic PTKs were able to substitute for IL-3, emphasizing their functional equivalence for growth regulation as mentioned in the introduction. Second, activated Raf expressed in 32D cells in the presence of IL-3 shortened the G1 phase of the cell cycle, leading to a more rapid rate of growth, and increased the saturation density (Troppmair et al., 1992b; Cleveland et al., 1993). In the absence of IL-3, v-Raf had survival activity similar to that of Bcl-2, a weak oncogene that contributes to the development of some natural B cell malignancies (Bakshi et al., 1985; Tsujimoto et al., 1985). Induction of survival involves suppression of apoptosis, which is the mode of death that 32D cells undergo in the absence of IL-3 (Askew et al., 1991). Bcl-2 and Raf work through separate pathways to achieve apoptosis suppression. This conclusion is based on the finding that

Bcl-2 does not activate Raf kinase, Raf kinase does not phosphorylate or induce expression of Bcl-2, and a combination of both genes leads to cooperative survival induction (Reed et al., 1991; Miyashita et al., 1993). Third, constitutive expression of Myc, an oncogene that acts synergistically with Raf in the transformation of cells from many lineages (Blasi et al., 1985; Rapp et al., 1985d; Cleveland et al., 1986a, 1986b; Morse and Rapp, 1988; Klinken et al., 1988, 1989; Pfeifer et al., 1989; Worland et al., 1990; Principato et al., 1990), does not lower or eliminate IL-3 dependence in 32D cells. In this regard 32D cells differ from certain other strains of the murine myeloid cell line, FDC-P1, where exogenous myc partially or completely substitutes for IL-3 (Rapp et al., 1985c; Dean et al., 1987; Morse and Rapp, 1988), perhaps because the Raf pathway is already activated. In fact, in 32D cells, ectopic Myc expression accelerates apoptotic cell death in the absence of IL-3 (Askew et al., 1991). However, in combination with v-Raf, v-Myc is as effective in abrogating IL-3 dependence as is ectopic expression of the most effective PTK oncogenes (Cleveland et al., 1989). Thus, activation of Raf and Myc is sufficient to account for most of the mitogenic activity of the IL-3 receptor. Based on these observations, we have proposed (Rapp et al., 1985b, 1986, 1988a, 1988b) that minimally two pathways are activated by growth-inducing PTKs, a Myc pathway and the Raf pathway, and these are jointly required (Figure 16-5). The data on TAM67 inhibition of partial transformation by *v-myc* may, at first glance, appear to be inconsistent with this model. However, in light of the data on apoptosis induction by constitutively expressed c-Myc in the absence of growth factors (Askew et al., 1991; Evan et al., 1992), and the survival activity of Raf (Troppmair et al., 1992b; Cleveland et al., 1993; Miyashita et al., 1993), it may be explained if we assume that Raf-1-induced AP-1 activity is mediating the *v-raf*-induced survival activity and was, therefore, selected in the constitutive *v-myc*-expressing cells. If AP-1 is in fact required for Raf survival activity it presumably includes forms of AP-1 that do not contain c-Jun, as this gene is not expressed in v-Raf-expressing 32D cells (Cleveland et al., 1993). Whether in NIH3T3 cells secondary changes were, indeed, responsible for the increased c-Jun levels or whether *v-myc* expression somehow contributed directly to the induction of c-Jun, will require experiments with inducible myc expression vectors.

Dominant negative mutants have also been derived from c-Myc (Sawyers et al., 1992), and it was recently shown that the PTK oncogene *abl*, which was TAM67 sensitive (Rapp et al., 1993), was *myc* dependent for transformation (Sawyers et al., 1992). We have previously demonstrated that the *abl* PTK, like other PTK oncogenes, induces not only Raf kinase activity (Carroll et al., 1990) but also constitutive *myc* expression (Cleveland et al., 1989). Thus, the combined data with dominant negative mutants of c-Myc and c-Jun are consistent with our model (Figure 16-6), since they emphasize the essential nature of the Raf pathway that modulates Jun activity in NIH3T3 cells in addition to the Myc pathway.

We have also studied the effect of Myc and Raf, alone and in combination, on cell systems with the potential to differentiate. From experiments in newborn mice with retroviruses expressing these oncogenes we knew that the preferred target cells for transformation differed between *v-myc* and *v-raf*, although the range of tissues affected by either gene was broad and partially overlapping (Rapp et al., 1985d; Morse et al., 1986; Frederickson et al., 1988; Morse and Rapp, 1988; Pfeifer et al., 1989; Worland et al., 1990). We inferred from this pattern that Myc is normally rate limiting in lymphoid and epithelial tissues, whereas Raf is rate limiting in mesenchymal (erythroid, fibroblastic) and epithelial cells (Rapp et al., 1986). Several of these lineages were further examined in cell culture, including the erythroid, B/myeloid, and mast cell systems. With bone marrow cells grown in the presence of 2-ME, either *v-myc* or *v-raf* alone immortalized cells *in vitro* at the preB cell stage (Principato et al., 1990). Combination of *v-raf* with *v-*

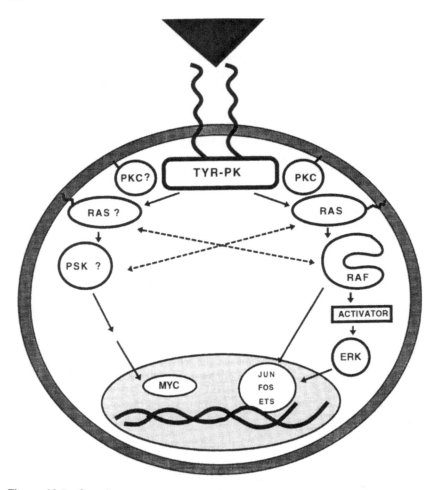

Figure 16-5 Signaling pathways originating from PTK receptor. The Raf signaling pathway is well characterized (Rapp, 1991; Kyriakis et al., 1992). Components in the Myc pathway are unknown.

myc led to synergistic induction of differentiation of B lineage cells *in vitro*, including the mature B cell stage (Klinken et al., 1988; Principato et al., 1990; Chen et al., 1992). These mature B cells appeared to switch lineage and give rise to mature macrophages at low frequency (Klinken et al., 1988; Principato et al., 1990). If in the same experiments cells were cultured in the absence of 2-ME, exclusive immortalization of mature CSF-1-independent macrophages was observed when v-Raf plus v-Myc were jointly expressed but not with either v-Raf or v-Myc alone (Blasi et al., 1985). In order to evaluate the effect of v-Raf and v-Myc on the mast cell lineage, fetal liver cultures were maintained in the presence of IL-3 (Ihle et al., 1987). In this system *v-raf* but not *v-myc* was an immortalizing oncogene, without blocking differentiation or abrogating IL-3 requirements (Cleveland et al., 1986b). Finally, when bone marrow cultures from phenylhydrazine-treated mice were plated in methylcellulose in the presence of suboptimal amounts of erythropoietin, v-Raf was very effective in inducing colonies of well-differentiated hemoglobin-synthesizing

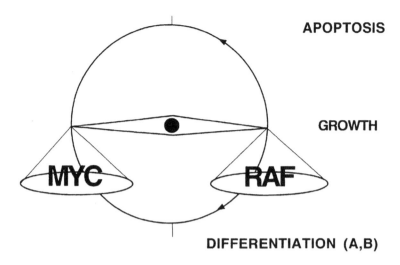

APOPTOSIS

GROWTH

DIFFERENTIATION (A,B)

Figure 16-6   Model for Raf/Myc effects on cell fates.

erythroid cells (Klinken et al., 1989). In contrast, *v-myc* was unable to stimulate the formation of erythroid colonies. The combination of *v-myc* with *v-raf* did not increase the yield of erythroid colonies. However, the cells in the colonies grew much more rapidly and were inhibited from terminally differentiating. This effect of Myc in the erythroid system is reminiscent of v-erbA, a dominant negative mutant of the thyroid receptor, which cooperates with v-erbB, an oncogenic form of the EGF receptor in transformation of erythroid lineage cells (Beug and Vennström, 1991). We, therefore, speculate that there is convergence in the pathways in which Myc and the thyroid receptor function.

From the above findings we conclude that v-Raf is compatible with or induces differentiation in a wide range of cell types *in vivo* and *in vitro*, whereas Myc by itself either is ineffective (apoptotic effects were not monitored before 1991), immortalizes, or blocks differentiation in a lineage-specific fashion. The reprogramming of early response gene expression that was observed in *v-raf* transformed fibroblast cell lines (fos downregulated, jun constitutive) (Kolch et al., 1993b; Rapp et al., 1993), may be a cryptic form of differentiation in these immortalized cells and indicate a critical role for the jun homodimer form of AP-1 in this process.

Based on these conclusions, we hypothesize that the ratio of Raf and Myc determines cell fate in terms of apoptosis, growth, and differentiation and that their level determines the rate at which these processes develop. This hypothesis makes several predictions that can be readily tested in cell culture. For example, growth and differentiation factors should be interconvertable by mutations or conditions of stimulation that affect their Myc and Raf activating potential. Alternatively, manipulation of the level of receptor-stimulated Myc and Raf activity in cells by expression of transforming versions or dominant inhibitory mutants of Raf and Myc should allow conversion of a growth receptor into a differentiation receptor and vice versa. We have undertaken to test this hypothesis in the PC12 cell system in which Myc and Raf are regulated by both growth (EGF) and differentiation (NGF) factors, and have obtained data that support such a model. The following experiments were carried out. First, PC12 cells grown in serum-containing medium were infected with viruses expressing *v-myc*, dominant negative *c-myc*, *v-raf*, or *v-raf* together with *v-myc* (Rapp et al., unpublished data). The results clearly demonstrated that *v-myc* did not antagonize *v-raf* and, instead, accelerated the induction of differentiation, as does *trk* overexpression in combination with NGF (Hempstead et al., 1992). *v-raf* alone had activities similar to NGF, while *v-myc* alone did not induce differentiation and instead had cytotoxic effects. Expression of inhibitory mutants of Myc altered the response of PC12 cells to EGF. which then behaved like a differentiation factor. The latter findings are compatible with an inhibitory role of Myc on differentiation, although the data with the *v-raf /v-myc* combination also indicate that Myc is not dominant and that the response depends on the strength of the differentiation signal. Nevertheless, the phenotype of differentiated cells differed somewhat between the various induction conditions (*v-myc* plus *v-raf* versus *v-raf* and dominant negative *myc* versus *v-raf* or NGF), suggesting that Myc/Raf ratios might be instructive as to the type of differentiation that is induced (Figure 16-6).

Work with EGF and NGF receptors supports the idea that signal transducers common to both receptors, such as Raf and Myc, regulate differentiation as well as growth. Thus, autocrine expression of TGFα (Derynck, 1992), an alternative ligand for the EGF receptor, by infection of PC12 cells with a retroviral expression vector led to the induction of differentiation within hours, as described above for Raf/Myc or Trk (NGF receptor) overexpression. This effect can be blocked by EGF and presumably results from a stronger long-term activation of the EGF receptor by endogenously produced TGFα than is seen with EGF. Moreover, expression of a chimeric form of the human EGF receptor with the cytoplasmic part of an oncogenic form of the EGF receptor, v-erbB, rendered PC12 cells sensitive to induction of differentiation with EGF (Rapp et al., unpublished data). These experiments show that the EGF receptor can be activated in such a way that it behaves like the NGF receptor, suggesting that differences between these receptors are quantitative rather than qualitative. They also raise the possibility that one and the same receptor may normally be used for regulation of either process (growth, differentiation) through developmentally controlled changes in autocrine versus paracrine production of ligand. The data on Raf and Myc demonstrate that these are the relevant downstream targets for growth- and differentiation-related receptor effects, as Myc and Raf by themselves are sufficient to regulate cell death, growth, and differentiation (Figure 6). We believe that for an understanding of the major components of this decision making process we can now set aside other known receptor activities and focus on those events that are downstream of and perhaps synergistically affected by Raf and Myc. It is likely that of the multiple targets of an activated Raf pathway, the induction of AP-1 as well as Ets activity will mediate some of its differentiation effects. In fact, overexpression of *c-jun* as well as *c-fos* has been reported to induce differentiation in F9 embryonal carcinoma cells (Müller and Wagner, 1984; Yamaguchi-Iwai et al., 1990). Other data obtained with the F9 cell system are also consistent with our model. Specifically, by use of *myc* expression vectors it was shown that retinoic acid-induced differentiation could be blocked, whereas *myc*-antisense expression-induced differentiation (Griep and Westphal, 1988) and similar findings were made in other cell systems (Cleveland et al., 1988; Holt et al., 1988). In addition to Fos and Jun, an oncogenic form of Ets, v-Ets, has also been associated with differentiation processes (Metz and Graf, 1991; Macleod et al., 1992). If it is indeed true that the differentiated phenotypes induced at different Myc/Raf ratios in PC12 cells are distinct, as suggested by their morphology, this might represent a general mechanism for regulation of differentiation in multipotential stem cells. It will be interesting to test whether agents that induce different types of differentiation of the same precursor cell, such as TPA and retinoic acid with HL-60 (Rovera et al., 1979; Breitman et al., 1980), affect the Myc/Raf balance differentially. There is little information that would aid speculation on the mechanism involved in Myc/Raf-mediated differentiation induction. To bring to light more of the machinery that mediates this fundamental process it will be important to identify the nature of relevant Myc- and AP-1/ets-regulated target genes. There is a good chance that such genes will be jointly regulated by Myc and AP-1/ets transcription factor complexes.

ADDENDUM

Following completion of this manuscript a series of papers appeared that address receptor—Ras (Buday and Downward, 1993; Egan et al., 1993; Rozakis-Adcock et al., 1993) and Ras–Raf interaction (Van Aelst et al., 1993; Zhang et al., 1993), as well as UV activation of Raf–1 kinase (Radler-Pohl et al., 1993). In addition, activation of Raf-1 through G protein coupled receptors has been reported by several groups (Kyriakis et al., 1993; Siegel et al., 1993; Winitz et al., 1993; Howe and Marshall, 1993; Cook et al., 1993). The hypotheses contained in this review have been presented by U.R.Rapp at the Nato Conference in Spetsai, Greece, in September 1-10, 1992.

NOTE ADDED DURING PROOF

The predicted role of ras in the membrane translocation of raf has been demonstrated recently (Leevers et al., 1994, Stokoe et al., 1994)

REFERENCES

Adler, V., Franklin, C. C., and Kraft, A. S., Phorbol esters stimulate the phosphorylation of c-Jun but not v-Jun: Regulation by the N-terminal δ domain, *Proc. Natl. Acad. Sci. U.S.A.*, 89, 5341, 1992.

Alexandropoulos, K., Qureshi, S. A., Bruder, J. T., Rapp, U. R., and Foster, D. A., The induction of Egr-1 expression by v-Fps is via a protein kinase C-independent intracellular signal that is sequentially dependent upon HaRas and Raf-1, *Cell Growth Differ.*, 3, 731, 1992.

Alvarez, E., Northwood, I. C., Gonzalez, F. A., Latour, D. A., Seth, A., Abate, C., Curran, T., and Davis, R. J., Pro-Leu-Ser/Thr-Pro is a consensus primary sequence for substrate protein phosphorylation, *J. Biol. Chem.*, 266, 15277, 1991.

Ambrosio, L., Mahowald, A. P., and Perrimon, N., Requirement of the Drosophila raf homolog for torso function, *Nature*, 342, 288, 1989.

Armelin, H. A., Armelin, M. C., Kelly, K., Stewart, T., Leder, P., Cochran, B. H., and Stiles, C. D., Functional role for c-myc in mitogenic response to platelet-derived growth factor, *Nature*, 310, 655, 1984.

Ashworth, A., Nakielny, S., Choen, P., and Marshall, C., The amino acid sequence of a mammalian MAP kinase kinase, *Oncogene*, 7, 2555, 1992.

Askew, D. S., Ashmun, R. A., Simmons, B. C., and Cleveland, J. L., Constitutive c-myc expression in IL-3 dependent myeloid cell line suppresses cell cycle arrest and accelerates apoptosis, *Oncogene*, 6, 1915, 1991.

Auwerx, J. and Sassone-Corsi, P., IP-1: A dominant inhibitor of Fos/Jun whose activity is modulated by phosphorylation, *Cell*, 64, 983, 1991.

Baccarini, M., Sabatini, D. M., App, H., Rapp, U. R., and Stanley, E. R., Colony stimulating factor-1 (CSF-1) stimulates temperature dependent phosphorylation and activation of the Raf-1 proto-oncogene product, *EMBO J.*, 9, 3649, 1990.

Baichwal, V. R. and Tjian, R., Control of c-Jun activity by interaction of a cell-specific inhibitor with regulatory domain δ: Differences between v- and c-Jun, *Cell*, 63, 815, 1990.

Baichwal, V. R., Park, A., and Tjian, R., v-Src and EJ Ras alleviate repression of c-Jun by a cell-specific inhibitor, *Nature*, 352, 165, 1991.

Baker. S. J., Kerppola, T. K., Luk. D., Vandenberg. M. T., Marshak. D. R., Curran, T., and Abate, C., Jun is phosphorylated by several protein kinases at the same sites that are modified in serum-stimulated fibroblasts, *Mol. Cell. Biol.,* 12. 4694, 1992.

Bakshi, A., Jensen. J. P., Goldman. P., Wright. J. J., McBride. O. W., Epstein. A. L., and Korsemeyer. S. J., Cloning the chromosomal translocation of t(14;18) human lymphoma: Clustering around JH on chromosome 14 and near a transcriptional unit on 18, *Cell,* 4, 899, 1985.

Ballou, L. M., Luther, H., and Thomas. G., MAP2 kinase and 70K S6 kinase lie on distinct signaling pathways, *Nature,* 349. 348, 1991.

Barbacid, M., ras genes, *Annu. Rev. Biochem.,* 56, 779. 1987.

Beug. H. and Vennström, B., in *Nuclear Hormone Receptors,* Parker, M. D., Ed.. Academic Press, London, p. 355, 1991.

Binétruy, B., Smeal, T., and Karin. M., Ha-Ras augments c-Jun activity and stimulates phosphorylation of its activation domain, *Nature,* 351, 122, 1991.

Blasi, E., Mathieson, B. J., Varesio. L., Cleveland. J. L., Borchert. P. A., and Rapp. U. R., Selective immortalization of murine macrophages from fresh bone marrow by a raf/myc recombinant murine retrovirus, *Nature,* 318. 667, 1985.

Bonner, T. I., Kerby, S. B., Sutrave, P., Gunnell. M. A., Mark. G., and Rapp. U. R., Structure and biological activity of human homologs of the raf/mil oncogene. *Mol. Cell. Biol.,* 5. 1400. 1985.

Bonner, T. I., Oppermann. H., Seeburg. P., Kerby, M. A., Gunnell, M. A., Young. A. C., and Rapp. U. R., The complete coding sequence of the human raf oncogene and the corresponding structure of the c-raf-1 gene, *Nucleic Acids Res.,* 14. 1009. 1986.

Breitman, T. R., Selonick, S. E., and Collins. J., Induction of differentiation of the human promyelocytic leukemia cell line (HL-60) by retinoic acid, *Proc. Natl. Acad. Sci. U.S.A.,* 77. 2936, 1980.

Bruder, J. T., Heidecker, G., and Rapp. U. R., Serum TPA and Ras induced expression from Ap-1/Ets driven promoters requires Raf-1 kinase, *Genes Dev.,* 6, 545, 1992.

Bruder, J. T., Heidecker, G., Tan, T.-H., Weske, J. C., Derse. D., and Rapp. U. R., Oncogene activation of HIV-LTR-driven expression via the NF-κB binding sites, *Nucl. Acids Res.,* 21. 5229, 1993.

Buday, L. and Downward, J., Epidermal growth factor regulates p21ras through the formation of a complex of receptor, Grb2 adaptor protein, and Sos nucleotide exchange factor, *Cell,* 73. 611, 1993.

Cai, H., Szeberenyi, J., and Cooper, G. M., Effect of a dominant inhibitory Ha-ras mutation on mitogenic signal transduction in NIH3T3 cells, *Mol. Cell. Biol.,* 10. 5314. 1990.

Carin, B. R., Ramer, S. W., and Kornberg, R. D., Order of action of components in the yeast pheromone response pathway revealed with a dominant allele of the STE11 kinase and multiple phosphorylation of the STE7 kinase. *Genes Dev.,* 6. 1305, 1992.

Carroll, M. P., Clark-Lewis, I., Rapp. U. R., and May. W. S., Interleukin-3 and granulocyte-macrophage colony stimulating factor mediate rapid phosphorylation of cytosolic c-Raf, *J. Biol. Chem.,* 265. 19812. 1990.

Carroll, M. P., McMahon, M., Weich, N., Spivak, J. L., Rapp. U. R., and May. W. S., Erythropoietin induces Raf-1 activation and Raf-1 is required for erythropoietin-mediated proliferation, *J. Biol. Chem.,* 266. 14964, 1991.

Chen, R. H., Sarnecki, C., and Blenis. J., Nuclear localization and regulation of erk- and rsk-encoded protein kinases, *Mol. Cell. Biol.,* 12. 915, 1992.

Chen, Y. W., Vora, K. A., and Lin. M. S., Generation and differentiation potential of retrovirus-immortalized mature murine B cells, *Curr. Top. Microbiol. Immunol.,* 182. 337, 1992.

Chung, J., Pelech, S. L., and Blenis, J., Mitogen-activated Swiss mouse 3T3 RSK kinase I and II are related to pp44^{mpk} from sea star oocytes and participate in the regulation of pp90^{rsk} activity, *Proc. Natl. Acad. Sci. U.S.A.*, 88, 4981, 1991.

Clark, S. G., Stern, M. J., and Horvitz, H. R., C. elegans cell-signaling gene sem-5 encodes a protein with SH2 and SH3 domains, *Nature*, 356, 285, 1992.

Cleveland, J. L., Morse, H. C., Ihle, J. N., and Rapp, U. R., Interaction between raf and myc oncogenes in transformation in vitro and in vivo, *J. Cell. Biochem.* 30, 195, 1986a.

Cleveland, J. L., Weinstein, Y., Ihle, J. N., Askew, D. S., and Rapp, U. R., Transformation and insertional mutagenesis in vitro of primary hematopoietic stem cell cultures, *Curr. Top. Microbiol. Immunol.*, 132, 44, 1986b.

Cleveland, J. L., Morse, H. C., III, and Rapp, U. R., Myc oncogenes and tumor induction, in *ISI Atlas of Science: Biochemistry 1*, Institute for Scientific Information, Inc., Philadelphia, 1988, p. 93.

Cleveland, J. L., Dean, M., Rosenberg, N., Wang, J. Y., and Rapp, U. R., Tyrosine kinase oncogenes abrogate interleukin-3 dependence of murine myeloid cells through signaling pathways involving c-myc: Conditional regulation of c-myc transcription by temperature sensitive v-abl, *Mol. Cell. Biol.*, 9, 5685, 1989.

Cleveland, J. L., Troppmair, J., Packham, G., Askew, D. S., Lloyd, P., Gonzolez-Garcia, M., Nunez, G., Ihle, J. N., and Rapp, U. R., *Oncogene*, in press.

Cohen P., Signal integration at the level of protein kinases, protein phosphatases and their substrates, *TIBS*, 17, 408, 1992.

Cook, S. J., Rubinfeld, B., Albert, I., and McCormick, F., RapV12 antagonizes Ras-dependent activation of ERK1 and ERK2 by LPA and EGF in Rat-1 fibroblasts, *EMBO J.*, 12, 3475, 1993.

Cortner, J., Chen, S., Flanagan, M. A., and Farnham, P. J., v-raf stimulates expression of the immediate early growth response gene egr-2 and of the delayed early response gene ornithine decarboxylase, 1993, submitted.

Courchesne, W. E., Kunisawa, R., and Thorner, J., A putative protein kinase overcomes pheromone-induced arrest of cell cycling in S. cerevisiae, *Cell*, 58, 1107, 1989.

Crues, C. M., Alessandrini, A., and Erickson, R. L., The primary structure of MEK, a protein kinase that phosphorylates the ERK gene product, *Science*, 258, 478, 1992.

Cuadrado, A., Bruder, J. T., Heidaran, M. A., App, H., Rapp U. R., and Aaronson, S. A., H-ras and raf-1 cooperate in transformation of NIH3T3 fibroblasts, *Oncogene*, 8, 2443, 1993.

Curran, T. and Franza, B. R., Fos and Jun. The AP-1 connection, *Cell*, 55, 395, 1988.

Dalton, S. and Treisman, R., Characterization of SAP-1 a protein recruited by serum response factor to the c-fos serum response element, *Cell*, 68, 597, 1992.

Davis, R. J. and Czech, M. P., Tumor-promoting phorbol ester cause the phosphorylation of epidermal growth factor receptors in normal human fibroblasts at threonine-654, *Proc. Natl. Acad. Sci. U.S.A.*, 82, 1974, 1985.

Dean, M., Cleveland, J. L., Rapp, U. R., and Ihle, J. N., Role of myc in the abrogation of IL-3 dependence of myeloid FDC-P1 cells, *Oncogene Res.*, 1, 279, 1987.

Dent, P., Haser, W., Haystead, T. A., Vincet, L. A., Roberts, T. M., and Sturgill, T. W., Activation of mitogen-activated protein kinase kinase by v-raf in NIH3T3 cells in vitro, *Science*, 257, 1404, 1992.

Derynck, R., The physiology of transforming growth factor-alpha, *Adv. Cancer Res.*, 58, 27, 1992.

de Vries-Smits, A. M., Burgering, B. M., Leevers, S. J., Marshall, C. J., and Bos, J. L., Involvement of p21ras in activation of extracellular signal-related kinase 2, *Nature*, 357, 602, 1992.

Dickson, B., Sprenger, F., Morrison, D., and Hafen, E., Raf functions downstream of Ras1 in sevenless signal transduction, *Nature*, 360, 600, 1992.

Egan, S. E., Giddings, B. W., Brooks, M. W., Buday, L., Sizeland, A. M., and Weinberg, R.A., Association of Sos ras exchange protein with Grb2 is implicated in tyrosine kinase signal transduction and transformation, *Nature*, 363, 45, 1993.

Eilers, M., Schirm, S., and Bishop, J. M., The Myc protein activates transcription of the alpha-prothymosin gene, *EMBO J.*, 10, 133, 1991.

Elion, E. A., Grisafi, P. L., and Fink, G. R., FUS3 encodes a cdc2+/CDC28-related kinase required for the transition from mitosis into conjugation, *Cell*, 60, 649, 1990.

Evan, G. I., Wyllie, A. H., Gilbert, C. S., Littlewood, T. D., Land, H., Brooks, M., Waters, C. M., Penn, L. Z., and Hancock, D. C., Induction of apoptosis in fibroblasts by c-myc protein, *Cell*, 69, 119, 1992.

Farnham, P. J. and Kollmar, R., Characterization of the 5′ end of the growth-regulated syrian hamster CAD gene, *Cell Growth Differ.*, 1, 179, 1990.

Farnham, P. J. and Means, A. L., Sequences downstream of the transcriptional initiation site modulate the activity of the murine dihydrofolate reductase promoter, *Mol. Cell. Biol.*, 10, 1390, 1990.

Farnsworth, C. L. and Feig, L. A., Dominant inhibitory mutations in the Mg^{2+} binding site of RasH prevent its activation by GTP, *Mol. Cell. Biol.*, 11, 4822, 1991.

Feig, L. A. and Cooper, G. M., Inhibition of NIH3T3 cell proliferation by a mutant ras protein with preferential affinity for GDP, *Mol. Cell. Biol.*, 8, 3235, 1988.

Force, T., Bonventre, J. V., Heidecker, G., Rapp, U. R., Avruch, J., and Kyriakis, J. M., Enzymatic characteristics of the c-Raf-1 protein kinase, *Proc. Natl. Acad. Sci. U.S.A.*, 91, 1270, 1994.

Frederickson, T. N., Hartley, J. W., Wolford, N. K., Resau, J. H., Rapp, U. R., and Morse, H. C., III, Histogenesis and clonality of pancreatic tumors induced by v-*myc* and v-*raf* oncogenes in NFS/N mice, *Am. J. Pathol.*, 131, 444, 1988.

Fujii, M., Shalloway, D., and Verma, I. M., Gene regulation by tyrosine kinases: src protein activates various promoters, including c-fos, *Mol. Cell. Biol.*, 9, 2493, 1989.

Fukumoto, Y., Kaibuchi, K., Oku, N., Hori, Y., and Taka, Y., Activation of the c-fos serum-response element by the activated c-Ha-ras protein in a manner independent of protein kinase C and cAMP-dependent protein kinase, *J. Biol. Chem.*, 265, 774, 1990.

Gallego, C., Gupta, S. K., Heasley, L. E., Quian, N. X., and Johnson, G. L., Mitogen-activated protein kinase activation resulting from selective oncogene expression in NIH3T3 and Rat 1a cells, *Proc. Natl. Acad. Sci. U.S.A.*, 89, 7355, 1992.

Gille, H., Sharrocks, A. D., and Shaw, P. E., Phosphorylation of transcription factor $p62^{TCF}$ by MAP kinase stimulates ternary complex formation at the c-fos promoter, *Nature*, 358, 414, 1992.

Graham, R. and Gilman, M., Distinct protein targets for signals acting at the c-fos serum response element, *Science*, 251, 189, 1991.

Griep, A. E. and Westphal, H., Antisense myc sequences induce differentiation of F9 cells, *Proc. Natl. Acad. Sci. U.S.A.*, 85, 5539, 1988.

Grugel, S., Bruder, J. T., Lee, J. E., Sithanandam, G., Beck, T. W., Troppmair, J., and Rapp, U. R., Identification of A-Raf protein and regulation of its kinase activity by a mechanism common to all three Raf isozymes, 1993, submitted.

Gutman, A., Wasylyk, C., and Wasylyk, B., Cell-specific regulation of oncogene-responsive sequences of the c-fos promoter, *Mol. Cell. Biol.*, 11, 5381, 1991.

Hall, A., Signal transduction through small GTPases-a tale of two GAPs, *Cell*, 69, 389, 1992.

Han, M., Golden, A., Han, Y., and Sternberg, P. W., C. elegans lin-45 raf gene participates in let-60 ras-stimulated vulval induction, *Nature*, 363, 133, 1993.

Heidecker, G., Huleihel, M., Cleveland, J. L., Beck, T. W., Lloyd, P., Pawson, T., and Rapp, U. R., Mutational activation of c-raf-1 and definition of the minimal transforming sequence, *Mol. Cell. Biol.*, 10, 2503, 1990.

Heidecker, G., Kolch, W., Morrison, D. K., and Rapp, U. R., Role of Raf-1 phosphorylation in signal transduction, *Adv. Cancer Res.*, 58, 53, 1992.

Heldin, C. H., SH2 domains: Elements that control protein interaction during signal transduction, *TIBS*, 16, 450, 1991.

Hempstead, B. L., Rabin, S. J., Kaplan, L., Reid, S., Parada, L. F., and Kaplan, D. R., Overexpression of the trk tyrosine kinase rapidly accelerates nerve growth factor-induced differentiation, *Neuron*, 9, 883, 1992.

Hipskind, R. A., Rao, V. N., Mueller, C. G. F., Reddy, E. S. P., and Nordheim, A., Ets-related protein Elk-1 is homologous to the c-fos regulatory factor p62^{TCF}, *Nature*, 354, 531, 1991.

Holt, J. T., Redner, R. L., and Nienhuis, A. W., An oligomer complementary to c-myc mRNA inhibits proliferation of HL-60 promyelocytic cells and induces differentiation, *Mol. Cell. Biol.*, 8, 963, 1988.

Howe, L. R. and Marshall, C. J., Lysophosphatidic acid stimulates mitogen-activated protein kinase activation via a G-protein coupled pathway requiring p21 ras and p74 raf-1, *J. Biol. Chem.*, 268, 20717, 1993.

Howe, L. R., Leevers, S. J., Gomez, N., Nakielny, S., Cohen, P., and Marshall, C. J., Activation of the MAP kinase pathway by the protein kinase raf, *Cell*, 7, 335, 1992.

Hunter, T., Ling, N., and Cooper, J. A., Protein kinase C phosphorylation of the EGF receptor at a threonine residue close to the cytoplasmatic face of the plasma membrane, *Nature*, 311, 480, 1984.

Ihle, J. N., Weinstein, Y., Cleveland, J. L., Dean, M. C., and Rapp, U. R., Regulation of normal and transformed hemopoietic stem cell growth by interleukin 3. UCLA Symposia on Molecular and Cellular Biology, in *Recent Advances in Leukemia and Lymphoma*, Gale, R. P. and Golde, D. W., Eds., Alan R. Liss, New York, p. 293, 1987.

Ikawa, S., Fukui, M., Ueyama, Y., Tamaoki, N., Yamamoto, T., and Toyoshima, K., B-raf, a new member of the raf family, is activated by DNA rearrangement, *Mol. Cell. Biol.*, 8, 2651, 1988.

Jamal, S. and Ziff, E., Transactivation of c-fos and ß-actin genes by raf as a step in early response to transmembrane signals, *Nature*, 344, 463, 1990.

Kaibuchi, K., Fukumoto, Y., Oku, N., Hori, Y., Yamamoto, T., Toyoshima, K., and Takai, Y., Activation of the serum response element and 12-O-tetradecanoylphorbol-13-acetate response element by the activated c-raf-1 protein in a manner independent of protein kinase C, *J. Biol. Chem.*, 264, 20855, 1989.

Karin, M. and Smeal, T., Control of transcription factors by signal transduction pathways: The beginning of the end, *TIBS*, 17, 418, 1992.

Klinken, S. P., Alexander, W. S., and Adams, J. M., Hemopoietic lineage switch: v-raf oncogene converts Eµ-myc transgenic B cells into macrophages, *Cell*, 53, 857, 1988.

Klinken, S. P., Rapp, U. R., and Morse, H. C., III, Raf/myc-infected erythroid cells are restricted in their ability to terminally differentiate, *J. Virol.*, 63, 1489, 1989.

Kolch, W., Heidecker, G., Lloyd, P., and Rapp, U. R., Raf-1 protein kinase is required for growth of induced NIH/3T3 cells, *Nature*, 349, 426, 1991.

Kolch, W., Heidecker, G., Kochs, G., Hummel, R., Vahidi, H., Mischak, H., Finkenzeller, G., Marme, D., and Rapp, U. R., PKCα is a Raf-1 kinase kinase and activates by phosphorylation of serine 499 and 259, *Nature*, 364, 249, 1993a.

Kolch, W., Heidecker, G., Troppmair, J., Yanagihara, K., Bassin, R. H., and Rapp, U. R., Raf revertant cells resist transformation by non-nuclear oncogenes and are deficient in induction of early genes by TPA and serum, *Oncogene*, 8, 361, 1993b.

König, H., Ponta, H., Rahmsdorf, U., Büscher, M., Schönthal, A., Rahmsdorf, H. J., and Herrlich, P., Autoregulation of Fos: The dyad symmetry element as the major target of repression, *EMBO J.*, 89, 2559, 1989.

Kremer, N. E., D'Arcangelo, G., Thomas, S. M., DeMarco, M., Brugge, J. S., and Halegoua, S., Signal transduction by nerve-growth factor and fibroblast growth factor in PC12 cells requires a sequence of src and ras actions. *J. Cell. Biol.*, 115, 809, 1992.

Kypta, R. M., Goldberg, Y., Ulug, E. T., and Courtneidge, S. A., Association between the PDGF receptor and members of the src family of tyrosine kinases. *Cell*, 62, 481, 1990.

Kyriakis, J. M., App, H., Zhang, X., Banerjee, P., Brautigan, D. L., Rapp, U. R., and Avruch, J., Raf-1 activates MAP kinase-kinase. *Nature*, 358, 417, 1992.

Kyriakis, J. M., Force, T. L., Rapp, U. R., Bonventre, J. V., and Avruch, J., Mitogen regulation of c-Raf-1 protein kinase activity toward mitogen-activated protein kinase-kinase. *J. Biol. Chem.*, 268, 16009, 1993.

Langer, S. J., Bortner, D. M., Roussel, M. F., Sherr, C., and Ostrowski, M. C., Mitogenic signaling by colony-stimulating factor 1 and ras is suppressed by the ets-2 DNA-binding domain and restored by myc overexpression. *Mol. Cell. Biol.*, 12, 5355, 1992.

Lee R., Rapp, U. R., and Blackshear, P. J., Evidence for one or more Raf-1 kinases activated by insulin and polypeptide growth factors. *J. Biol. Chem.*, 266, 10351, 1991.

Leevers, S. J. and Marshall, C. J., Activation of extracellular signal-regulated kinase, erk 2, by p21ras. *EMBO J.*, 11, 569, 1992.

Leevers, S. J., Paterson, H. F., and Marshall, C. J., Requirement for Ras in Raf activation is overcome by targeting Raf to the plasma membrane. *Nature*, 369, 411, 1994.

Lemaire, P., Vesque, C., Schmidt, J., Stunnenberg, H., Frank, R., and Charnay, P., The serum-inducible mouse gene Krox-24 encodes a sequence-specific transcriptional activator. *Mol. Cell. Biol.*, 10, 3456, 1990.

Lloyd, A., Yancheva, N., and Wasylyk, B., Transformation suppressor activity of a jun transcription factor lacking its activation domain. *Nature*, 352, 635, 1992.

Lowenstein, E. J., Daly, R. J., Batzer, A. G., Li, W., Margolis, B., Lammers, R., Ullrich, A., Skolnik, E. Y., Bar-Sagi, D., and Schlessinger, J., The SH2 and SH3 domain-containing protein GRB2 links receptor tyrosine kinases to ras signaling. *Cell*, 70, 431, 1992.

Lüscher, B. and Eisenman, R. N., New light on Myc and Myb. Part I. Myc. *Genes Dev.*, 4, 2025, 1990.

Macleod, K., Leprince, D., and Stehelin, D., The ets gene family. *TIBS*, 17, 251, 1992

Marsh, L., Neiman, A. M., and Herskowitz, I., Signal transduction during pheromone response in yeast. *Annu. Rev. Cell Biol.*, 7, 699, 1991.

Metz, T. and Graf, T., v-myc and v-ets transform chicken erythroid cells and cooperate both in trans and cis to induce distinct differentiation phenotypes. *Genes Dev.*, 5, 369, 1991.

Miyashita, T., Hovey, L., Torigoe, T., Krajewsky, S., Troppmair, J., Rapp, U. R., and Reed, J. C., Novel form of oncogene cooperation: Synergistic suppression of apoptosis by combination of *bcl*-2 and *raf* oncogenes, submitted.

Moodie, S. A., Willumsen, B. M., Weber, M. J., and Wolfman, A., Complexes of Ras.GTP with Raf-1 and mitogen-activated protein kinase kinase. *Science*, 260, 1658, 1993.

Morse, H. C., III. and Rapp, U. R., Tumorigenic activity of artificially activated c-oncogenes, in *Cellular Oncogene Activation*, Klein, G., Ed., Marcel Dekker, New York, 1988, 335.

Morse, H. C., III, Hartley, J. W., Frederickson, T. N., Yetter, R. A., Majumdar, C., Cleveland, J. L., and Rapp, U. R., Recombinant murine retroviruses containing avian v-myc induce a wide spectrum of neoplasms in newborn mice. *Proc. Natl. Acad. Sci. U.S.A.*, 83, 6868, 1986.

Müller, R. and Wagner, E. F., Differentiation of F9 teratocarcinoma stem cells after transfer of c-fos proto-oncogenes. *Nature*, 311, 438, 1984.

Nori, M., L'Allemain, G., and Weber, M., Regulation of TPA induced responses in NIH3T3 cells by GAP, the GTPase activating protein associated with p21c-ras. *Mol. Cell. Biol.*, 12, 936, 1992.

Nose, K., Itami, M., Satake, M., Ito, Y., and Kuroki, T., Abolishment of c-fos inducibility in ras-transformed mouse osteoblast cell lines, *Mol. Carcinogenesis*, 2, 208, 1989.

Oshima, M., Sithanandam, G., Rapp, U. R., and Guroff, G., The phosphorylation and activation of B-raf in PC12 cells stimulated by nerve growth factor, *J. Biol. Chem.*, 266, 23753, 1991.

Owaki, H., Varma, R., Gillis, B., Bruder, J. T., Rapp, U. R., Davis, L. S., and Geppert, T. D., Raf-1 is required for T cell IL-2 production, *EMBO J.*, 12, 4367, 1993.

Parker, P. and Waterfield, M. D., Phosphatidylinositol 3-kinase: A novel effector, *Cell Growth Differ.*, 3, 747, 1992.

Patwardhan, S., Gashler, A., Siegel, M. G., Chang, L. C., Joseph, L. J., Shows, T. B., LeBeau, M., and Sukhatme, V. P., Egr-3, a novel member of the Egr family of genes encoding immediate-early transcription factors, *Oncogene*, 6, 917, 1991.

Pawson, T. and Gish, G. D., SH2 and SH3 domains: From structure to function, *Cell*, 71, 359, 1992.

Perkins, L. A., Larsen, I., and Perrimon, N., Corkscrew encodes a putative protein tyrosine phosphatase that functions to transduce the terminal signal from the receptor tyrosine kinase torso, *Cell*, 70, 225, 1992.

Pfeifer, A. M. A., Mark, G. E., III, Malan-Shibley, L., Graziano, S., Amstal, P., and Harris, C. C., Cooperation of c-raf-1 and c-myc protooncogenes in the neoplastic transformation of simian virus 40 large tumor antigen-immortalized human bronchial epithelial cells, *Proc. Natl. Acad. Sci. U.S.A.*, 86, 10075, 1989.

Principato, M., Cleveland, J. L., Rapp, U. R., Holmes, K. L., Pierce, J. H., Morse H. C., III, and Klinken, S. P., Transformation of murine bone marrow cells with combined v-raf/v-myc oncogenes yields clonally related mature B cells and macrophages, *Mol. Cell. Biol.*, 10, 3562, 1990.

Propst, F., Storm, S. M., and Rapp, U. R., Oncogenes coding for protein-serine/threonine kinases: mos and raf, in *The Molecular Basis of Human Cancer*, Barbacid, M. and Kumar, A., Eds., 1993.

Pulverer, B. J., Kyriakis, J. M., Avruch, J., Nikolakaki, E., and Woodgett, J. R., Phosphorylation of c-jun mediated by MAP kinases, *Nature*, 353, 670, 1991.

Qureshi, S. A., Cao, X. M., Sukhatme, V. P., and Foster, D. A., v-Src activates mitogen-responsive transcription factor Egr-1 via serum response elements, *J. Biol. Chem.*, 266, 10802, 1991a.

Qureshi, S. A., Joseph, C. K., Rim, M., Maroney, A., and Foster, D. A., v-Src activates both protein kinase C-dependent and independent signaling pathways in murine fibroblasts, *Oncogene*, 6, 995, 1991b.

Qureshi, S. A., Rim, M., Bruder, J. T., Kolch, W., Rapp, U. R., Sukhatme, V. P., and Foster, D. A., An inhibitory mutant of c-Raf-1 blocks v-Src induced activation of the Egr-1 promoter, *J. Biol. Chem.*, 266, 20594, 1991c.

Qureshi, S. A., Alexandropoulos, K., Rim, M., Joseph, C. K., Bruder, J. T., Rapp, U. R., and Foster, D. A., Evidence that Ha-Ras mediates two distinguishable intracellular signals activated by v-Src, *J. Biol. Chem.*, 267, 17635, 1992.

Radler-Pohl, A., Sachsenmaier, C., Gebel, S., Auer, H.-P., Bruder, J. T., Rapp, U. R., Angel, P., Rahmsdorf, H. J., and Herrlich, P., UV-induced activation of AP-1 involves obligatory extranuclear steps including Raf-1 kinase, *EMBO J.*, 12, 1005, 1993.

Rapp, U. R., Role of Raf-1 serine/threonine protein kinase in growth factor signal transduction, *Oncogene*, 6, 495, 1991.

Rapp U. R., Goldsborough M. D., Mark G. E., Bonner T. I., Groffen J., Reynolds, F. H., Jr., and Stephenson, J., Structure and biological activity of v-raf, a unique oncogene transduced by a retrovirus, *Proc. Natl. Acad. Sci. U.S.A.*, 80, 4218, 1983.

Rapp, U. R., Bonner, T. I., and Cleveland, J. L., The raf oncogenes, in *Retroviruses and Human Pathology*, Gallo, R. C., Stehelin, D., and Varnier, O. E., Eds., Humana Press, Clifton, NJ, pp. 449–472, 1985a.

244

Rapp, U. R., Bonner, T. I., Moelling, K., Jansen, H. W., Bister, K., and Ihle, J., Genes and gene products involved in growth regulation of tumor cells, *Recent Results Cancer Res.*, 99, 221, 1985b.

Rapp, U. R., Cleveland, J. L., Brightman, K., Scott, A., and Ihle, J. N., Abrogation of IL-3 and IL-2 dependence by recombinant murine retroviruses expressing v-myc oncogenes, *Nature*, 317, 434, 1985c.

Rapp, U. R., Cleveland, J. L., Frederickson, J. L., Holmes, K. L., Morse, H. C., III, Jansen, H. W., Patschinsky, T., and Bister, K., Rapid induction of hemopoietic neoplasms in newborn mice by a raf(mil)/myc recombinant murine retrovirus, *J.Virol.*, 55, 23, 1985d.

Rapp, U. R., Cleveland, J. L., Storm, S. M., Beck, T. M., and Huleihel, M., Transformation by *raf* and *myc* oncogenes, in *Oncogenes and Cancer*, Proceedings of the 17th International Symposium of the Princess Takamatsu Cancer Research Fund, Tokyo, 1986, Aaronson, S. A., Bishop, M. J., Sugimura, T., Terada, M., Toyoshima, K., and Vogt, P. K., Eds., Japan Scientific Press, Tokyo, 1987, 55.

Rapp, U. R., Storm, S. M., and Cleveland, J. L., Oncogenes: Clinical relevance, *Haematol. Bluttransfus.*, 31, 450, 1987.

Rapp, U. R., Cleveland, J. L., Bonner, T. I., and Storm S. M., The raf oncogenes, in *The Oncogene Handbook*, Curran, T., Reddy, E. P., and Skalka, A., Eds., p. 213, 1988a.

Rapp, U. R., Heidecker, G., Huleihel, M., Cleveland, J. L., Choi, W. C., Pawson, T., Ihle, J. N., and Anderson, W. B., raf family serine/threonine protein kinases in mitogen signal transduction, *Cold Spring Harbor Symp. Quanti. Biol.*, 53, 173, 1988b.

Rapp, U. R., Huleihel, M., Pawson, T., Linnoila, I., Minna, J. D., Heidecker, G., Cleveland, J. L., Beck, T., Forchhammer, J., and Storm, S. M., Role of raf oncogenes in lung carcinogenesis, *J. Int. Assoc. Study Lung Cancer*, 4, 162, 1988c.

Rapp, U. R., Troppmair, J., Carroll, M., and May, S., The role of Raf-1 protein kinase in IL-3 and GM-CSF-mediated signal transduction, *Curr. Top. Microbiol. Immunol.*, 166, 129, 1990.

Rapp, U. R., Troppmair, J., and Birrer, M., Differential sensitivity of transformed cells to growth inhibition by a dominant negative mutant of c-Jun, *Oncogene*, in press.

Reed, J. C., Yum, S., Cuddy, M. P., Turner, B. C., and Rapp, U. R., Differential regulation of the p72-74 Raf-1 kinase in 3T3 fibroblasts expressing ras or src oncogenes, *Cell Growth Differ.*, 2, 235, 1991.

Rim, M., Qureshi, S. A., Gius, D., Nho, J., Sukhatme, V. P., and Foster, D. A., Evidence that activation of the Egr-1 promoter by v-Raf involves serum response elements, *Oncogene*, 7, 2065, 1992.

Roussel, M. F., Cleveland, J. L., Shurtleff, S. A., and Sherr, C., Myc rescue of a mutant CSF-1 receptor impaired in mitogenic signaling, *Nature*, 353, 361, 1991.

Roux, P., Blanchard, J., Fernandez, A., Lamb, N., Jeanteur, P., and Peichaczyk, M., Nuclear localization of c-Fos, but not v-Fos proteins is controlled by extracellular signals, *Cell*, 63, 341, 1990.

Rovera, G., Santoli, D., and Damsky, C., Human promyeolcytic leukemia cells in culture differentiate into macrophage-like cells when treated with a phorbol diester, *Proc. Natl. Acad. Sci. U.S.A.*, 76, 2779, 1979.

Rozakis-Adcock, M., McGlade, J., Mbamalu, G., Pelicci, G., Daly, R., Li, W., Thomas, S., Brugge, J., Pelicci, P. G., Schlessinger, J., and Pawson, T., Association of the Shc and Grb2/Sem5 SH2-containing proteins is implicated in activation of the Ras pathway by tyrosine kinases, *Nature*, 360, 689, 1992.

Rozakis-Adcock, M., Fernely, R., Wade, J., Pawson, T., and Bowtell, D., The SH2 and SH3 domains of mammalian Grb2 couple the EGF receptor to the Ras activator mSos1, *Nature*, 363, 83, 1993.

Sassone-Corsi, P., Sisson, J. C., and Verma, I. M., Transcriptional autoregulation of the proto-oncogene Fos, *Nature*, 334, 314, 1988.

Sassone-Corsi, P., Der, C. J., and Verma, I. M., ras-induced neuronal differentiation of PC12 cells: Possible involvement of fos and jun, *Mol. Cell. Biol.*, 9, 3174, 1989.

Sawyers, C., Callahan, W., and Witte, O. N., Dominant negative myc blocks transformation by abl oncogenes, *Cell*, 70, 901, 1992.

Seth, A., Gonzalez, F. A., Gupta, S., Raden, D. L., and Davis, R. J., Signal transduction within the nucleus by mitogen-activated protein kinase, *J. Biol. Chem.*, 3267, 24796, 1992.

Settleman, J., Naraimhan, V., Foster, L. C., and Weinberg, R. A., Molecular cloning of cDNAs encoding the GAP-associated protein p190: Implication for a signaling pathway from Ras to the nucleus, *Cell*, 69, 539, 1992.

Shaw, P. E., Schroter, H., and Nordheim, A., The ability of a ternary complex to form over the serum response element correlates with serum inducibility of the human c-fos promoter, *Cell*, 56, 563, 1989.

Shou, C., Farnsworth, C. L., Neel, B. G., and Feig, L. A., Molecular cloning of cDNAs encoding a guanine-nucleotide-releasing factor for Ras p21, *Nature*, 358, 351, 1992.

Siegel, J. N., June, C. H., Yama, H., Rapp, U. R., and Samelson, L. E., Rapid activation of c-Raf-1 after stimulation of the T-cell receptor or the muscarinic receptor type i in resting T cells, *J. Immunol.*, 151, 4116, 1993.

Siegfried, Z. and Ziff, E. B., Altered transcriptional activity of c-fos promoter plasmids in v-raf transformed cells, *Mol. Cell. Biol.*, 10, 6073, 1990.

Sithanandam, G., Kolch, W., Duh, F. M., and Rapp, U. R., Complete coding sequence of human B-raf cDNA and detection of B-Raf protein kinase with isozyme specific antibodies, *Oncogene*, 5, 1775, 1990.

Smeal, T., Binétruy, B., Mercola, D. A., Birrer, M., and Karin, M., Oncogenic and transcriptional cooperation with Ha-Ras requires phosphorylation of c-Jun on serines 63 and 73, *Nature*, 354, 494, 1991.

Smeal, T., Binétruy, B., Mercola, D., Grover-Bardwick, A., Heidecker, G., Rapp, U. R., and Karin, M., Oncoprotein-mediated signaling cascade stimulates c-Jun activity by phosphorylation of serines 63 and 73, *Mol. Cell. Biol.*, 12, 3507, 1992.

Smith, M. R., Heidecker, G., Rapp, U. R., and Kung, H. F., Induction of transformation and DNA synthesis after microinjection of raf proteins, *Mol. Cell. Biol.*, 7, 3828, 1990.

Sozeri, O., Vollmer, K., Liyanage, M., Frith, D., Kour, G., Mark, G., III, and Stabel, S., Activation of the c-Raf protein kinase by protein kinase C, *Oncogene*, 7, 2259, 1992.

Stephens, R. M., Sithanandam, G., Copeland, T. D., Kaplan, D. R., Rapp, U. R., and Morrison, D. K., 95-kilodalton B-Raf serine/threonine kinase: Identification of the protein and its major autophosphorylation site, *Mol. Cell. Biol.*, 12, 3733, 1992.

Stern, D. F., Robberts, A. B., Roche, N. S., Sporn, M. B., and Weinberg, R. A., Differential responsiveness of myc- and ras-transformed cells to growth factors: Selective stimulation of myc-transformed cells by epidermal growth factor, *Mol. Cell. Biol.*, 6, 870, 1986.

Stokoe, D., Campbell, D. G., Nakielny, S., Hidaka, H., Leevers, S. J., Marshall, C., and Cohen, P., MAPKAP kinase-2: A novel protein kinase activated by mitogen-activated protein kinase, *EMBO J.*, 11, 3985, 1992.

Stokoe, D., Macdonald, S. G., Cadwallader, K., Symons, M., and Hancock, J. F., Activation of Raf as a result of recruitment to the plasma membrane. *Science*, 264, 1463, 1994.

Storm, S. M. and Rapp, U. R., Oncogene activation: c-raf-1 gene mutations in experimental and naturally occurring tumors, *Toxicol. Lett.*, 67, 201, 1993.

Storm, S. M. and Rapp, U. R., Consistent raf-1 point mutations in ENU induced lung tumors and lymphomas, submitted.

Storm, S. M., Brennscheidt, U., Sithanandam, G., and Rapp, U. R., Raf oncogenes in carcinogenesis, *CRC Crit. Rev. Carcinogenesis*, 2, 1, 1990.

Sturgill, T. W., Ray, L. B., Erickson, E., and Maller, J. L., Insulin-stimulated MAP-2 kinase phosphorylates and activates ribosomal protein S6 kinase II, *Nature*, 334, 715, 1988.

Sukhatme. V. P.. Kartha. S., Toback. F. G.. Taub. R.. Hoover. R. G.. and Tsai-Morris, C.-H.. A novel early gene rapidly induced in fibroblasts. epithelial. and lymphocytes mitogens. *Oncogene Res.,* 1, 343, 1987.

Sukhatme. V. P.. Cao. X.. Chang. L. C.. Tsai-Morris. C.-H.. Stamenkovich. D.. Ferreira. P. C. P.. Choen. D. R.. Edwards. S. A.. Curran. T.. Lebeau. M. M.. and Adamson. E. D.. A zinc-finger encoding gene coregulated with c-Fos during growth and differentiation and after depolarization. *Cell,* 53. 37. 1988.

Szeberenyi. J., Cai. H.. and Cooper. G. M.. Effect of a dominant inhibitory Ha-ras mutation on neuronal differentiation in PC12 cells. *Mol. Cell. Biol.,* 10. 5324. 1990.

Teague. M. A.. Chaleff. D. T.. and Errede. B.. Nucleotide sequence of the yeast regulatory gene STE7 predicts a protein homologous to protein kinases. *Proc. Natl. Acad. Sci. U.S.A.,* 83, 7371. 1986.

Thomas. S. M.. DeMarco. M.. D'Arcangelo. G., Halegoua. S., and Brugge. J. S.. Ras is essential for nerve growth factor- and phorbol ester-induced tyrosine phosphorylation of MAP kinase. *Cell,* 68. 1031. 1992.

Toda. T.. Shimanuki. M., and Yanagida. M., Fission yeast genes that confer resistance to staurosporine encode an AP-1 like transcription factor and a protein kinase related to the mammalian ERK/MAP2 and budding yeast FUS3 and KSS1 kinases. *Genes Dev.,* 5. 60. 1991.

Trahey. M.. Wong. G.. Halenbeck. R.. Rubinfeld. B.. Martin. G. A.. Ladner. M.. Long. C. M.. Crosier. W. J.. Watt. K.. Koths. K.. and McCormick. F.. Molecular cloning of two types of GAP complementary DNA from human placenta. *Science,* 242. 1697. 1988.

Treisman. R.. The serum response element. *TIBS,* 17. 423. 1992.

Troppmair. J.. Potter. M.. Wax. J. S.. and Rapp. U. R.. An altered v-raf is required in addition to v-myc in J3V1 virus for acceleration of murine plasmacytomagenesis. *Proc. Natl. Acad. Sci. U.S.A.,* 86. 9941. 1989.

Troppmair. J.. Bruder. J. T.. App. H.. Cai. H.. Liptak. L.. Szeberenyi. J.. Cooper. G. M.. and Rapp. U. R.. Ras controls coupling of growth factor receptors and protein kinase C in the membrane to Raf-1 and B-Raf protein serine kinases in the cytosol. *Oncogene,* 7. 1867. 1992a.

Troppmair. J.. Cleveland. J. L.. Askew. D. S.. and Rapp. U. R.. v-Raf/v-Myc synergism in abrogation of IL-3 dependence: v-Raf suppresses apoptosis. *Curr. Top. Microbiol. Immunol.,* 182. 453. 1992b.

Troppmair. J.. Bruder. J. T.. Munoz. H.. Lloyd. P. A.. Kyriakis. J. M.. Banerjee. P.. Avruch. J.. and Rapp. U. R.. Mitogen-activated protein kinase/extracellular signal-regulated protein kinase activation by oncogenes. serum. and 12-O-tetradecanoylphorbol-13-acetate requires Raf and is necessary for transformation. *J. Biol. Chem.,* 269. 7030. 1994.

Tsujimoto. Y.. Cossman. J.. Jaffe. E.. and Croce. C. M.. Involvement of the bcl-2 gene in human follicular lymphoma. *Science,* 228. 1440. 1985.

Turner. B. C.. Rapp. U. R.. App. H.. Green. M.. Dobashi. K.. and Reed. J. C.. Interleukin-2 (IL-2) induces tyrosine phosphorylation and activation of p72-74 Raf-1 kinase in T cell line lymphocytes. *Proc. Natl. Acad. Sci. U.S.A.,* 88. 1227. 1991.

Van Aelst. L.. Barr. M.. Marcus. S.. Polverino. A.. and Wigler. M.. Complex formation between RAS and RAF and other protein kinases. *Proc. Natl. Acad. Sci. U.S.A.,* 90. 6213. 1993.

Vogel. U. S.. Dixon. R. A. F.. Schaber. M. D.. Diehl. R. E.. Marshall. M. S.. Sconick. E. M.. Sigal. I. S.. and Gibbs. J. B.. Cloning of bovine GAP and its interaction with oncogenic ras p21, *Nature,* 335. 90. 1988.

Vogt. P. K. and Bos. T. J.. jun: Oncogene and transcription factor, *Adv. Cancer Res.,* 55. 1. 1990.

Wang, Y., Xu, H., Riggs, M., Rodgers, L., and Wigler, M., byr2, a Schizosaccharomyces pombe gene encoding a protein kinase capable of partial suppression of ras1 mutant phenotype, *Mol. Cell. Biol.*, 11, 3554, 1991.

Wasylyk, C., Flores, P., Gutman, A., and Wasylyk, B., PEA3 is a nuclear target for transcription activation by non-nuclear oncogenes, *EMBO J.*, 8, 3371, 1989a.

Wasylyk, C., Wasylyk, B., Heidecker, G., Huleihel, M., and Rapp, U. R., Expression of raf oncogenes activates the PEA1 transcription factor motif, *Mol. Cell. Biol.*, 9, 2247, 1989b.

Wasylyk, B., Wasylyk, C., Flores, P., Begue, A., Leprince, D., and Stehelin, D., The c-ets proto-oncogenes encode transcription factors that cooperate with c-Fos and c-Jun for transcriptional activation, *Nature*, 346, 191, 1990.

Wilson, T. and Treisman, R., Fos C-terminal mutations block downregulation of c-Fos transcription following serum stimulation, *EMBO J.*, 7, 4193, 1988.

Winitz, S., Russell, M., Quian, N., Gardner, A., Dwyer, L., and Johnson, G. L., Involvement of Ras and Raf in the Gi-coupled acetylcholine muscarinic m2 receptor activation of mitogen-activated protein (MAP) kinase kinase and MAP kinase, *J. Biol. Chem.*, 268, 19196, 1993.

Wong, G., Müller, O., Clark, R., Conroy, L., Moran, M. F., Polakis, P., and McCormick, F., Molecular cloning and nucleic acid binding properties of the GAP-associated tyrosine phosphoprotein p62, *Cell*, 69, 551, 1992.

Wood, K. W., Sarnecki, C., Roberts, T. M., and Blenis, J., Ras mediates nerve growth factor receptor modulation of three signal-transducing protein kinases: MAP kinase, Raf-1, and RSK, *Cell*, 68, 1041, 1992.

Worland, P. L., Hampton, L. L., Thorgeissen, S. S., and Huggett, A. C., Development of an in vitro model of tumor progression using v-raf and v-raf/v-myc transformed rat liver epithelial cells: Correlations of tumorigenicity with the downregulation of specific proteins, *Mol. Carcinogenesis*, 3, 20, 1990.

Yamaguchi-Iwai, Y., Satake, M., Murakami, Y., Sakai, M., Muramatsu, M., and Ito, Y., Differentiation of F9 embryonal carcinoma cells induced by the c-jun and activated c-Ha-ras oncogenes, *Proc. Natl. Acad. Sci. U.S.A.*, 87, 8670, 1990.

Zhang, X., Settleman, J., Kyriakis, J. M., Suzuki, E. T., Elledge, S. J., Marshall, M. S., Bruder, J., Rapp, U. R., and Avruch, J., p21 ras binds to the amino terminal regulatory region of c-Raf-1, *Nature*, 364, 308, 1993.

Chapter 17

Oncogene and Tumor Suppressor Gene Products Regulating *c-fos* Expression

Axel Schönthal and James R. Feramisco

CONTENTS

SIGNIFICANCE OF c-*fos* IN SIGNAL TRANSDUCTION

Control of cellular growth and differentiation is a prerequisite for the proper development of higher eucaryotic organisms. Extracellular molecules, such as hormones or growth factors, are important agents in determining this control. The genetic response of cells to these molecules often requires signal receptors, signal transduction to the nucleus (second and third messengers), and usually alterations in transcription factor activity, which activates or represses the respective target genes. The importance of adequate regulation of these signal transduction pathways has been emphasized by the finding that many protein products of proto-oncogenes are components of this network. If mutated or inappropriately expressed they may become activated oncoproteins that no longer underlie the fine-tuned external control mechanisms but are able to cause unrestricted cellular growth and tumor formation (Cooper, 1990; Franks and Teich, 1991).

Great effort is being made to identify components of these signal transduction pathways, and their hierarchical order and interactions are targets for intense scrutiny (Watson, 1988). The characterization of promoter elements regulated by these pathways is important because it allows studies to trace backward from the gene to the original input signal; e.g., the promoter element aids in identifying transcription factors binding to this element, then the regulation of transcription factor activity by other, more distal components can be studied, and so forth. On the other end, at the signal input side, studies are proceeding downstream; e.g., *sis*/PDGF binds to and activates its receptor; the PDGF receptor, in turn, associates with other molecules, such as *src*, *raf*, or GTPase activating protein (GAP), which interacts with ras, and these components transduce the signal further to the nucleus (Cross and Dexter, 1991). Ultimately, both research directions will meet and reveal the complete cascade of events in cellular signal transduction. Undoubtedly, the discovery of the *fos* gene was extremely helpful in these studies not only because this gene is the target of so many different external stimuli that are transduced by various signal transduction pathways, but also the promoter of this gene has proven to be a rich source of response elements that are and will be valuable tools in the retrograde analysis of signal transduction. Since c-*fos* gene expression is rapidly stimulated in response to external stimuli, and since Fos/Jun proteins (AP-1) act as specific transcription factors to regulate expression of "late" genes, it is generally assumed that the product of the c-*fos* gene is one of the components that play a decisive role in converting incoming short-term signals

0-8493-4573-1/94/$0.00+$.50
© 1994 by CRC Press, Inc.

into long-term responses such as proliferation or differentiation. In this chapter we will focus on known (proto-)oncogene and tumor suppressor gene products that regulate c-*fos* gene expression, and describe promoter elements that are responsive to their activity.

POSITIVE REGULATORY PROTEINS

Several approaches are being used to identify proteins that are able to regulate c-*fos* gene expression. For example, purified growth factors or cytokines are added into the culture medium of cells and then c-*fos* expression is analyzed. In this way, platelet-derived growth factor (PDGF) (Cochran et al., 1984; Kruijer et al., 1984; Stumpo and Blackshear, 1986; Siegfried and Ziff, 1989; Salhany et al., 1992), fibroblast growth factor (FGF) (Stumpo and Blackshear, 1986; Ito et al., 1990), epidermal growth factor (EGF) (Bravo et al., 1985; Greenberg et al., 1985; Fisch et al., 1987; Ito et al., 1990; McDonnell et al., 1990; Sagar et al., 1991), nerve growth factor (NGF) (Greenberg et al., 1985; Visvader et al., 1988; Bartel et al., 1989; Ito et al., 1990), growth hormone (Doglio et al., 1989; Slootweg et al., 1990, 1991), insulin (Stumpo and Blackshear, 1986; Stumpo et al., 1988; Messina, 1990; Messina et al., 1992), colony stimulating factor 1 (CSF-1) (Nakamura et al., 1991), and several interleukins (ILs) (Fagarasan et al., 1990; Trouche et al., 1991; Körholz et al., 1992) have been found to stimulate c-*fos* expression. The discovery of the PDGF-2 gene as the cellular counterpart of the v-*sis* oncogene of simian sarcoma virus (SSV) provided early evidence of oncoprotein-regulated gene expression (Doolittle et al., 1983; Waterfield et al., 1983). Other examples are inferred from the homology of the CSF-1 receptor to the fms proto-oncogene product (Sherr et al., 1985), from the close similarity of *erbB*-2 to the EGF receptor (Downward et al., 1984; Yamamoto et al., 1986), or from the identification of int-2, hst, and fgf-5 proto-oncogenes as FGF family members (Bovi et al., 1987; Yoshida et al., 1987; Delli-Bovi et al., 1988; Zhan et al., 1988). In addition, IL-2 and CSF-1 have exhibited oncogenic potential in certain cells (Chen et al., 1985; Lang et al., 1985; Baumbach et al., 1988).

Co-transfection has been widely used to analyze potential c-*fos* regulation by oncoproteins, in more detail. Usually, an (oncogene) expression vector is transfected together with a plasmid bearing c-*fos* promoter sequences fused to a reporter gene, such as chloramphenicol-acetyltransferase (CAT). The level of CAT expression reflects the activity of the transfected c-*fos* promoter in the presence of the cotransfected oncogene. This method allows for a variety of manipulations that help to analyze signal transduction in more detail. On the upstream, signal side, the expression plasmid can be mutated to evaluate the importance of various domains of the oncoprotein for the observed effects. On the downstream, receiving side, mutations of the c-*fos* promoter allow the definition of response elements that may be regulated by binding of specific transcription factors.

A multitude of these studies established the regulation of c-*fos* gene expression by a variety of proto-oncoproteins. These include cellular proto-oncoproteins and their retroviral counterparts, for example, sis/PDGF, *erbB*-2, Ha-ras, *src, fps, raf,* and *mos* (Schönthal et al., 1988a, 1988b; Fujii et al., 1989; Kaibuchi et al., 1989; Sassone-Corsi et al., 1989; Siegfried and Ziff, 1989; Jamal and Ziff, 1990; Fujimoto et al., 1991; Jähner and Hunter, 1991; Jehn et al., 1992); the tax-1 protein of human T cell leukemia virus type-1 (HTLV-1) (Fujii et al., 1988, 1991, 1992; Alexandre and Verrier, 1991); and oncoproteins of DNA tumor viruses, such as E1A of adenovirus or middle T antigen of polyomavirus (Py-mT) (Sassone-Corsi and Borrelli, 1987; Offringa et al., 1990; Simon et al., 1990; de Groot et al., 1991; Kitabayashi et al., 1991; Schönthal et al., 1992b; Glenn and Eckhart, 1993).

Multiple regulatory elements in the c-*fos* promoter have been identified, most notably the serum response element (SRE), the major mediator of growth factor stimulation (Deschamps et al., 1985; Treisman, 1985, 1986; Gilman et al., 1986; Gilman, 1988;

Leung and Miyamoto, 1989). Immediately downstream of the SRE there is a *fos* AP-1 (FAP) site that is highly homologous to a consensus 12-*O*-tetradecanoylphorbol-13-acetate (TPA) responsive element (TRE) and to a cyclic AMP (cAMP) responsive element (CRE) (Fisch et al., 1989a; Velcich and Ziff, 1990). Occupation of the TRE and the SRE seem mutually exclusive (König et al., 1989). The region further downstream includes several stretches of sequences that are homologous to the AP-2 binding site of the human metallothionein IIA gene (Imagawa et al., 1987; Fisch et al., 1989). Their significance for c-*fos* regulation, however, has not yet been studied in detail. Another regulatory site is the *sis* inducible element (SIE), which responds to conditioned medium from v-*sis* transformed cells and PDGF (Hayes et al., 1987; Wagner et al., 1990). In addition, there are multiple elements regulating the response to cAMP (CREs) (Fisch et al., 1987, 1989a; Sassone-Corsi et al., 1988b; Berkowitz et al., 1989; Härtig et al., 1991) or estrogen (EREs) (Weisz and Rosales, 1990; Hyder et al., 1992). Another element, the retinoblastoma control element (RCE), is regulated by the product of the retinoblastoma (Rb) tumor suppressor gene (Robbins et al., 1990; Kim et al., 1991). Recent reports provide evidence of at least nine different elements, including the SRE, a CRE, and a NF-1/CTF consensus sequence, that figure in basal level transcription of the c-*fos* promoter (Lucibello et al., 1991; Runkel et al., 1991). It is important to note that several of the above-mentioned elements act in a cell type-specific manner, and some exhibit positive and negative effects (Lucibello et al., 1991). The precise function of these sites appears to depend on the activity of neighboring elements, indicating some form of interaction between different regulatory sites (see below).

Which c-*fos* promoter elements are regulated by oncoproteins? Not surprisingly, the SRE appears to be a prime target. It is activated by *sis*/PDGF (Gilman, 1988; Wagner et al., 1990), *erbB*-2 (Fujimoto et al., 1991), Ha-ras (Fukumoto et al., 1990; Gauthier-Rouvièr et al., 1990; Gutman et al., 1991), *src* (Fujii et al., 1989), *raf* (Kaibuchi et al., 1989; Jamal and Ziff, 1990), *mos* (Schönthal and Feramisco, 1990), tax-1 (Alexandre and Verrier, 1991; Fujii et al., 1992) and Py-mT (Glenn and Eckhart, 1993). Usually, the viral or activated cellular oncoproteins are more potent activators than the normal cellular proto-oncoproteins, indicating a stronger effect on signal transduction pathways. As mentioned before, there appear to be cell type-specific requirements for induction. For example, in NIH 3T3 cells the inner core of the SRE, the CArG box, is sufficient for transactivation by Ha-ras, whereas in HeLa cells the complete SRE is required (Gutman et al., 1991). Moreover, by transient cotransfection analysis it has been shown that TPA, an activator of protein kinase C (PKC) pathways, activates the SRE in HeLa, but not in NIH 3T3 cells (Siegfried and Ziff, 1989). However, using stably transfected NIH 3T3 cells, another group did demonstrate TPA inducibility of the SRE in these cells (Büscher et al., 1988). These differences may be accounted for by the different methods used: in HeLa cells it has been shown that TPA activates the FAP site of stably integrated *fos*–CAT fusion genes, whereas in transient transfection assays this element has been reported to be nonresponsive (Fisch et al., 1989a).

In addition to the SRE, other elements of the c-*fos* promoter appear to play a role in its regulation by oncoproteins. The tax-1 protein acts through at least four regulatory elements, the SRE, SIE, CRE, and the octanucleotide direct repeat element (DR) (Alexandre and Verrier, 1991). In the SRE the CArG box by itself is sufficient for this transactivation (Fujii et al., 1992). In the case of E1A-dependent transactivation of the c-*fos* promoter, the primary target has been identified as the TATA element (Simon et al., 1990). Other experiments suggest some inducibility of a promoter region containing the SRE as well as cooperation with the AMP pathway through the CRE (Müller et al., 1989; Velcich and Ziff, 1990; Engel et al., 1991). In addition to transactivating the SRE, *sis*/PDGF also stimulates the SIE. For efficient induction of the SIE, however, cooperation with other sequences located between −100 and −57 seems to be required. Similarly, the combination of the SIE with the SRE results in additive transactivation in response to *sis*/PDGF (Wagner et al., 1990). Multiple response elements have also been described as targets for Py-mT (Glenn and Eckhart, 1993), *src* (Fujii et al., 1989) and *raf* (Jamal and Ziff, 1990; Siegfried and Ziff, 1990) oncoprotein activity.

In general, it appears as though no single response element mediates the full response of c-*fos* to a defined stimulus. Although some isolated promoter elements by themselves are activated in response to a stimulus, in the context of the full-length promoter the interaction of multiple elements may be required for the appropriate level of c-*fos* expression. Furthermore, this analysis is complicated by the presence of positive as well as negative elements. For example, c-*fos* promoter sequences from −700 to +42 are stimulated approximately 23-fold by cotransfected v-*raf* expression plasmid (Jamal and Ziff, 1990). A shorter c-*fos* construct from −225 to +42 is even more inducible (37-fold), although the deleted sequences harbor the SRE (at −319 to −297) that by itself is strongly inducible by viral and activated c-*raf* (Jamal and Ziff, 1990).

Most studies of c-*fos* promoter activity in response to certain stimuli analyze isolated fragments of the promoter. While this approach is certainly helpful for an approximate assessment of the contribution of the respective response element, more precise information could be obtained using full-length promoter constructs where only the element(s) of interest is inactivated by point mutations. This would reveal the importance of a defined element in the background of the other regulatory sites. Although constructs of this type have been analyzed, the size of the c-*fos* promoter adds another constraint. The majority of published papers studied c-*fos* sequences between −700 and +42 and revealed at least 12 regulatory promoter elements. The recent identification of an ERE located between −1300 and −1060, and of another ERE located 1500 bp downstream of the poly(A) signal, suggests that there may be many more response elements contributing to c-*fos* gene expression (Weisz and Rosales, 1990; Hyder et al., 1992).

NEGATIVE REGULATORY PROTEINS

Repression of c-*fos* promoter activity appears to be a complex mechanism as well (Sassone-Corsi and Verma, 1987). First, the gene product itself, c-*fos* protein as part of transcription factor AP-1, downregulates elevated c-*fos* transcription (Sassone-Corsi et al., 1988a; Schönthal et al., 1988b, 1989; Wilson and Treisman 1988; Lucibello et al., 1989; Ofir et al., 1990). This mechanism most likely contributes to the transient kinetics of c-*fos* expression observed in response to various stimuli such as serum or overexpression of oncoproteins. The CArG box of the SRE appears to be sufficient for the negative autoregulatory effect (König et al., 1989; Shaw et al., 1989a; Subramaniam et al., 1989; Gius et al., 1990; Rivera et al., 1990). Curiously, it has also been shown that serum induction of the SRE can be repressed by the CRE binding (CREB) protein (Ofir et al., 1991).

In addition, negative regulation of c-*fos* expression by the protein products of tumor suppressor genes has been described. One such protein, wild-type p53, but not a transforming mutant of p53, has been shown to repress c-*fos* promoter activity via a region between nucleotides −53 and +42 (Ginsberg et al., 1991; Kley et al., 1992). Another protein of this class, Rb, has been found to downregulate c-*fos* expression via the RCE (Robbins et al., 1990). However, this latter effect appears to be highly cell type specific. Whereas in some mouse fibroblast cell lines Rb acts inhibitory, it stimulates c-*fos* expression in mink lung epithelial or human lung adenocarcinoma cells (Kim et al., 1991).

By virtue of their capacity to downregulate c-*fos* expression, other proteins are being considered as negative regulators of cellular growth. For example, the CRE modulator (CREM) protein has been shown to bind to the CRE and downregulate basal and cAMP-stimulated c-*fos* expression (Foulkes et al., 1991a, 1991b). The use of drugs that specifically block certain cellular components allows the analysis of other potentially regulatory proteins. For instance, treatment of cells with okadaic acid (Kim et al., 1990; Schönthal et al., 1991a, 1991c, 1992a; Thévenin et al., 1991), an inhibitor of serine/threonine specific phosphoprotein phosphatases type 1 and 2A (PP-1, PP-2A), or with thapsigargin (Schönthal et al., 1991b), an inhibitor of intracellular membrane Ca^{2+}-ATPases, or with ouabain/amiloride (Shibanuma et al., 1987; Nakagawa et al., 1992), inhibitors of the Na/K-ATPase, resulted in elevated c-fos expression via the SRE. These studies provided indirect evidence that certain phosphatases and ion pumps may act as negative regulators of c-*fos* expression.

More direct evidence has been derived from experiments where purified proteins were microinjected into living cells. To study phosphatase involvement in c-*fos* regulation, Alberts et al. (1992) microinjected the catalytic subunit of PP-2A into cells that harbored an SRE reporter construct. Under these conditions, serum stimulation of the SRE was diminished, indicating a negative regulation by the serine/threonine protein phosphatase. This result complements other studies that showed induction of SRE activity in response to microinjection of casein kinase II (Gauthier-Rouvièr et al., 1991), a ubiquitous serine/threonine protein kinase that phosphorylates and, thereby, activates the serum response factor (SRF), the principal regulatory protein binding to the SRE (Prywes and Roeder, 1986; Treisman, 1986; Shaw et al., 1989b; Schröter et al., 1987; Hipskind and Nordheim, 1991). Previous experiments using the microinjection technique had also directly demonstrated c-*fos* induction by oncogenic ras protein (Stacey et al., 1987; Gauthier-Rouvièr et al., 1990).

CONCLUSION

In summary, the above-described studies have pointed to complex relationships of promoter elements in the regulation of c-*fos* expression, and have provided a framework within which the molecular pathways governed by oncoproteins regulate this expression. We expect that further refinement of the signal transduction components of these pathways will come from future studies and lead to an understanding of the rationale behind the complex regulatory networks and, in turn, of the full consequences of the expression of c-*fos*.

REFERENCES

Alberts, A. S., Deng, T., Lin, A., Meinkoth, J. L., Schönthal, A., Mumby, M. C., Karin, M., and Feramisco, J. R., Protein phosphatase 2A potentiates activity of promoters containing AP-1 binding elements, *Mol. Cell Biol.*, 13, 2104, 1993.

Alexandre, C. and Verrier, B., Four regulatory elements in the human c-*fos* promoter mediate transactivation by HTLV-1 Tax protein, *Oncogene*, 6, 543, 1991.

Bartel, D. P., Sheng, M., Lau, L. F., and Greenberg, M. E., Growth factors and membrane depolarization activate distinct programs of early response gene expression: Dissociation of *fos* and *jun* induction, *Genes Dev.*, 3, 304, 1989.

Baumbach, W. R., Colston, E. M., and Cole, M. D., Integration of the BALB/c ecotropic provirus into the colony-stimulating factor-1 growth factor locus in a myc retrovirus-induced murine monocyte tumor, *J. Virol.*, 62, 3151, 1988.

254

Berkowitz, L. A., Riabowol, K. T., and Gilman, M. Z., Multiple sequence elements of a single functional class are required for cyclic AMP responsiveness of the mouse c-*fos* promoter, *Mol. Cell. Biol.*, 9, 4272, 1989.

Bovi, P. D., Curatola, A. M., Kern, F. G., Greco, A., Ittmann, M., and Basilico, C., An oncogene isolated by transfection of Kaposi's Sarcoma DNA encodes a growth factor that is a member of the FGF family, *Cell*, 50, 729, 1987.

Bravo, R., Burckhardt, J., Curran, T., and Müller, R., Stimulation and inhibition of growth by EGF in different A431 cell clones is accompanied by the rapid induction of c-*fos* and c-myc proto-oncogenes, *EMBO J.*, 4, 1193, 1985.

Büscher, M., Rahmsdorf, H. J., Litfin, M., Karin, M., and Herrlich, P., Activation of the c-*fos* gene by UV and phorbol ester: Different signal transduction pathways converge to the same enhancer element, *Oncogene*, 3, 301, 1988.

Chen, S. J., Holbrook, N. J., Mitchell, K. F., Vallone, C. A., Greengard, J. S., Crabtree, G. R., and Lin, Y., A viral long terminal repeat in the interleukin 2 gene of a cell line that constitutively produces interleukin 2, *Proc. Natl. Acad. Sci. U.S.A.*, 82, 7284, 1985.

Cochran, B. H., Zullo, J., Verma, I. M., and Stiles, C. D., Expression of the c-*fos* oncogene and of *fos*-related gene stimulated by platelet-derived growth factor, *Science*, 266, 1080, 1984.

Cooper, J. M., *Oncogenes*, Jones and Bartlett, Boston, MA, 1990.

Cross, M. and Dexter, T. M., Growth factors in development, transformation, and tumorigenesis, *Cell*, 64, 271, 1991.

de Groot, R., Foulkes, N., Mulder, M., Kruijer, W., and Sassone-Corsi, P., Positive regulation of *jun*/AP-1 by E1A, *Mol. Cell. Biol.*, 11, 192, 1991.

Delli-Bovi, P., Curatola, A. M., Newman, K. M., Sato, Y., Moscatelli, D., Hewick, R. M., Rifkin, D. B., and Basilico, C., Processing, secretion, and biological properties of a novel growth factor of the fibroblast growth factor family with oncogenic potential, *Mol. Cell. Biol.*, 8, 2933, 1988.

Deschamps, J., Meijlink, F., and Verma, I. M., Identification of a transcriptional enhancer element upstream from the proto-oncogene *fos*, *Science*, 230, 1174, 1985.

Doglio, A., Dani, C., Grimaldi, P., and Ailhaud, G., Growth hormone stimulates c-*fos* gene expression by means of protein kinase C without increasing inositol lipid turnover, *Proc. Natl. Acad. Sci. U.S.A.*, 86, 1148, 1989.

Doolittle, R. F., Hunkapiller, M. W., Hood, L. E., Devare, S. G., Robbins, K. C., Aaronson, S. A., and Antoniades, H. N., Simian sarcoma virus Oncogene, v-*sis*, is derived from the gene (or genes) encoding a platelet-derived growth factor, *Science*, 221, 275, 1983.

Downward, J., Yarden, Y., Mayes, E., Scrace, G., Totty, N., Stockwell, P., Ullrich, A., Schlessinger, J., and Waterfield, M. D., Close similarity of epidermal growth factor receptor and v-erb-B oncogene protein sequences, *Nature*, 307, 521, 1984.

Engel, D. A., Muller, U., Gedrich, R. W., Eubanks, J. S., and Shenk, T., Induction of c-*fos* mRNA and AP-1 DNA-binding activity by cAMP in cooperation with either the adenovirus 243- or the adenovirus 289-amino acid E1A protein, *Proc. Natl. Acad. Sci. U.S.A.*, 88, 3957, 1991.

Fagarasan, M. O., Aiello, F., Muegge, K., Durum, S., and Axelrod, J., Interleukin 1 induces β-endorphin secretion via *Fos* and *Jun* in AtT-20 pituitary cells, *Proc. Natl. Acad. Sci. U.S.A.*, 87, 7871, 1990.

Fisch, T. M., Prywes, R., and Roeder, R. G., c-*fos* sequences necessary for basal expression and induction by epidermal growth factor, 12-O-tetradecanoylphorbol-13-acetate, and the calcium ionophore, *Mol. Cell. Biol.*, 7, 3490, 1987.

Fisch, T. M., Prywes, R., and Roeder, R. G., An AP1-binding site in the c-*fos* gene can mediate induction by epidermal growth factor and 12-O-tetradecanoyl phorbol-13-acetate, *Mol. Cell. Biol.*, 9, 1327, 1989a.

Fisch, T. M., Prywes, R., Simon, M. C., and Roeder, R. G., Multiple sequence elements in the c-*fos* promoter mediate induction by cAMP. *Genes Dev.*, 3, 198, 1989b.

Foulkes, N. S., Borrelli, E., and Sassone-Corsi, P., CREM gene: use of alternative DNA-binding domains generates multiple antagonists of cAMP-induced transcription. *Cell*, 64, 739, 1991.

Foulkes, N. S., Laoide, B. M., Schlotter, F., and Sassone-Corsi, P., Transcriptional antagonist cAMP-responsive element modulator (CREM) down-regulates c-*fos* cAMP-induced expression, *Proc. Natl. Acad. Sci. U.S.A.*, 88, 5448, 1991.

Franks, L. M. and Teich, N. M., *Introduction to the Cellular and Molecular Biology of Cancer*, Oxford University Press, Oxford, 1991.

Fujii, M., Sassone-Corsi, P., and Verma, I. M., c-*fos* promoter trans-activation by the tax₁ protein of human T-cell leukemia virus type I. *Proc. Natl. Acad. Sci. U.S.A.*, 85, 8526, 1988.

Fujii, M., Shalloway, D., and Verma, I. M., Gene regulation by tyrosine kinases: *src* protein activates various promoters, including c-*fos*, HIV-LTR, c-*myc*, *Mol. Cell. Biol.*, 9, 2493, 1989.

Fujii, M., Niki, T., Mori, T., Matsuda, T., Matsui, M., Nomura, N., and Seiki, M., HTLV-1 Tax induces expression of various immediate early serum responsive genes, *Oncogene*, 6, 1023, 1991.

Fujii, M., Tsuchiya, H., Chuhjo, T., Akizawa, T., and Seiki, M., Interaction of HTLV-1 Tax I with p67SRF causes the aberrant induction of cellular immediate early genes through CArG boxes, *Genes Dev.*, 6, 2066, 1992.

Fujimoto, A., Kai, S., Akiyama, T., Toyoshima, K., Kaibuchi, K., Takai, Y., and Yamamoto, T., Transactivation of the TPA-responsive element by the oncogenic c-*erbB*-2 protein is partly mediated by protein kinase C. *Biochem. Biophys. Res. Commun.*, 178, 724, 1991.

Fukumoto, Y., Kaibuchi, K., Oku, N., Hori, Y., and Takai, Y., Activation of the c-*fos* serum-response element by the activated c-Ha-*ras* protein in a manner independent of protein kinase C and cAMP-dependent protein kinase, *J. Biol. Chem.*, 265, 774, 1990.

Gauthier-Rouviér, C., Fernandez, A., and Lamb, N. J. C., *ras*-induced c-*fos* expression and proliferation in living rat fibroblasts involves C-kinase activation and the serum response element pathway, *EMBO J.*, 9, 171, 1990.

Gauthier-Rouviér, C., Basset, M., Blanchard, J.-M., Cavadore, J.-C., Fernandez, A., and Lamb, N. J. C., Casein kinase II induces c-*fos* expression via the serum response element pathway and p67SRF phosphorylation in living fibroblasts, *EMBO J.*, 10, 2921, 1991.

Gilman, M. Z., The c-*fos* serum response element responds to protein kinase C-dependent and -independent signals but not to cyclic AMP, *Genes Dev.*, 2, 394, 1988.

Gilman, M. Z., Wilson, R. N., and Weinberg, R. A., Multiple protein-binding sites in the 5′-flanking region regulate c-*fos* expression, *Mol. Cell. Biol.*, 6, 4305, 1986.

Ginsberg, D., Mechta, F., Yaniv, M., and Oren, M., Wild-type p53 can down-modulate the activity of various promoters, *Proc. Natl. Acad. Sci. U.S.A.*, 88, 9979, 1991.

Gius, D., Cao, X., Rauscher, F. J., III, Cohen, D. R., Curran, T., and Sukhatme, V. P., Transcriptional activation and repression by *fos* are independent functions: The C terminus represses immediate-early gene expression via CArG elements, *Mol. Cell. Biol.*, 10, 4243, 1990.

Glenn, G. M. and Eckhart, W., Polyomavirus middle T antigen activates transcription of the c-*fos* promoter through multiple response elements, submitted.

Greenberg, M. E., Greene, L. A., and Ziff, E. B., Nerve growth factor and epidermal growth factor induce rapid transient changes in proto-oncogene transcription in PC12 cells, *J. Biol. Chem.*, 260, 14101, 1985.

Gutman, A., Wasylyk, C., and Wasylyk, B., Cell-specific regulation of oncogene-responsive sequences of the c-*fos* promoter, *Mol. Cell. Biol.*, 11, 5381, 1991.

Härtig, E., Loncarevic, I. F., Büscher, M., Herrlich, P., and Rahmsdorf, H. J., A new cAMP response element in the transcribed region of the human c-fos gene, *Nucleic Acids Res.*, 19, 4153, 1991.

Hayes, T. E., Kitchen, A. M., and Cochran, B. H., Inducible binding of a factor to the c-*fos* regulatory region, *Proc. Natl. Acad. Sci. U.S.A.*, 84, 1272, 1987.

Hipskind, R. A. and Nordheim, A., Functional dissection *in vitro* of the human c-*fos* promoter, *J. Biol. Chem.*, 266, 19583, 1991.

Hyder, S. M., Stancel, G. M., Nawaz, Z., McDonnell, D. P., and Loose-Mitchell, D. S., Identification of an estrogen response element in the 3'-flanking region of the murine c-*fos* protooncogene, *J. Biol. Chem.*, 267, 18047, 1992.

Imagawa, M., Chiu, R., and Karin, M., Transcription factor AP-2 mediates induction by two different signal-transduction pathways: Protein kinase C and cAMP, *Cell*, 51, 251, 1987.

Ito, E., Sweterlitsch, L. A., Tran, P. B.-V., Rauscher, F. J., III, and Narayanan, R., Inhibition of PC-12 cell differentiation by the immediate early gene *fra*-1, *Oncogene*, 5, 1755, 1990.

Jähner, D. and Hunter, T., The stimulation of quiescent rat fibroblasts by v-*src* and v-*fps* oncogenic protein-tyrosine kinases leads to the induction of a subset of immediate early genes, *Oncogene*, 6, 1259, 1991.

Jamal, S. and Ziff, E., Transactivation of c-*fos* and β-actin genes by *raf* as a step in early response to transmembrane signals, *Nature*, 344, 463, 1990.

Jehn, B., Costello, E., Marti, A., Keon, N., Deane, R., Li, F., Friis, R. R., Burri, P. H., Martin, F., and Jaggi, R., Overexpression of *mos, ras, src,* and *fos* inhibits mouse mammary epithelial cell differentiation, *Mol. Cell. Biol.*, 12, 3890, 1992.

Kaibuchi, K., Fukumoto, Y., Oku, N., Hori, Y., Yamamoto, T., Toyoshima, K., and Takai, Y., Activation of the serum response element and 12-O-tetradecanoylphorbol-13-acetate response element by the activated c-*raf*-1 protein in a manner independent of protein kinase C, *J. Biol. Chem.*, 265, 20855, 1989.

Kim, S.-J., Lafyatis, R., Kim, K. Y., Angel, P., Fujiki, H., Karin, M., Sporn, M. B., and Roberts, A. B., Regulation of collagenase gene expression by okadaic acid, an inhibitor of protein phosphatases, *Cell Reg.*, 1, 269, 1990.

Kim, S.-J., Lee, H.-D., Robbins, P. D., Busam, K., Sporn, M. B., and Roberts, A. B., Regulation of transforming growth factor β1 gene expression by the product of the retinoblastoma-susceptibility gene, *Proc. Natl. Acad. Sci. U.S.A.*, 88, 3052, 1991.

Kitabayashi, I., Chiu, R., Gachelin, B., and Yokoyama, K., E1A dependent up-regulation of c-*jun*/AP-1 activity, *Nucleic Acids Res.*, 19, 649, 1991.

Kley, N., Chung, R. Y., Fay, S., Loeffler, J. P., and Seizinger, B. R., Repression of the basal c-*fos* promoter by wild-type p53, *Nucleic Acids Res.*, 20, 4083, 1992.

König, H., Ponta, H., Rahmsdorf, U., Büscher, M., Schönthal, A., Rahmsdorf, H. J., and Herrlich, P., Autoregulation of *fos:* The dyad symmetry element as the major target of repression, *EMBO J.*, 8, 2559, 1989.

Körholz, D., Gerdau, S., Enczmann, J., Zessack, N., and Burdach, S., Interleukin 6-induced differentiation of a human B cell line into IgM-secreting plasma cells is mediated by c-*fos*, *Eur. J. Immunol.*, 22, 607, 1992.

Kruijer, W., Cooper, J. A., Hunter, T., and Verma, I. M., Platelet-derived growth factor induces rapid but transient expression of the c-*fos* gene and protein, *Nature*, 312, 711, 1984.

Lang, R. A., Metcalf, D., Gough, N. M., Dunn, A. R., and Gonda, T. J., Expression of a hemopoietic growth factor cDNA in a factor-dependent cell line results in autonomous growth and tumorigenicity, *Cell*, 43, 531, 1985.

Leung. S. and Miyamoto, N. G., Point mutational analysis of the human c-*fos* serum response factor binding site, *Nucleic Acids Res.*, 17, 1177, 1989.

Lucibello, F. C., Lowag. C.. Neuberg, M., and Müller, R., Trans-repression of the mouse c-*fos* promoter: A novel mechanism of *fos*-mediated trans-regulation. *Cell*, 59, 999, 1989.

Lucibello, F. C., Ehlert, F., and Müller, R., Multiple interdependent regulatory sites in the mouse c-*fos* promoter determine basal level transcription: Cell type-specific effects, *Nucleic Acids Res.*, 19, 3583, 1991.

McDonnell, S. E., Kerr, L. D., and Matrisian. L. M., Epidermal growth factor stimulation of stromelysin mRNA in rat fibroblasts requires induction of proto-oncogenes c-*fos* and c-*jun* and activation of protein kinase C, *Mol. Cell. Biol.*, 10, 4284, 1990.

Messina. J. L., Insulin's regulation of c-*fos* gene transcription in hepatoma cells. *J. Biol. Chem.*, 265, 11700, 1990.

Messina. J. L., Standaert, M. L., Ishizuka, T., Weinstock, R. S., and Farese, R. V., Role of protein kinase C in insulin's regulation of c-*fos* transcription. *J. Biol. Chem.*, 267, 9223, 1992.

Müller, U., Roberts, M. P., Engel, D. A., Doerfler, W., and Shenk. T., Induction of transcription factor AP-1 by adenovirus E1A protein and cAMP, *Genes Dev.*, 3, 1991, 1989.

Nakagawa, Y., Rivera, V., and Larner, A. C.. A role for the Na/K-ATPase in the control of human c-*fos* and c-*jun* transcription, *J. Biol. Chem.*, 267, 8785, 1992.

Nakamura. T., Datta, R., Kharbanda. S., and Kufe. D., Regulation of *jun* and *fos* gene expression in human monocytes by the macrophage colony-stimulating factor. *Cell Growth Differ.*, 2, 267, 1991.

Offringa, R. Gebel. S., van Dam, H., Timmers, M., Smits. A., Zwart, R., Stein, B., Bos, L., van der Eb. A., and Herrlich, P., A novel function of the transforming domain of E1a: Repression of AP-1 activity, *Cell*, 62, 527, 1990.

Ofir, R., Dwarki, V. J., Rashid. D., and Verma. I. M., Phosphorylation of the C terminus of Fos protein is required for transcriptional transrepression of the c-*fos* promoter, *Nature*, 348, 80, 1990.

Ofir, R., Dwarki, V. J., Rashid. D., and Verma, I. M., CREB represses transcription of *fos* promoter: role of phosphorylation. *Gene Exp.*, 1, 55, 1991.

Prywes. R. and Roeder, R. G., Inducible binding of a factor to the c-*fos* enhancer, *Cell*, 47, 777, 1986.

Rivera, V. M., Sheng, M., and Greenberg, M. E., The inner core of the serum response element mediates both the rapid induction and subsequent repression of c-*fos* transcription following serum stimulation. *Genes Dev.*, 4, 255, 1990.

Robbins, P. D., Horowitz, J. M., and Mulligan, R. C., Negative regulation of human c-*fos* expression by the retinoblastoma gene product, *Nature*, 346, 668, 1990.

Runkel, L., Shaw, P. E., Herrera, R. E., Hipskind, R. A., and Nordheim, A., Multiple basal promoter elements determine the level of human c-*fos* transcription, *Mol. Cell. Biol.*, 11, 1270, 1991.

Sagar, S. M., Edwards, R. H., and Sharp, F. R., Epidermal growth factor and transforming growth factor α induce c-*fos* gene expression in retinal Muller cells *in vivo*, *J. Neurol. Res.*, 29, 549, 1991.

Salhany, K. E., Robinson-Benion, C., Candia, A. F., Pledger. W. J., and Holt, J. T., Differential induction of the c-*fos* promoter through distinct PDGF receptor-mediated signaling pathways, *J. Cell. Physiol.*, 150, 386, 1992.

Sassone-Corsi, P. and Borrelli, E., Promoter trans-activation of protooncogenes c-*fos* and c-*myc*, but not c-Ha-*ras*, by products of adenovirus early region 1A, *Proc. Natl. Acad. Sci. U.S.A.*, 84, 6430, 1987.

Sassone-Corsi, P. and Verma. I. M., Modulation of c-*fos* gene transcription by negative and positive cellular factors, *Nature*, 326, 507, 1987.

Sassone-Corsi, P., Sisson, J. C., and Verma, I. M., Transcriptional autoregulation of the proto-oncogene *fos*, *Nature*, 334, 314, 1988a.

Sassone-Corsi, P., Visvader, J., Ferland, L., Mellon, P. L., and Verma, I. M., Induction of proto-oncogene *fos* transcription through the adenylate cyclase pathway: Characterization of a cAMP-responsive element, *Genes Dev.*, 2, 1529, 1988b.

Sassone-Corsi, P., Der, C. J., and Verma, I. M., *ras*-induced neuronal differentiation of PC12 cells: Possible involvement of *fos* and *jun*, *Mol. Cell. Biol.*, 9, 3174, 1989.

Schönthal, A. and Feramisco, J. R., Different promoter elements are required for the induced expression of c-*fos* and c-*jun* proto-oncogenes by the v-mos oncogene product, *New Biol.*, 2, 143, 1990.

Schönthal, A., Gebel, S., Stein, B., Ponta, H., Rahmsdorf, H. J., and Herrlich, P., Nuclear oncoproteins determine the genetic program in response to external stimuli, *Cold Spring Harbor Symp. Quant. Biol.*, 53, 779, 1988a.

Schönthal, A., Herrlich, P., Rahmsdorf, H. J., and Ponta, H., Requirement for *fos* gene expression in the transcriptional activation of collagenase by other oncogenes and phorbol esters, *Cell*, 54, 325, 1988b.

Schönthal, A., Büscher, M., Angel, P., Rahmsdorf, H. J., Ponta, H., Hattori, K., Chiu, R., Karin, M., and Herrlich, P., The Fos and Jun/AP-1 proteins are involved in the downregulation of Fos transcription, *Oncogene*, 4, 629, 1989.

Schönthal, A., Alberts, A. S., Frost, J. A., and Feramisco, J. R., Differential regulation of jun family gene expression by the tumor promoter okadaic acid, *New Biol.*, 3, 977, 1991a.

Schönthal, A., Sugarman, J., Brown, J. H., Hanley, M. R., and Feramisco, J. R., Regulation of c-*fos* and c-jun protooncogene expression by the Ca^{2+}-ATPase inhibitor thapsigargin, *Proc. Natl. Acad. Sci. U.S.A.*, 88, 7096, 1991b.

Schönthal, A., Tsukitani, Y., and Feramisco, J., Transcriptional and posttranscriptional regulation of c-*fos* proto-oncogene expression by the tumor promoter okadaic acid, *Oncogene*, 6, 423, 1991c.

Schönthal, A., Alberts, A. S., Rahman, A., Meinkoth, J., Mumby, M., and Feramisco, J. R., Involvement of serine/threonine specific protein phosphatases in signal transduction pathways that regulate gene expression and cell growth, in *Recent Advances in Cellular and Molecular Biology*, vol. 2, Wegmann, R. J. and Wegmann, M. A., Eds., Peeters Press, Leuven, Belgium, 1992a, 67.

Schönthal, A., Srinivas, S., and Eckhart, W., Induction of c-*jun* protooncogene expression and transcription factor AP-1 activity by the polyoma virus middle-sized tumor antigen, *Proc. Natl. Acad. Sci. U.S.A.*, 89, 4972, 1992b.

Schröter, H., Shaw, P. E., and Nordheim, A., Purification of intercalator-released p67, a polypeptide that interacts specifically with the c-fos serum response element, *Nucleic Acids Res.*, 15, 10145, 1987.

Shaw, P. E., Frasch, S., and Nordheim, A., Repression of c-*fos* transcription is mediated through p67SRF bound to the SRE, *EMBO J.*, 8, 2567, 1989a.

Shaw, P. E., Schröter, H., and Nordheim, A., The ability of a ternary complex to form over the serum response element correlates with serum inducibility of the human c-*fos* promoter, *Cell*, 56, 563, 1989b.

Sherr, C. J., Rettenmier, C. W., Sacca, R., Roussel, M. F., Look, A. T., and Stanley, E. R., The c-fms proto-oncogene product is related to the receptor for the mononuclear phagocyte growth factor, CSF-1, *Cell*, 41, 665, 1985.

Shibanuma, M., Kuroki, T., and Nose, K., Inhibition of proto-oncogene c-*fos* transcription by inhibitors of protein kinase C and ion transport, *Eur. J. Biochem.*, 164, 15, 1987.

Siegfried, Z. and Ziff, E. B., Transcription activation by serum, PDGF, and TPA through the c-*fos* DSE: Cell type specific requirements for induction, *Oncogene*, 4, 3, 1989.

Siegfried, Z. and Ziff, E. B., Altered transcriptional activity of c-fos promoter plasmids in v-*raf*-transformed NIH 3T3 cells, *Mol. Cell. Biol.*, 10, 6073, 1990.

Simon, M. C., Rooney, R. J., Fisch, T. M., Heintz, N., and Nevins, J. R., E1A-dependent trans-activation of the c-*fos* promoter requires the TATAA sequence, *Proc. Natl. Acad. Sci. U.S.A.*, 87, 513, 1990.

Slootweg, M. C., van Genesen, S. T., Otte, A. P., Duursma, S. A., and Kruijer, W., Activation of mouse osteoblast growth hormone receptor: c-*fos* oncogene expression independent of phosphoinositide breakdown and cyclic AMP, *J. Mol. Endocrinol.*, 4, 265, 1990.

Slootweg, M. C., de Groot, R. P., Herrmann-Erlee, M. P. M., Koornneef, I., Kruijer, W., and Kramer, Y. M., Growth hormone induces expression of c-jun and jun B oncogenes and employs a protein kinase C signal transduction pathway for the induction of c-*fos* oncogene expression, *J. Mol. Endocrinol.*, 6, 179, 1991.

Stacey, D. W., Watson, T., Kung, H.-F., and Curran, T., Microinjection of transforming *ras* protein induces c-*fos* expression, *Mol. Cell. Biol.*, 7, 523, 1987.

Stumpo, D. J. and Blackshear, P. J., Insulin and growth factor effects on c-*fos* expression in normal and protein kinase C-deficient 3T3-L1 fibroblasts and adipocytes, *Proc. Natl. Acad. Sci. U.S.A.*, 83, 9453, 1986.

Stumpo, D. J., Stewart, T. N., Gilman, M. Z., and Blackshear, P. J., Identification of c-*fos* sequences involved in induction by insulin and phorbol esters, *J. Biol. Chem.*, 263, 1611, 1988.

Subramaniam, M., Schmidt, L. J., Crutchfield, C. E., III, and Getz, M. J., Negative regulation of serum-responsive enhancer elements, *Nature*, 340, 64, 1989.

Thévenin, C., Kim, S.-J., and Kehrl, J. H., Inhibition of protein phosphatases by okadaic acid induces AP1 in human T cells, *J. Biol. Chem.*, 266, 9363, 1991.

Treisman, R., Transient accumulation of c-*fos* RNA following serum stimulation requires a conserved 5′ element and c-*fos* 3′ sequences, *Cell*, 42, 889, 1985.

Treisman, R., Identification of a protein-binding site that mediates transcriptional response of the c-*fos* gene to serum factors, *Cell*, 46, 567, 1986.

Trouche, D., Robin, P., Robillard, O., Sassone-Corsi, P., and Harel-Bellan, A., c-*fos* transcriptional activation by IL-2 in mouse CTL-L2 cells is mediated through two distinct signal transduction pathways converging on the same enhancer element, *J. Immunol.*, 147, 2398, 1991.

Velcich, A. and Ziff, E. B., Functional analysis of an isolated *fos* promoter element with AP-1 site homology reveals cell type-specific transcriptional properties, *Mol. Cell. Biol.*, 10, 6273, 1990.

Visvader, J., Sassone-Corsi, P., and Verma, I. M., Two adjacent promoter elements mediate nerve growth factor activation of the c-*fos* gene and bind distinct nuclear complexes, *Proc. Natl. Acad. Sci. U.S.A.*, 85, 9474, 1988.

Wagner, B. J., Hayes, T. E., Hoban, C. J., and Cochran, B. H., The SIF binding element confers *sis*/PDGF inducibility onto the c-*fos* promoter, *EMBO J.*, 9, 4477, 1990.

Waterfield, M. D., Scrage, G. T., Whittle, N., Stroobant, P., Johnsson, A., Wasteson, A., Westermark, B., Heldin, C.-H., Huang, J. S., and Denel, T. F., Platelet-derived growth factor is structurally related to the putative transforming protein p28^{SIS} of simian sarcoma virus, *Nature*, 304, 35, 1983.

Watson, J. D., Molecular biology of signal transduction, in *Cold Spring Harbor Symposium on Quantitative Biology, Vol. 53*, Cold Spring Harbor, NY, 1988.

Weisz, A. and Rosales, R., Identification of an estrogen response element upstream of the human c-*fos* gene that binds the estrogen receptor and the AP-1 transcription factor, *Nucleic Acids Res.*, 18, 5097, 1990.

Wilson, T. and Treisman, R., Fos c-terminal mutations block down-regulation of c-*fos* transcription following serum stimulation, *EMBO J.*, 7, 4193, 1988.

Yamamoto, T., Ikawa, S., Akiyama, T., Semba, K., Nomura, N., Miyajima, N., Saito, T., and Toyoshima, K.. Similarity of protein encoded by the human *c-erb-B*-2 gene to epidermal growth factor receptor, *Nature*, 319, 230, 1986.

Yoshida, T., Miyagawa, K., Odagiri, H., Sakamoto, H., Little, P. F. R., Terada, M., and Sugimura, T.. Genomic sequence of *hst*, a transforming gene encoding a protein homologous to fibroblast growth factors and the *int-2*-encoded protein, *Proc. Natl. Acad. Sci. U.S.A.*, 84, 7305, 1987.

Zhan, X., Bates, B., Hu, X., and Goldfarb, M.. The human FGF-5 oncogene encodes a novel protein related to fibroblast growth factors, *Mol. Cell. Biol.*, 8, 3487, 1988.

Chapter 18

Regulation of AP-1 Activity by Adenovirus E1A: Dissection of Dimers Containing Jun, Fos, and ATF/CREB Members

Hans van Dam and Alex J. van der Eb

CONTENTS

INTRODUCTION

The proteins encoded by early region 1 A (E1A) of human adenoviruses (Ad) modulate the expression of both adenovirus genes and various host cell genes. These transcription-regulating properties enable the virus to redirect the metabolism of infected cells, allowing efficient production of virus progeny (Berk, 1986; Flint and Shenk, 1989). In non-permissive rodent cells, however, adenovirus infection is abortive and can lead to cell transformation, a process in which the E1A gene also plays a decisive role (van der Eb and Zantema, 1992). As the E1A proteins do not show significant DNA binding activity themselves, they apparently mediate their effects by altering the activity of cellular transcription factors, directly or indirectly. Since several promoters and/or enhancers whose activities are positively or negatively affected by E1A contain binding sites for the transcription factor AP-1, the various members of the AP-1 family appear to be candidates for E1A-dependent regulation. In the last few years it has indeed become clear that the E1A proteins can affect both the activity and expression levels of certain AP-1 components. These findings indicate that regulation of AP-1 activity plays a role in both adenovirus infection and tumorigenic transformation.

PUTATIVE STIMULATION OF AP-1 ACTIVITY VIA CONSERVED REGION 3 OF E1A PROTEIN

In the case of Ad serotype 5, the adenovirus E1A gene codes for two main protein products of 289 and 243 amino acids (aa), which are generated through alternative splicing and differ only in an internal stretch of 46 residues (Figure 18-1). These two proteins show essential differences in their transcription-regulatory properties. The 289-

0-8493-4573-1/94/$0.00+$.50

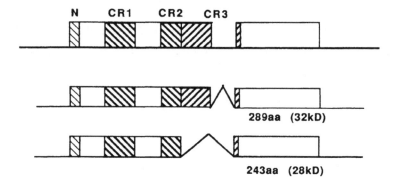

Figure 18-1 Schematic representation of the Ad 5 E1A gene and its two major protein products. Two mRNAs, 13S and 12S, are generated through alternative splicing and encode two related proteins. The domains conserved between the different Ad serotypes (CR1: aa 41–73, CR2: aa 121–136, CR3: aa 143–188) are indicated by the hatched boxes. The *N*-terminal region, which, in addition to CR1 and CR2, is essential for E1A-induced transformation, is indicated as well.

aa E1A protein (289R) carries a strong transactivation domain that is essential for activation of the adenovirus early promoters (e.g., the AdE3 and AdE4 promoters), and, therefore, plays a crucial role in the viral life cycle (Flint and Shenk, 1989). This transactivation domain, which is absent in the 243-aa E1A protein (243R), is encoded by conserved region 3 (CR3), one of the three regions conserved among different Ad serotypes (Figure 1). It has been proposed that CR3 mediates its effects on the AdE2, AdE3, and AdE4 promoters at least in part by stimulating the activity of AP-1 transcription factors, which would activate the promoters via their AP-1 (E3) and/or closely related ATF/CREB (E2, E3, and E4) binding sites. This model is based on the fact that some of the Ad early genes can also be induced by cAMP, and that adenovirus infection enhances cAMP-induced synthesis of two AP-1 components, JunB and cFos, in mouse S49 cells in an E1A-dependent fashion (Müller et al., 1989). This suggests that the induction of the Ad early genes by E1A could be mediated via increased de novo synthesis of AP-1. In a subsequent study, however, the stimulatory effect of E1A on the synthesis of JunB and cFos in S49 cells was found to be strictly dependent on cAMP, whereas the induction of the AdE2 and AdE4 promoters was not. Moreover, 243R–E1A, which lacks CR3, was also able to enhance the induction of AP-1 synthesis by cAMP, but could not induce the AdE4 promoter (Engel et al., 1991). Thus, only the synergistic effect of 289R–E1A and cAMP on the Ad early promoters, as observed in S49 cells, may be established via the induction of AP-1 synthesis. CR3-mediated transactivation of the Ad early promoters in the absence of cAMP, however, does not involve induction of de novo synthesized AP-1 activity. This conclusion is also supported by the finding that CR3-dependent activation can, in fact, occur in the absence of any de novo protein synthesis (Green et al., 1988).

A putative role for preexisting AP-1 in CR3-dependent transactivation was suggested by a study performed in HeLa cells. In these cells the AP-1 binding site of the AdE3 promoter seems to be the main target for induction by E1A. Moreover, the phorbol ester TPA, a well-known inducer of AP-1 activity, was also found to stimulate E3 promoter activity via this element, and TPA and E1A were found to activate AdE3 in a synergistic manner (Buckbinder et al., 1989). This result is somewhat surprising since, in various other promoters, such as that of the human collagenase I gene, AP-1 sites are not targets for CR3-dependent transactivation, but, in contrast, mediate E1A-dependent repression (see below). An explanation for this difference might be that E1A activation and inhibition involve different AP-1 components (i.e., the different members of the Jun and Fos families). Another explanation could be that activation of the AdE3 promoter is actually mediated via members of the ATF/CREB family, some of which also can bind to AP-1 or AP-1-related sites when dimerized to AP-1 components* (Hai and Curran, 1991). It should be noted that transactivation by CR3 is rather promiscuous; not only can it be established via AP-1 or ATF binding sites, but also via binding sites for E2F, Sp-1, and CTF, and via octamer motifs. In some cases the presence of only a TATA element appears to be sufficient (Wu et al., 1987; Pei and Berk, 1989; Weintraub and Dean, 1992). Since only certain members of the ATF and octamer protein families have been found to mediate CR3-dependent transactivation, and since the E1A proteins have been shown to be able to bind to the TATA binding protein (TBP) directly, at least *in vitro*, E1A has been proposed to act as a molecular bridge between certain gene-specific transcription factors and TBP (Liu and Green, 1990; Horikoshi et al., 1991; Lee et al., 1991). These interactions would allow CR3 to interact directly with the transcription initiation complex and, subsequently, enhance the activity of this complex. Since the ATF sites in the E4 promoter mediate the activation by E1A, and at least two different ATF family members seem to be CR3-inducible (ATF-2 and E4F) (Liu and Green, 1990; Rooney et al., 1990; Bondesson et al., 1992), it is surprising that in the AdE3 promoter the AP-1 site, rather than the ATF site, is required for CR3-dependent stimulation. This may indicate that only the AP-1 site is properly located in this promoter to allow E1A to function as a bridging factor between an AP-1 or ATF member and TBP. An alternative explanation may be that ATF-2 and E4F do not or only weakly bind to the AdE3 ATF site (Rooney et al., 1990; van Dam, unpublished results).

In summary, it appears that in the case of the AdE2 and AdE4 promoters, E2F, ATF-2, and E4F transcription factors, rather than Jun or Fos family members, mediate transactivation by CR3. Transactivation through the AP-1 site in the AdE3 promoter, however, might involve either a Jun or Fos family member, or an ATF family member present in an AP-1/ATF heterodimer. E1A-induced *de novo* synthesis of specific AP-1 components may in some cases contribute to or enhance CR3-induced transactivation. cFos might be of particular importance in this respect, since the c-fos promoter has been reported to be inducible by E1A via CR3 (Sassone-Corsi et al., 1987).

INHIBITION OF AP-1 (cJUN/cFOS AND cJUN/cJUN) ACTIVITY VIA CONSERVED REGION 1 OF E1A

Studies of the expression of growth factor- and TPA-inducible genes in Ad-transformed cells have identified the collagenase I, stromelysin, and CD44 genes as AP-1-responsive genes that are repressed by E1A (Offringa et al., 1989, 1990; Frisch et al., 1990; Hofmann et al., 1993). By transient cotransfection and/or stable cell line analysis, it was demonstrated that both 289R– and 243R–E1A repress stromelysin and collagenase promoter activity, although repression by 289R is less efficient, and that this repression requires CR1, but not CR2 (Offringa et al., 1990; van Dam et al., 1989 and unpublished results). The TPA-responsive element of the collagenase promoter (collTRE: TGAGTCA), which is a consensus AP-1 binding site also present in the stromelysin promoter, was shown to be the target for repression. The inhibitory effect of E1A on the collTRE is specific and not restricted to the collagenase promoter; in contrast to the c-fos serum-responsive

---

* ATF-2, ATF-3, and ATF-4, but not ATF-1 and CREB, are able to form heterodimers with cJun. ATF-4, but not ATF-2 and ATF-3, can bind to cFos and Fra-1. JunB cannot bind to ATF-2 (Ivashkiv et al., 1990; Benbrook and Jones, 1990; Hai and Curran, 1991).

element (SRE) and the TPA-responsive elements of the c-jun promoter (jun1- and jun2TRE), both the basal and TPA-induced activity of the collTRE were found to be repressed by E1A when placed in front of a minimal promoter that only contains a TATA box (Offringa et al., 1990).

Since CR1 is one of the regions essential for E1A-dependent transformation, together with CR2 and the amino-terminus (Lillie et al., 1986; Schneider et al., 1987; Jelsma et al., 1989; see also Figure 18-1), E1A-induced transformation is linked to repression of collTRE activity. However, as is evident from the data mentioned above, not all promoters containing AP-1 binding sites are inhibited by E1A. Requirements for repression appear to be (1) the nature of the AP-1 components that actually bind to a particular site, (2) the contribution of the AP-1 site to the overall promoter activity, and (3) the promoter context. The AP-1 subtypes cJun/cJun and cJun/cFos belong to the AP-1 factors that are subject to E1A-specific inhibition. These two dimers appear to be the natural factors acting on the collagenase promoter, since inhibition of their expression by the use of antisense constructs blocks the activation of the collagenase promoter by TPA (Schönthal et al., 1988). Using extracts of HeLa- and E1A-transformed 3T3 cells it was shown that the collTRE binding complexes *in vitro*, indeed, contain cFos and/or cJun (van Dam et al., 1993). More evidence for selective inhibition of transactivation by cJun/cJun and cJun/cFos is provided by experiments using cFos and cJun expression vectors. Stimulation of either the collagenase promoter or minimal promoters driven by the collTRE or junTREs via cotransfection of these expression vectors is inhibited by E1A in all cases (Offringa et al., 1990, van Dam et al., 1990 and unpublished results). The fact that E1A does not inhibit the TPA-induced activity of either the c-jun promoter or the two junTREs (TGACATCA and TTACCTCA), even though these elements can be bound by cJun/cJun or cJun/cFos, indicates that they are normally regulated by different AP-1(-related) factors that are not sensitive to E1A inhibition. *In vitro* binding studies, indeed, have shown that the junTREs bind multiple AP-1 complexes that differ in their electrophoretic mobilities and relative affinities from the complexes binding to the collTRE (Stein et al., 1992; van Dam, 1992; Herr et al., 1994).

Although transactivation of the collTRE by cJun/cFos is inhibited by E1A, not all promoters regulated by these factors need to be repressed. The effect of E1A on promoters that, in addition to E1A-repressable AP-1 sites, also contain elements that can be activated by E1A (e.g., by the strong transactivation domain of CR3), will depend on the relative contributions of repression and activation. An example of this is provided by a synthetic promoter in which one or multiple copies of the collTRE are placed in front of the −105/+51 herpes simplex virus TK promoter. In HeLa cells the collTRE is inhibited through CR1, but the TK promoter itself is induced by CR3. The net effect observed is reduced activation, as compared to the TK promoter alone, but no repression (Offringa et al., 1990). It should also be taken into account that a promoter that contains AP-1 site(s) that do not contribute significantly to promoter activity will not be inhibited by E1A either. Such a situation might occur, for example, when the AP-1 components occupying these elements have a low intrinsic transactivation potential, or when the localization of the AP-1 site in the promoter does not allow these factors to interact efficiently with the transcription initiation machinery. The latter case is exemplified by the c-fos promoter, in which the SRE and AP-1 sites partially overlap.

STIMULATION OF AP-1 ACTIVITY VIA CONSERVED REGION 1

Although not as strong as transactivation through CR3, the 243R–E1A protein is also able to stimulate promoter activity. One of the 243R-inducible genes is c-jun, the activation of which requires CR1 but apparently not CR2 (van Dam et al., 1990; Kitabayashi et al.,

1991). Elevated expression of c-jun mRNA on E1A-induced transformation has been observed in a variety of cell types. In transient transfection assays, stimulation of c-jun promoter activity by 289R is stronger than by 243R, indicating that, although 243R is already sufficient, CR3 may play an additional role (van Dam et al., 1990; Kitabayashi et al., 1991; de Groot et al., 1991). At first glance, the activation of c-jun expression appears paradoxical: why should E1A induce the expression of cJun when it also inhibits the function of this protein? However, analysis of the mechanism of activation of the c-jun promoter provides more insight. Transactivation of the c-jun promoter was found to require the two AP-1(-like) binding sites in this promoter, the jun1TRE and the jun2TRE, which can confer TPA inducibility on heterologous promoters (Stein et al., 1992; van Dam et al., 1993). The jun2TRE is the decisive element in c-jun promoter activation by E1A, since, in contrast to the jun1TRE, it can efficiently confer 243R–E1A-responsiveness on heterologous promoters (van Dam et al., 1993). In vitro binding studies with extracts from various cell types revealed that, in contrast to the jun1TRE, the jun2TRE binds only one particular protein complex in significant amounts. This complex also binds with high affinity to the jun1TRE and the consensus ATF/CREB binding site (TGACGTCA), but binds with low affinity to the collTRE. By the use of UV cross-linking and treatment with specific antibodies, this jun2TRE complex was found to be composed of cJun and ATF-2 and/or an ATF-2-related factor, which might be ATF-a (Gaire et al., 1990). Thus, in contrast to cJun/cJun and cJun/cFos, cJun/ATF-2 seems to be an AP-1 (or AP-1/ATF) subtype that is activated by 243R–E1A, supporting the hypothesis that activation and repression of AP-1 activity by E1A are mediated via different factors. In agreement with this model, stimulation of the jun2TRE by overexpression of cJun, forcing binding of cJun homodimers to this element in spite of their low affinity, was found to be inhibited be E1A (van Dam, 1992).

PUTATIVE MECHANISMS FOR E1A–CR1-DEPENDENT REGULATION OF cJUN ACTIVITY

To study the mechanism by which E1A inhibits cJun/cJun and cJun/cFos, hybrid transcription factors have been analyzed in which the DNA binding domain of cJun is fused to the transactivation domain of the transcription factor GHF1/pit1, and vice versa. These studies revealed that it is the DNA binding domain of cJun that is the target for E1A inhibition. In vivo footprinting analysis showed that in E1A-expressing cells the collagenase AP-1 site is not occupied by protein, in contrast to untransformed cells. This suggests that E1A inhibits the binding of cJun to the collTRE (Hagmeyer et al., 1993). The two AP-1 sites in the c-jun promoter involved in E1A activation show similar occupation in E1A-expressing cells and control cells in vivo, confirming the differential regulation of the collTRE and c-junTREs by E1A as observed in transient cotransfection. Thus, it appears that heterodimerization of cJun to ATF-2 has at least two effects on the cJun DNA binding domain: (1) its relative affinity for AP-1 motifs is altered, resulting in preferential binding to ATF/CREB motifs, and (2) it becomes insensitive to inhibition by E1A. Since several studies have shown that the DNA binding activity of cJun can be regulated by differential phosphorylation of four serine and threonine residues located close to the basic region, Hagmeyer and co-workers (1993) examined whether the inhibitory effect of E1A involved differences in the phosphorylation pattern of these sites. However, this appeared not to be the case, because transactivation by a cJun mutant in which all four phosphorylation sites were mutated was still found to be inhibited by E1A (Hagmeyer et al., 1993). Another possibility is that E1A inhibits the binding of cJun/cJun and cJun/cFos to DNA by causing a third protein to bind to these dimers. This interaction would be unstable, since the in vitro binding activity to the collTRE does not appear to be significantly affected in extracts of E1A-transformed cells, and in some cell types is in fact increased (Offringa et al., 1990; van Dam et al., 1993).

By studying the effects of E1A on cJun hybrid genes in which the transactivation domain of cJun is fused to the DNA binding domain of GHF1, it was found that 243R–E1A can enhance the transactivation activity of cJun through CR1. This result suggests that the 243R–E1A-induced activation of cJun/ATF-2 is at least in part mediated by cJun. The activity of ATF-2 in this dimer may be increased as well, as was shown for a cMyb–ATF-2 fusion protein in which the activity of ATF-2 was increased about threefold by 243R–E1A (Maekawa et al., 1991). In contrast to the inhibitory effect, the stimulatory effect of E1A on cJun appears to be caused by alterations in its phosphorylation state. cJun proteins immunoprecipitated from E1A-transformed cells show an altered phosphorylation pattern, as compared to proteins precipitated from untransformed cells. These differences include increased phosphorylation of serines 63 and 73, two amino acids located in the transactivation domain. Phosphorylation of these sites has been shown to be associated with increased transactivating activity of cJun after treatment with TPA, ultraviolet light, and various membrane-associated oncogene products, as described in Chapter 5. It, therefore, appears that E1A is able to activate the cJun transactivation function via a mechanism similar to some growth-factor-inducible signal transduction routes. How E1A activates this cJun domain, and whether the same or similar kinases or phosphatases are involved, remains to be established. Important in this respect may be that several of the functions of 243R–E1A appear to be mediated via direct binding to a set of cellular proteins implicated in cell cycle regulation. These proteins include p105-RB, p107, p58-cyclin A, and p33-cdk2, proteins that interact with protein kinases or represent the kinases themselves (Hunter, 1991; Nevins, 1992; and references therein). The effects of E1A on these proteins are believed to affect the relative amounts and activities of various kinase complexes in the cell. For the E1A-dependent regulation of cJun activity, the association of E1A to p300, a nuclear phosphoprotein with a molecular weight of 300 kDa, might be instrumental. p300 does not bind to E1A via CR1 and CR2, but via CR1 and the N-terminus, in contrast to the previously mentioned E1A-associating proteins. The binding of p300, which is differentially phosphorylated during the cell cycle (Yaciuk and Moran, 1991), might trigger alterations in kinase and/or phosphatase activities that lead to the enhanced phosphorylation of the cJun transactivation domain. It remains to be established, however, whether the effects of E1A on cJun also require the N-terminus of E1A. A possibility is that this putative p300-dependent regulation of cJun phosphorylation is mediated by altering the expression levels of cyclin A and cyclin D. E1A-transformed cells were found to express increased levels of cyclin A, but decreased levels of cyclin D, which is also dependent on CR1, but not on CR2 (Buchou et al., 1993).

Enhancement of the cJun transactivation function by E1A, thus, does not lead to activation of genes controlled by cJun/cJun or cJun/cFos, but appears to lead to constitutive expression of genes controlled by cJun/ATF-2 and possibly other cJun-containing dimers. However, the presence of a cJun/ATF-2 binding site may not be sufficient for promoter activation. Although the cJun/ATF-2 binding site in the c-jun promoter (jun2TRE) is able to confer 243R–E1A inducibility on heterologous promoters, additional transcription factor binding sites are required for efficient E1A induction. When cloned as a single element in front of a minimal TATA promoter, the jun2TRE was found to be only weakly induced, and maximal activation of the c-jun promoter also required the jun1TRE, a CTF binding site, and a region downstream of the transcription initiation site. However, as a multimer the jun2TRE is efficiently inducible, even in a minimal TATA promoter (van Dam et al., 1993). These data suggest that cJun/ATF-2 needs to cooperate with itself or other factors to mediate the maximal response to 243R–E1A. This cooperation might involve the formation of a multiprotein complex, which is required for the efficient interaction of cJun/ATF-2 with the transcription initiation machinery. The enhanced phosphorylation of the cJun molecules in this multiprotein complex would then lead to enhanced activation of the transcription initiation complex, for example, by triggering a conformational change in this complex that facilitates the interaction of DNA-bound cJun or another member of the complex with the TBP or TBP-associating factors.

POSSIBLE FUNCTIONS OF ALTERED cJUN ACTIVITY IN ADENOVIRUS-INDUCED PROCESSES

Since CR1 is required for both 243R–E1A-dependent regulation of cJun/AP-1 and E1A-induced cell transformation, cJun may have a crucial role in adenovirus transformation. As reviewed in chapters 11, 14, 15, 16, 17, and 19, cJun is believed to play a key function in the regulation of cell proliferation and differentiation. Multiple studies suggest that transformation via cJun involves a constitutively activated transactivation domain (Bos et al., 1990; Lloyd et al., 1991; Smeal et al., 1991). Interestingly, in E1A-transformed cells, various cJun/cJun- or cJun/cFos-controlled genes are downregulated, whereas a cJun/ATF-2-controlled gene is induced. E1A is not unique in this respect, since the synthetic glucocorticoid dexamethasone, which enhances polyomavirus- and ras oncogene-mediated transformation (Martens et al., 1988; Marshall et al., 1991), also downregulates collagenase and stromelysin expression, but induces c-jun (Offringa et al., 1989, 1990; Jonat et al., 1990). Moreover, by analysis of v-jun and c-jun mutants in chicken embryo fibroblasts, Vogt and collaborators found an inverse correlation between transformation and transcriptional activation of collagenase and stromelysin (Håvarstein et al., 1992; Hartl and Vogt, personal communication). These data may, therefore, indicate that, at least in certain cell types, constitutive activation of cJun/ATF-2 rather than of cJun/cJun or cJun/cFos is required for transformation.

Although inhibition of cJun/cJun or cJun/cFos by E1A or dexamethasone might have a stimulatory effect on cell growth in certain cell types or under certain conditions, reduced expression of the secreted proteases stromelysin and collagenase I is unlikely to contribute to E1A-induced oncogenic transformation. In fact, it could have a negative effect on cancer development during the later stages by inhibiting metastasis. In this regard, it is of interest that adenovirus-transformed cells are generally nonmetastatic. Increased expression of stromelysin and collagenase, for example, induced by the ras oncogene, is associated with enhanced invasive and metastatic properties, whereas E1A represses the metastatic properties of ras-transformed rodent cells and certain human tumor cells (Pozzatti et al., 1986; Frisch et al., 1990; Hofmann et al., 1993). Recent findings support the hypothesis that the inhibitory effect of E1A on cJun/cJun or cJun/cFos is responsible for the antimetastatic effect of E1A. E1A has also been found to repress the ras-induced expression of CD44, a gene whose specific splice variants can confer metastatic potential on tumor cells (Günthert et al., 1991; Hofmann et al., 1993). Both ras-dependent activation and E1A-dependent inhibition of CD44 were found to be mediated via a consensus AP-1 binding site in the CD44 promoter.

It should be added that cJun-dependent functions other than the ones described above may also be affected by E1A. cJun can influence the function of transcription factors other than AP-1 and ATF members, for example, and may play a role in DNA replication. This aberrant functioning of cJun-dependent processes in the presence of E1A may play a role not only in adenovirus-induced transformation, but also in lytic adenovirus infection. Target cells of adenovirus infection *in vivo* are usually in a quiescent state, and both 289R– and 243R–E1A can induce cell cycle progression in these cells, thereby creating an environment in which DNA replication is more efficient (Stabel et al., 1985; Zerler et al., 1987). Induction of DNA synthesis and subsequent mitosis is dependent on the presence of CR1 and CR2. Since CR1 is able to trigger DNA synthesis in the absence of CR2, the regulation of cJun activity by CR1 might play a role in this process.

DIFFERENTIAL EFFECTS OF Ad5- AND Ad12E1A ON THE EXPRESSION OF *junB*

In addition to transactivation, CR3 of the E1A gene of Ad serotype 12 is also involved in inhibition of transcription. In Ad12-transformed derivatives of certain primary cell types, transcription of several cellular genes is downregulated in a CR3-dependent manner. One of these genes encodes the AP-1 component junB, while others code for the class I major histocompatibility complex (MHC) antigens (Schrier et al., 1983; Bernards et al., 1983; van Dam et al., 1990; Meijer et al., 1991). The inhibition is not observed when CR3 of Ad12E1A is replaced by the corresponding region of Ad5. The mechanism by which Ad12E1A inhibits junB transcription is not known. Although CR3 is essential, it cannot be excluded that the integrity of the transforming domains located in CR1 and CR2 is also required. Some experiments, in fact, suggest that repression by Ad12CR3 is the result of the Ad12-specific transformation process and, therefore, is not a direct effect of Ad12E1A (Meijer et al., unpublished data). Moreover, CR3 of Ad12E1A is essential for the ability of Ad12-transformed primary cells to grow as tumors in immunocompetent animals, an ability that Ad5-transformed cells lack (Schrier et al., 1983; Bernards et al., 1983). This may imply that reduced expression of junB, a gene that has been suggested to exhibit tumor-suppressor properties (Schütte et al., 1989), contributes to the tumorigenic character of Ad12-transformed cells.

Differential regulation of junB expression by Ad5- and Ad12E1A is also observed in Ad-transformed derivatives of established cell lines. In this case the presence of Ad5E1A leads to enhanced expression, whereas Ad12E1A has no effect (van Dam et al., 1990; Meijer et al., 1991). Ad5E1A-induced junB expression in established cells is dependent on CR1, and may, therefore, also be linked to cell transformation. However, since the effects of E1A on junB are cell type and serotype dependent, regulation of junB expression does not appear to be essential for E1A-induced transformation in general, which, in contrast to tumor induction in immunocompetent animals, does not show a similar cell and serotype dependence.

CONCLUSIONS AND PERSPECTIVES

Regulation of AP-1 activity by adenovirus E1A is a complex process because multiple AP-1 components are affected and different E1A domains, which can elicit opposite effects, are involved. Transactivation of promoter activity via CR3, which plays a crucial role in adenovirus lytic infection, might, in some cases, be mediated via interaction with specific AP-1 family members. Alternatively, it may involve AP-1 members only indirectly, as components of AP-1/ATF heterodimers. Transformation by E1A is accompanied by multiple changes in AP-1 activity, some of which are cell type or Ad serotype specific. The major target of E1A appears to be cJun, whose activity in certain types of dimers is inhibited, whereas in others it is increased (Figure 18-2). By affecting cJun activity E1A interferes in certain signal transduction pathways. It blocks the activation of various genes regulated by cJun/cJun or cJun/cFos (and, presumably, also other Jun/Fos members), apparently by inhibiting binding of these factors to their recognition sites. This inhibition might involve the interaction of E1A or an E1A-dependent protein with the DNA binding domain of Jun. Binding of cJun/ATF-2 is not inhibited by E1A. The activity of this complex is, in fact, increased, since E1A stimulates the transactivation function of cJun

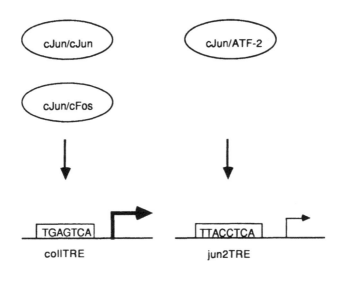

+ **E1A**

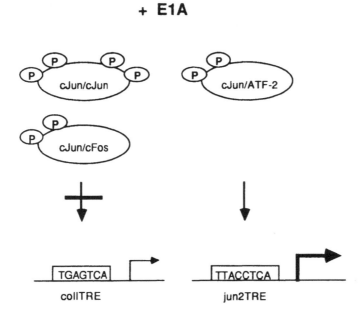

Figure 18-2 Model describing the putative mechanism for the differential regulation of AP-1-dependent gene expression as observed in E1A-transformed cells. Enhanced phosphorylation of the cJun protein leads to increased expression of genes controlled by cJun–ATF-2. Various genes that are controlled by cJun–cJun and cJun–cFos are repressed, however, since these dimers are unable to bind to their recognition sites.

by inducing hyperphosphorylation of the transactivation domain. Activation of cJun/ ATF-2 leads to constitutive activation of c-jun expression and, presumably, other cJun/ ATF-2-dependent genes. Resolving the mechanism of inhibition of cJun DNA binding and the identification of the putative kinases and phosphatases involved in E1A-dependent cJun activation will be challenging projects for future research. Most important, continuing the search for cJun target genes that are induced by E1A might lead to the identification of cJun-dependent genes that are directly involved in cell transformation.

ACKNOWLEDGMENTS

We would like to thank J. P. Sleeman and L. Sherman for critical reading of the manuscript. H. v. D. is supported by an EMBO long-term fellowship.

REFERENCES

Benbrook. D. M. and Jones. N. C., Heterodimer formation between CREB and JUN proteins. *Oncogene*, 5. 295. 1990.

Berk. A. J.. Adenovirus promoters and E1A transactivation, *Annu. Rev. Genet.*, 20. 45. 1986.

Bernards. R.. Schrier. P. I.. Houweling. A.. Bos. J. L.. van der Eb. A. J.. Zijlstra. M., and Melief. C. J. M., Tumorigenicity of cells transformed by adenovirus type 12 by evasion of T-cell immunity. *Nature*, 305. 776. 1983.

Bondesson. M.. Svensson. C.. Linder. S.. and Akusjärvi. G., The carboxy-terminal exon of the adenovirus E1A protein is required for E4F-dependent transcription activation. *EMBO J.*, 11. 3347. 1992.

Bos. T. J.. Monteclaro. F. S.. Mitsunobu. F.. Ball. A. R.. Jr.. Chang. C. H. W.. Nishimura. T.. and Vogt. P. K.. Efficient transformation of chicken embryo fibroblasts by c-Jun requires structural modification in coding and noncoding sequences. *Genes Dev.*, 4. 1677. 1990.

Buchou. T.. Kranenburg. O.. van Dam. H.. Roelen. D.. Zantema. A.. Hall. F. L.. and van der Eb. A.. Increased cyclin A and decreased cyclin D levels in adenovirus 5 E1A-transformed rodent cell lines, *Oncogene*, 8. 1765. 1993.

Buckbinder. L., Miralles. V. J.. and Reinberg. D.. TPA can overcome the requirement for E1a and together act synergistically in stimulating expression of the adenovirus EIII promoter. *EMBO J.*, 8. 4239. 1989.

de Groot. R. P.. Foulkes. N.. Mulder. M.. Kruyer. W.. and Sassone-Corsi. P.. Positive regulation of jun/AP-1 by E1A. *Mol. Cell. Biol.*, 11. 192. 1991.

Engel. D. A.. Muller. U.. Gedrich. R. W.. Eubanks. J. S.. and Shenk. T., Induction of c-fos mRNA and AP-1 DNA binding activity by cAMP in cooperation with either the adenovirus 243- or the adenovirus 289-amino acid E1A protein, *Proc. Natl. Acad. Sci. U.S.A.*, 88. 3957. 1991.

Flint. K. J. and Shenk. T., Adenovirus E1A protein paradigm viral transactivator. *Annu. Rev. Genet.*, 23. 141. 1989.

Frisch. S. M.. Reich. R.. Collier. I. E.. Genrich. L. T.. Martin. G.. and Goldberg. G. I., Adenovirus E1A represses protease gene expression and inhibits metastasis of human tumor cells. *Oncogene*, 5. 75. 1990.

Gaire. M.. Chatton. B.. and Kedinger. C., Isolation and characterization of two novel. closely related ATF cDNA clones from HeLa cells. *Nucleic Acids Res.*, 18. 3467. 1990.

Green. M.. Loewenstein. P. M.. Pusztai. R.. and Symington. J. S., An adenovirus E1A protein domain activates transcription in vivo and in vitro in the absence of protein synthesis. *Cell*, 53. 921. 1988.

Günthert, U., Hofmann, M., Rudy, W., Reber, S., Zöller, M., Haußmann, I., Matzku, S., Wenzel, A., Ponta, H., and Herrlich, P., A new variant of glycoprotein CD44 confers metastatic potential to rat carcinoma cells, *Cell*, 65, 13, 1991.

Hagmeyer, B., König, H., Herr, I., Offringa, R., Zantema, A., van der Eb, A. J., Herrlich, P., and Angel, P., Adenovirus E1A negatively and positively modulates transcription of AP-1 dependent genes by dimer-specific regulation of the DNA binding and transactivation activities of Jun, *EMBO J.*, 1993, in press.

Hai, T. and Curran, T., Cross-family dimerization of transcription factors Fos/Jun and ATF/CREB alters DNA binding specificity, *Proc. Natl. Acad. Sci. U.S.A.*, 88, 3720, 1991.

Håvarstein, L. S., Morgan, I. M., Wong, W.-Y., and Vogt, P. K., Mutations in the Jun Delta region suggest an inverse correlation between transformation and transcriptional activation, *Proc. Natl. Acad. Sci. U.S.A.*, 89, 618, 1992.

Herr, I., van Dam, H., and Angel, P., Binding of promoter associated AP-1 is not altered during induction and subsequent repression of the *c-jun* promoter by TPA and UV irradiation. *Carcinogenesis*, 15, 1105, 1994.

Hofmann, M., Rudy, W., Günthert, U., Zimmer, S. G., Zawadzki, V., Zöller, M., Lichtner, R. B., Herrlich, P., and Ponta, H., A link between RAS and metastatic behavior of tumor cells: ras induces CD44 promoter activity and leads to low-level expression of metastasis-specific variants of CD44 in CREF cells. *Cancer Res.*, 53, 1516, 1993.

Horikoshi, N., Maguire, K., Kralli, A., Maldonado, E., Reinberg, D., and Weinmann, R., Direct interaction between adenovirus E1A protein and the TATA box binding transcription factor IID, *Proc. Natl. Acad. Sci. U.S.A.*, 88, 5124, 1991.

Hunter, T., Cooperation between oncogenes, *Cell*, 64, 249, 1991.

Ivashkiv, L. B., Liou, H.-C., Kara, C. J., Lamph, W. W., and Verma, I. M., mXBP/CRE-BP2 and c-Jun form a complex which binds to the cyclic AMP, but not to the 12-O-tetradecanoylphorbol-13-acetate, response element, *Mol. Cell. Biol.*, 10, 1609, 1990.

Jelsma, T. N., Howe, J. A., Mymryk, J. S., Evelegh, C. M., Cunniff, N. F., and Bayley, S. T., Sequences in E1A proteins of human adenovirus 5 required for cell transformation, repression of a transcriptional enhancer, and induction of proliferating cell nuclear antigen, *Virology*, 171, 120, 1989.

Jonat, C., Rahmsdorf, H. J., Park, K.-K., Cato, A. C. B., Gebel, S., Ponta, H., and Herrlich, P., Anti-tumor promotion and antiinflammation: Down-modulation of AP-1 (Fos/Jun) activity by glucocorticoid hormone, *Cell*, 62, 1189, 1990.

Kitabayashi, I., Chiu, R., Gachelin, G., and Yokoyama, K., E1A dependent up-regulation of c-jun/AP-1 activity, *Nucleic Acids Res.*, 19, 649, 1991.

Lee, W. S., Kao, C. C., Bryant, G. O., Liu, X., and Berk, A. J., Adenovirus E1A activation domain binds the basic repeat in the TATA box transcription factor, *Cell*, 67, 365, 1991.

Lillie, J. W., Green, M., and Green, M. R., An adenovirus E1a protein region required for transformation and transcriptional repression, *Cell*, 46, 1043, 1986.

Liu, F. and Green, M. R., A specific member of the ATF transcription factor family can mediate transcription activation the adenovirus E1a protein, *Cell*, 61, 1217, 1990.

Lloyd, A., Yancheva, N., and Wasylyk, B., Transformation suppressor activity of a Jun transcription factor lacking its activation domain, *Nature*, 352, 635, 1991.

Maekawa, T., Matsuda, S., Fujisawa, J.-I., Yoshida, M., and Ishii, S., Cyclic AMP response element-binding protein, CRE-BP1, mediates the E1A-induced but not the Tax-induced trans-activation, *Oncogene*, 6, 627, 1991.

Marshall, H., Martens, I., Svensson, C., Akusjärvi, G., and Linder, S., Glucocorticoid hormones may partially substitute for adenovirus E1A in cooperation with ras, *Exp. Cell Res.*, 194, 35, 1991.

Martens, I., Nilsson, M., Magnusson, G., and Linder, S., Glucocorticoids facilitate the stable transformation of embryonal rat fibroblasts by a polyomavirus large tumor antigen-deficient mutant, *Proc. Natl. Acad. Sci. U.S.A.*, 85, 5571, 1988.

Meijer, I., van Dam, H., Boot, A. J. M., Bos, J. L., Zantema, A., and van der Eb, A. J., Co-regulated expression of junB and MHC class I genes in adenovirus-transformed cells, *Oncogene*, 6, 911, 1991.

Müller, U., Roberts, M. P., Engel, D. A., Doerfler, W., and Shenk, T., Induction of transcription factor AP-1 by adenovirus E1A protein and cAMP, *Genes Dev.*, 3, 1991, 1989.

Nevins, J. R., A closer look at E2F, *Nature*, 358, 375, 1992.

Offringa, R., Smits, A. M. M., Houweling, A., Bos, J. L., and van der Eb, A., Similar effects of adenovirus E1A and glucocorticoid hormones on the expression of the metalloprotease stromelysin, *Nucleic Acids Res.*, 16, 10973, 1989.

Offringa, R., Gebel, S., van Dam, H., Timmers, M., Smits, A., Zwart, R., Stein, B., Bos, J. L., van der Eb, A., and Herrlich, P., A novel function of the transforming domain of E1a: Repression of AP-1 activity, *Cell*, 62, 527, 1990.

Pei, R. and Berk, A. J., Multiple transcription factor binding sites mediate adenovirus E1A transactivation, *J. Virol.*, 63, 3499, 1989.

Pozzatti, R., Muschel, R., Williams, J., Padmanabhan, R., Howard, B., Liotta, L., and Khoury, G., Primary rat embryo cells transformed by one or two oncogenes show different metastatic potentials, *Science*, 232, 223, 1986.

Rooney, R. J., Raychaudhuri, P., and Nevins, J. R., E4F and ATF, two transcription factors that recognize the same site, can be distinguished both physically and functionally: A role for E4F in E1A trans activation, *Mol. Cell. Biol.*, 10, 5138, 1990.

Sassone-Corsi, P. and Borrelli, E., Promoter trans-activation of protooncogenes c-fos and c-myc, but not c-Ha-ras, by products of adenovirus early region 1A, *Proc. Natl. Acad. Sci. U.S.A.*, 84, 6430, 1987.

Schneider, J. F., Fisher, F., Goding, C. R., and Jones, N. C., Mutational analysis of the adenovirus E1a gene: The role of transcriptional regulation in transformation, *EMBO J.*, 6, 2053, 1987.

Schönthal, A., Herrlich, P., Rahmsdorf, H. J., and Ponta, H., Requirement for fos gene expression in the transcriptional activation of collagenase by other oncogenes and phorbol esters, *Cell*, 54, 325, 1988.

Schrier, P. I., Bernards, R., Vaessen, R. T. M. J., Houweling, A., and van der Eb, A. J., Expression of class I major histocompatibility antigens switched off by highly oncogenic adenovirus 12 in transformed rat cells, *Nature*, 305, 771, 1983.

Schütte, J., Viallet, J., Nau, M., Segal, S., Fedorko, J., and Minna, J., jun-B inhibits and c-fos stimulates the transforming and trans-activating activities of c-jun, *Cell*, 59, 987, 1989.

Smeal, T., Binétruy, B., Mercola, D. A., Birrer, M., and Karin, M., Oncogenic and transcriptional cooperation with Ha-ras requires phosphorylation of c-Jun on serines 63 and 73, *Nature*, 354, 494, 1991.

Stabel, S., Argos, P., and Philipson, L., The release of growth arrest by microinjection of adenovirus E1A DNA, *EMBO J.*, 4, 2329, 1985.

Stein, B., Angel, P., van Dam, H., Ponta, H., Herrlich, P., van der Eb, A., and Rahmsdorf, H. J., Ultraviolet-radiation induced c-jun gene transcription: Two AP-1 like binding sites mediate the response, *Photochem. Photobiol.*, 55, 409, 1992.

van Dam, H., Functions of conserved region 1 in transformation by the adenovirus E1A oncogene, Ph.D. thesis, University of Leiden, 1992.

van Dam, H., Offringa, R., Smits, A. M. M., Bos, J. L., Jones, N. C., and van der Eb, A. J., The repression of the growth factor-inducible genes JE, c-myc and stromelysin by adenovirus E1A is mediated by conserved region 1, *Oncogene*, 4, 1207, 1989.

van Dam, H., Offringa, R., Meijer, I., Stein, B., Smits, A. M., Herrlich, P., Bos, J. L., and van der Eb, A. J., Differential effects of the adenovirus E1A oncogene on members of the AP-1 transcription factor family, *Mol. Cell. Biol.*, 10, 5857, 1990.

van Dam, H., Duyndam, M., Rottier, R., Bosch, A., de Vries-Smits, L., Herrlich, P., Zantema, A., Angel, P., and van der Eb, A. J., Heterodimer formation of cJun and ATF-2 is responsible for induction of c-jun by the 243 amino acid adenovirus E1A protein, *EMBO J.*, 12, 479, 1993.

van der Eb, A. J. and Zantema, A., Adenovirus oncogenesis, in *Malignant Transformation by DNA Viruses*, Doerfler, W. and Bohm, P., Eds., VCH Publishers, Weinheim, Germany, pp. 115-140, 1992.

Weintraub, S. J. and Dean, D. C., Interaction of a common factor with ATF, Sp1, or TATAA promoter elements is required for these sequences to mediate transactivation by the adenoviral oncogene E1a, *Mol. Cell. Biol.*, 12, 512, 1992.

Wu, L., Rosser, D. S. E., Schmidt, M. C., and Berk, A., A TATA box implicated in E1A transcriptional activation of a simple adenovirus 2 promoter, *Nature*, 326, 512, 1987.

Yaciuk, P. and Moran, E., Analysis with specific polyclonal antiserum indicates that the E1A-associated 300-kDa product is a stable nuclear phosphoprotein that undergoes cell cycle phase-specific modification, *Mol. Cell. Biol.*, 11, 5389, 1991.

Zerler, B., Roberts, R. J., Mathews, M. B., and Moran, E., Different functional domains of the adenovirus E1A gene are involved in regulation of host cell cycle products, *Mol. Cell. Biol.*, 7, 821, 1987.

Chapter 19

Participation of Fos and Jun in Multistage Carcinogenesis

Adam Glick and Stuart H. Yuspa

CONTENTS

INTRODUCTION

The discovery that transforming genes of oncogenic retroviruses are transduced variants of eucaryotic genes launched the current era of molecular carcinogenesis. The interpretation of that discovery and the molecular insights that followed, however, depended on the foundation established by four prior decades of carcinogenesis research. Studies in animals had established that certain chemicals could cause malignant tumors, that cancer induction is a multistage process, and that carcinogenesis is organ specific in humans and other mammals (Yuspa and Poirier, 1988). Subsequent studies in cell culture proved that neoplastic transformation by chemicals is inductive rather than selective, and that somatic mutations are an important component of cancer pathogenesis. Cellular enzymes were identified that activated or detoxified chemical carcinogens, and DNA was recognized as the critical target for activated carcinogens (Miller, 1978). The development of methods to evaluate carcinogenesis in cultured human cells provided the foundation for studies of comparative carcinogenesis at the molecular level (Yuspa and Poirier, 1988; Harris, 1991). While recent genetic analyses have defined intimate details of the cellular pathways that are altered in response to carcinogenic stimuli and revealed the structural basis for many of these alterations, they must be evaluated in the context of cancer's multiple stages: initiation, promotion, and malignant conversion (Yuspa and Poirier, 1988; Harris, 1991).

Initiation of carcinogenesis is a consequence of carcinogen-induced genetic changes in a single cell. The interaction of a particular carcinogen with specific genetic sites

0-8493-4573-1/94/$0.00+$.50
© 1994 by CRC Press, Inc.

results, in part, from selectivity of metabolically activated carcinogens for particular nucleosides or DNA sequences and the capacity to repair damage accurately at these sites. In turn, modification of the molecular structure at specific genetic loci has tissue- and species-specific consequences dependent on the expression of a particular gene, its sequence, and the function of the gene product in the target cell. At the genetic level, alterations of coding and regulatory regions, genomic rearrangements, point mutations, and changes in gene expression all can result in an altered cell phenotype. Whatever genetic mechanisms are involved, initiated cells have an altered program of terminal differentiation, are resistant to cytotoxic substances, or show altered requirements for specific growth factors or nutrients. These cells would have a selective growth advantage in cytostatic or cytotoxic situations or under conditions favoring terminal differentiation or proliferation (Farber, 1984a,b; Yuspa and Poirier, 1988; Harris, 1991).

Tumor promoters, some acting through specific cellular receptors, produce a tissue environment conducive to the selective clonal outgrowth of the initiated cell population, resulting in a clinically evident premalignant lesion (Yuspa, 1987; Yuspa and Poirier, 1988; Harris, 1991). The tissue specificity for most promoters depends on the ability of a particular agent to produce the selective conditions required for the initiated phenotype of that organ.

Malignant conversion of benign tumors requires further genetic changes in the tumor cell. Such changes could result from inherent instability in the genome of initiated cells, resulting in spontaneous mutations or chromosomal changes, more likely to occur in the expanding population of proliferating benign tumor cells, or by genetic changes produced by additional exposure to exogenous or endogenously generated genotoxic agents. Malignant conversion is associated with marked deviation in cellular phenotype and autonomy, and with the acquisition of the ability to invade surrounding normal or benign tissues.

The contribution of specific genetic changes to multistage carcinogenesis has been the focus of considerable study. In general, three approaches have been used to evaluate the relevance of a gene for contributing to neoplastic transformation: (1) the pattern of expression or the structural and functional integrity of the gene and protein product have been analyzed in spontaneous and induced tumors and in normal cells treated with carcinogens or tumor promoters; (2) recombinant vectors and transfection methods have been used to test the oncogenic potential of certain genes that were isolated from a particular human or animal cancer; and (3) transgenic mice, overexpressing or lacking a particular gene, are created to evaluate an intact organism with modified cancer risk. Together, data obtained by these strategies are evaluated in the context of the function of the gene product to synthesize a model that defines both the relevance and mechanism of action of a particular oncogenic sequence. This chapter will evaluate data on the *c-fos* and *c-jun* gene products in the context of multistage carcinogenesis.

ALTERATIONS IN *fos/jun* DETECTED IN TUMORS

The *v-fos* oncogene was first identified as the sequence responsible for the production of osteosarcomas in mice by the FBJ-MSV and FBR-MSV retroviruses (Finkel et al., 1966; Curran et al., 1982). Further studies revealed its transforming potential for cultured cells and defined the critical structural features that are required for transformation (Miller et al., 1984; Lee et al., 1988). Changes in *c-fos* that contribute to carcinogenesis involve alterations in the amount of *fos* mRNA or protein. To date, structural changes in the *fos* gene that qualitatively influence the function of the gene product have not been described in spontaneous or carcinogen-induced tumors.

The *v-jun* oncogene was identified as the transforming sequence of a helper dependent avian retrovirus, ASV17, isolated from a chicken fibrosarcoma (Maki et al., 1987). ASV17 causes fibrosarcomas in chickens at the site of injection after a latency period, and transforms embryonic chicken fibroblasts, neuroretina cells, and myoblasts *in vitro* (Garcia and Samarut, 1990; Vogt and Bos, 1990; Su et al., 1991). Overexpression of the *v-jun* cellular homologue, *c-jun*, also transforms chicken embryo cells and may cooperate with other oncogenes to transform some mammalian cell lines (Schütte et al., 1989a; Garcia and Samarut, 1990). The essential components of *v-jun* can be replaced by *c-jun* sequences for transformation of cultured cells, but several mutations in the coding region of *v-jun* may be essential for carcinogenicity *in vivo* (Vogt and Bos, 1990).

Homodimers of c-Jun or heterodimers of c-Fos and c-Jun form the AP-1 complex that regulates gene expression through a short recognition sequence (AP-1 site or TRE) in the promoter of many genes (Vogt and Bos, 1990). c-Fos and c-Jun are each members of a larger family of homologous proteins (Fra-1, Fra-2, FosB) (Cohen and Curran, 1988; Zerial et al., 1989; Nishina et al., 1990) (JunB and JunD) (Chiu et al., 1989, Ryder et al., 1989) that can form similar homo- or heterodimers. Although some evidence exists for functional differences between AP-1 complexes composed of different Fos and Jun family members (Chiu et al., 1989; Schutte et al., 1989; Nakabeppu and Nathans, 1991), the specific role of these proteins *in vivo* and in carcinogenesis has not been defined.

OVEREXPRESSION OF *c-fos* IN OSTEOSARCOMAS

The biological basis for the remarkable specificity of v-Fos to induce osteosarcomas in mice has not been defined. After injection of *v-fos* retroviral supernatants, bone tumors in mice arise rapidly from periosteal cells, appearing first as cortical thickenings or dense areas of soft tissue adjacent to bone (Ward and Young, 1976). The tumors are composed of primitive mesenchymal cells with subcellular characteristics of osteoblasts and chondroblasts containing abundant alkaline phosphatase (Ward and Young, 1976). A number of experimental studies suggest that Fos has a fundamental function in bone morphogenesis and carcinogenesis. Analysis of *c-fos* transcripts in human long bones by *in situ* hybridization showed expression in chondrocytes bordering the joint space and in osteoblasts of developing bone (Sandberg et al., 1988). Similar studies in embryonic mice localized *c-fos* mRNA to perichondrial blastemal growth regions (Dony and Gruss, 1987), while protein in embryonic rat bones was detected in nuclei of chrondroblasts of skeletal bone and on the inner aspect of the periosteum (De Togni et al., 1988). Together, these studies suggest that Fos expression is important in bone development and bone growth during embryogenesis. However, c-Fos is not essential for bone morphogenesis in mice because c-Fos knockout mice develop a normal skeletal system (Johnson et al., 1992; Wang et al., 1992). Nevertheless, c-Fos null mice develop osteopetrosis, a disturbance of remodeling attributed to osteoblast function (Johnson et al., 1992; Wang et al., 1992), indicating that postnatal bone remodeling is also influenced by c-Fos.

Evidence for a causal relationship between high c-Fos expression and bone abnormalities, including tumors, comes from *c-fos* transgenic mice constructed with the metallothionein promoter and the FBJ 3'LTR (MT-*c-fos* LTR), which was required to stabilize the exogenous *fos* mRNA (Rüther et al., 1987, 1989). These mice frequently develop swellings at the end of long bones consisting of differentiated bone synthesizing cells and increased bone deposition. The changes begin postnatally when expression of the exogenous *fos* mRNA is detected (Rüther et al., 1987). A recent study has shown that osteoblasts are the specific target cell for the oncogenic action of *fos*, and that the effects are specific to fos,since transgenic mice expressing either fosB or c-jun in bone tissues do not develop any pathology (Grigoriadis et al., 1993). Swellings are limited in duration, but during the lifetime of these transgenic mice, 18% develop osteosarcomas, independent of the initial bone lesions (Rüther et al., 1989). Both bone hyperplasias and osteosarcomas occur predominantly (85%) in males (Rüther et al., 1987, 1989). These results directly implicate overexpression of c-Fos in sarcomagenesis but suggest secondary events are required and may be influenced by sex hormones or, possibly, the duration of postnatal bone growth, which is longer in males (Silverberg and Silverberg, 1971). Additionally, the c-Fos transgenic mice develop hyperplasia of the thymic epithelia, but tumors do not develop (Rüther et al., 1988). In contrast to osteosarcomas, chondrosarcomas developed in chimeric mice constructed by introducing MT-*c-fos* LTR into embryonic stem cells (Wang et al., 1991). In this example, the exogenous *fos* mRNA is detected in embryos, and the target cell is a more primitive chondroblast precursor (Wang et al., 1991). However, the tumors develop postnatally, confirming a need for additional changes to produce malignancy, although the latency is quite short, only 3 to 4 weeks.

The relevance of these animal experiments to human osteosarcomagenesis was supported by immunohistochemical analysis of expression of Fos in human osteosarcomas (Wu et al., 1990). This study revealed that 61% (16/30) of tumors had 1.5 to 2-fold higher nuclear immunoreactivity compared to benign or normal tissue, an increase comparable to tumors produced from v-*fos* transfected cells. When compared to the 39% of human osteosarcomas that had a normal staining pattern, the Fos overexpressing tumors did not differ in histology or clinical course.

OVEREXPRESSION OF *c-fos* IN OTHER HUMAN TUMORS

The concurrence of human and animal data indicates that overexpression of c-Fos contributes to the development of osteosarcomas and related bone tumors probably at an early stage of carcinogenesis. Surveys of other human tumors indicate that *c-fos* transcripts are frequently detected (Slamon et al., 1984), and several studies suggest that a causal relationship may exist. A survey of human breast tumor RNA for proto-oncogene overexpression revealed that the *c-fos* transcript was most frequently elevated (Biunno et al., 1988). AP-1 activity was consistently high in nuclear extracts of human breast carcinomas evaluated in a gel retardation assay and compared to normal breast tissue. However, this type of assay does not allow one to determine whether enhanced AP-1 activity is mediated by c-Fos or by Fos-related proteins, e.g., FosB, Fra-1, or Fra-2. It is also important to note that AP-1 activity, in normal breast tissue, is also high during breast involution (Linardopoulos et al., 1990). A large study comparing normal breast, benign and malignant breast lesions for nuclear Fos staining (Walker and Cowl, 1991) concluded that Fos overexpression was frequent in carcinomas and correlated to histological grading, with more cells positive in advanced lesions. Fos expression was not related to increases in estrogen growth factor (EGF) receptor or a high proliferation rate and did not correlate to estrogen receptor status. These findings suggest that the increase in c-Fos expression precedes changes in estrogen receptor expression but is not an early event in mammary carcinogenesis since it was not detected in benign tumors. Other human tumors associated with *c-fos* mRNA upregulation are thyroid adenomas and carcinomas (Terrier et al., 1988), pancreatic carcinomas (Wakita et al., 1992), neuromas and meningiomas (Riva and Larizza, 1992), melanomas (Yamanishi et al., 1991), and certain leukemias (Pinto et al., 1987). For the latter tumor type, overexpression was detected only in acute leukemias that expressed monocytic markers (Pinto et al., 1987) consistent with the experimental association of c-Fos expression and the induction of monocytic differentiation *in vitro* (Müller et al., 1985; Hass et al., 1991). However, too little data are available to suggest a causal role for Fos in these tumors and to specify a stage of carcinogenesis where Fos may be important.

UNDEREXPRESSION OF c-fos IN CARCINOGENESIS

While gain of function is generally associated with the activation of an oncogene, loss of gene function is also important in carcinogenesis, as exemplified by tumor suppressor genes. A recent study has associated c-fos expression with programmed cell death in a number of cell types (Smeyne et al., 1993), suggesting that reduced expression of fos could contribute to neoplasia through an alteration in the balance between proliferation and differentiation. Loss of c-fos gene expression has been consistently noted in human colon tumors, where c-fos mRNA and protein were lower in neoplastic lesions than in normal colon (Sugio et al., 1988; Klimpfinger et al., 1990; Nagai et al., 1992). Malignant variants of rat colon cell lines also underexpressed c-fos mRNA when compared to nontumorigenic control cells (Caignard et al., 1988). However, increased Fos protein expression was detected in morphological studies of variant crypts of rat colons after exposure to azoxymethane (Stopera et al., 1992). Such crypts are believed to be precursors of neoplastic lesions in this model. In normal colon, c-Fos expression is limited to the upper 1/3 of the crypt in the differentiated colonic population. Early expression of Fos in experimental colon carcinogenesis is manifested by aberrant expression in the lower 2/3 of the crypt, where proliferation is high (Lipkin and Deschner, 1976). Likewise, a decrease of fos mRNA in carcinomas may reflect a loss of the differentiating compartment as tumors diverge from the normal phenotype (Lipkin and Deschner, 1976).

EXPRESSION OF c-jun AND CARCINOGENESIS

Overexpressing c-jun or v-jun in mammalian cells generally does not result in neoplastic transformation unless the cells are immortalized or altered by other oncogenes such as ras (Schütte et al., 1989a). Transgenic mice overexpressing v-Jun driven by an H2-K MHC class 1 antigen gene promoter do not develop tumors unless subjected to full thickness skin wounding (Schuh et al., 1990). Wounded animals develop fibrosarcomas at the wound site after a long latency period, and progression through benign fibrous hyperplasia is documented (Schuh et al., 1990). Expression of v-Jun is exceptionally high in wound sarcomas, suggesting that both high v-Jun expression and additional genetic changes are required for carcinogenesis. The cell of origin of wound sarcomas may be a primitive mesenchymal cell since tumors frequently contain regions of muscle differentiation that are detected after malignant conversion has occurred (Schuh et al., 1992). Although high v-Jun expression was detected in thymus, spleen, and testes of transgenic mice, these sites did not develop tumors (Schuh et al., 1990). Few reports are available regarding c-Jun changes in human tumors. High levels of c-jun transcripts have been detected in human renal cell cancer and human lung cancer cell lines, but the gene is also highly expressed in the corresponding normal tissue (Minna et al., 1989; Koo et al., 1992). In contrast, c-jun was underexpressed in human metastatic melanoma cell lines when compared to neonatal melanocytes (Yamanishi et al., 1991), while junB and c-fos transcripts were elevated. Expression of c-Jun is increased after treating human B cell leukemia lymphocytes with phorbol esters and in some Hodgkin's lymphoma cell lines with the Reed-Sternberg phenotype (Gignac et al., 1990; Hsu et al., 1992). The relationship between c-Jun expression and cancer development cannot be documented in any of these examples, and studies of human sarcomas, comparable to the transgenic mouse model, have not been reported.

MODIFICATIONS OF FOS AND JUN
WHICH COULD CONTRIBUTE TO CARCINOGENESIS

The focus of most studies relating c-Fos and c-Jun to the process of carcinogenesis involve examination of levels of expression, since experiments with v-fos and v-jun indicate that increased expression can be carcinogenic. In the mouse skin model of chemically induced squamous cell cancer, expression of c-Fos or c-Jun is not altered at the mRNA level in either benign or malignant tumors compared to normal skin (Toftgard et al., 1985; Hashimoto et al., 1990). However, both benign and malignant tumors have increased expression of genes responsive to AP-1 levels, such as urokinase and metallothionein (Hashimoto et al., 1990), suggesting that AP-1 activity is elevated. Additionally, the expression of another AP-1-responsive gene, stromelysin, is significantly upregulated in carcinomas compared to benign papillomas (Ostrowski et al., 1988), suggesting that increases in AP-1 activity occur during skin tumor progression without changes in Fos or Jun expression. Recent biochemical studies have clearly shown that AP-1 activity can be regulated posttranslationally by direct modification of the proteins, for example, by phosphorylation, ADP ribosylation, or redox reactions; the latter one, however, has been demonstrated only by *in vitro* experiments (Abate et al., 1990; Amstad et al., 1990; Binétruy et al., 1991; Boyle et al., 1991; Smeal et al., 1991, 1992). Certain coexpressed dominant oncogenes, such as $c\text{-}ras^{Ha}$, $v\text{-}src$, $v\text{-}mos$, $v\text{-}raf$, E1A, and PyMT, can increase AP-1 activity without altering levels of Fos or Jun through some of these mechanisms (Herrlich and Ponta, 1989). Coexpression of other members of the *fos* and *jun* gene family can alter the activity of AP-1 and the transforming function of Jun and Fos (Chiu et al., 1989; Schütte et al., 1989b; Nakabeppu and Nathans, 1991). Many oncogenes may use AP-1 as a means to increase signal flow. Future studies must consider these mechanisms in evaluating the contribution of AP-1 transcription factors to carcinogenesis.

FOS/JUN (AP-1) AND BIOCHEMICAL PATHWAYS RELEVANT TO MULTISTAGE CARCINOGENESIS

INITIATION

The discovery that Fos interacts with Jun to regulate transcription has stimulated interest in the role of Fos, Jun, and AP-1 activity in specific stages of carcinogenesis. Expression of the *c-fos* and *c-jun* proto-oncogenes appears to be particularly relevant for the immediate interaction of carcinogens with target cells and the processing of consequent DNA damage. Thus, AP-1 activity may be important in initiation of carcinogenesis. A variety of DNA-damaging agents, including X-rays, ultraviolet light, chemical mutagens, oxidants, and cancer chemotherapeutic agents, cause a prolonged increase in c-Fos and c-Jun expression in a variety of mammalian cells, including cells of human origin (Stein et al., 1989, 1992; Crawford et al., 1989; Hollander and Fornace, 1989; Amstad et al., 1990; Sherman et al., 1990; Devary et al., 1991). X-rays and ultraviolet light are particularly potent and selective inducers of *c-jun* transcription and stabilize the *c-jun* mRNA (Sherman et al., 1990; Devary et al., 1991). Increased AP-1 activity after DNA damage may not require de novo protein synthesis, suggesting that posttranslational modification is also important in the response (Datta et al., 1992). The differential response of c-Jun and c-Fos to agents that have specific spectra of cellular targets suggests that *c-jun* may be induced by DNA-damaging agents, while *c-fos* may respond to damage to other cellular macromolecules as well as DNA (Holbrook and Fornace, 1991). Both protein kinase C-dependent and -independent pathways have been described for the induction of *c-fos* and *c-jun* by genotoxic compounds (Lin et al., 1990; Datta et al., 1992), and reactive oxygen intermediates may regulate the induction by X-rays (Datta et al., 1992). AP-1 transcriptional or DNA binding activity is generally increased in cells exposed to genotoxic agents (Stein et al., 1989; Devary et al., 1991), and AP-1-dependent secondary response genes (e.g., collagenase, urokinase) are expressed as a consequence of AP-1 induction. However, certain genes responsive to DNA damage, such as gadd 45, are independent of AP-1 activity (Papathanasiou et al., 1991).

The reproducible induction of c-Fos and c-Jun in many cell types after a genotoxic insult implies that these factors are important for the cellular response, but the function remains obscure. Assuming that AP-1 activity is protective, several targets for transactivation can be considered as functionally important in this response. Glutathione transferase catalyzes the conjugation of electrophilic compounds with glutathione. Overexpression of this enzyme protects cells from the influence of a number of genotoxic and cytotoxic agents (Nakagawa et al., 1990). The placental isoform of glutathione transferase, GT-P, is inducible by either phorbol ester exposure or *c-jun* transfection, suggesting that AP-1 activity is important for upregulating this enzyme (Sakai et al., 1992). The promoter region of the rat GT-P contains an enhancer element composed of two imperfect TREs that may be the AP-1 responsive site (Okuda et al., 1990). An AP-1 binding site has also been described within the antioxidant response element (ARE) of the NAD(P)H:quinone oxidoreductase (NQO) gene, and mutations in this region abolish basal and inducible activity (Li and Jaiswal, 1992). This enzyme is induced by a number of carcinogens with quinone intermediates and protects against free radical toxicity (Li and Jaiswal, 1992). The AP-1 region of NQO binds c-Fos and JunD, and the ARE is stimulated by carcinogens and antioxidants. While data are not available on induction by ionizing radiation, this pathway would be expected to play a protective role against oxidative intermediates generated internally from physical carcinogens, and supports a protective role for AP-1 induction. Interestingly, inhibition of tyrosine kinases involved in UV-induced activation of AP-1 potentiates cell killing by UV (Devary et al., 1992).

The involvement of oxidation reduction mechanisms in transcriptional regulation by Fos and Jun has been demonstrated recently with the discovery of Ref-1, a redox factor that modifies a conserved cysteine on Fos and Jun and regulates binding with the AP-1 site on DNA, at least *in vitro* (Xanthoudakis and Curran, 1992). Copurifying with AP-1 proteins, Ref-1 has been shown to have apurinic/apyrimidinic endonuclease DNA repair activity, thus linking AP-1 activity to DNA repair (Xanthoudakis et al., 1992). This association suggests that Fos and Jun and, perhaps, other AP-1 factors may be important in the repair of DNA damage and provides an additional mechanism of protection from the damaging agent.

So often in carcinogenesis research, paradoxical responses complicate interpretations of seemingly relevant phenomena. The induction of c-Fos and c-Jun by genotoxic agents may protect cells from DNA damage, but other data indicate that detrimental consequences can also be expected. For example, overexpression of *c-fos* in 3T3 cells results in an increase in chromosomal aberrations and point mutations (van den Berg et al., 1991), while overexpression in a human osteosarcoma cell line causes an increase in recombination of a transfected HSV-TK plasmid (van den Berg et al., 1993). Furthermore, the induction of mutations by ultraviolet light and dominant oncogenes is reduced by antisense *c-fos* oligonucleotides (van den Berg et al., 1991). Multiple reports have documented the induction of AP-1 activity and the formation of chromosomal aberrations after phorbol ester exposure (Petrusevska et al., 1988; Fürstenberger et al., 1989). These studies indicate that under certain conditions, Fos and, perhaps, Jun overexpression could enhance the carcinogenic process. These influences may contribute to the tumor formation in bone or cartilage seen so reproducibly in human osteosarcomas or c-Fos transgenic/chimeric mice.

TUMOR PROMOTION

The essential process of tumor promotion is the clonal outgrowth of initiated cells within the context of a normal tissue. The general mechanism of promotion in a number of epithelial tissues is a differential response of initiated and normal cells to a specific stimulus, which effects the proliferation and differentiation of the target tissue (Yuspa and Poirier, 1988; Harris, 1991). Changes in AP-1 could alter gene expression within initiated cells, producing a phenotype that responds differently to physiological signals. Alternatively, changes in AP-1 activity could alter the response of cells to external stimuli, such as exogenous tumor promoters, and indirectly contribute to tumor promotion.

AP-1 and Exogenous Tumor Promoters

Liver carcinogenesis often involves the action of cytotoxic promoting agents to which initiated cells are relatively resistant (Farber, 1980, 1984). Several AP-1-regulated detoxification enzymes may be involved in resistance. Spontaneous and chemically induced preneoplastic liver foci, as well as adenomatous and carcinomatous lesions of the lung and colon, have elevated constitutive levels of glutathione transferase isoforms (Sato, 1989). Overexpression of GT-P in 3T3 cells protects against the cytotoxic effects of adriamycin and ethacrinic acid (Nakagawa et al., 1990). NQ01 is also elevated in liver tumor cells (Schlager and Powis, 1990; Cresteil and Jaiswal, 1991) and is AP-1 regulated. Introduction of the *ras* oncogene into liver cells, upregulates GT-P (Power et al., 1987; Li et al., 1988). Elevated constitutive activity of AP-1 must occur early in initiated liver cells since resistance to cytotoxicity is essential to tumor promotion in this model. Some environmental/industrial compounds may cause cancer in humans by acting both as genotoxic agents and tumor promoters. A recent study has shown that the active compounds in asbestos fibers cause a persistent induction of c-*fos* and c-*jun* expression in lung mesothelial cells, and AP-1 activity in mesothelial and tracheal epithelium cells (Heintz et al., 1993). Thus in this case constitutive AP-1 activity results from continuous exposure to external signals.

In skin carcinogenesis, activation of c-*ras*[Ha] is an initiating genetic change (Balmain and Pragnell, 1983; Balmain et al., 1984; Roop et al., 1986). Initiated cells differ from normal epidermal cells in that they are resistant to the induction of terminal differentiation by phorbol esters and, thus, have a growth advantage *in situ* when skin is promoted by 12-*O*-tetradecanoylphorbol-13-acetate (Yuspa and Poirier, 1988; Yuspa, 1987). As seen for liver cells, the introduction of v-*ras*[Ha] into epidermal cells is associated with an increase in AP-1 activity (Glick, unpublished results), and benign tumors demonstrate overexpression of AP-1 regulated genes (Hashimoto et al., 1990), even though c-Fos and c-Jun are not elevated (Toftgard et al., 1985; Hashimoto et al., 1990). This suggests that posttranscriptional or posttranslational changes in Fos and Jun are constitutive in *ras* oncogene-initiated keratinocytes. It is possible that constitutive activation of AP-1 reprograms the response to phorbol esters from differentiation to proliferation. In the normal epidermis c-fos is expressed in both the basal and granular cell layers suggesting a role in both cell proliferation and differentiation (Fisher et al., 1991). Since TPA induces a transient increase in c-fos expression in cultured keratinocytes (Dotto et al., 1986), *fos* may play a role in tumor promotion of the epidermis. When expression of the v-*fos* oncogene was targeted to the epidermis of transgenic mice using a keratin 1 promoter construct, hyperplastic and hyperkeratotic lesions developed at sites of wounding, which with long latency led to benign squamous papillomas (Greenhalgh et al., 1993a). This suggests that v-*fos* requires additional signals such as those elicited by wounding to induce a constituitively hyperplastic epidermis characteristic of tumor promotion.

Phorbol ester-sensitive (P+) and -resistant (P-) subclones of the JB6 mouse epidermal cell line have provided some insight into potential mechanisms by which AP-1 activity contributes to resistance to phorbol esters in keratinocytes. P⁺ cells, but not P⁻ cells, are anchorage independent when exposed to phorbol esters. P⁺ clones respond to TPA with a significant increase in *c-jun* mRNA (Ben-Ari et al., 1992), and transfection of P⁺ cells with a dominant negative *c-jun* mutant blocked TPA induced transformation (Dong et al., 1994). Both P⁺ and P⁻ clones express a 35- and 46-kDa protein immunoprecipitated by anti-Fra-1 antibody. TPA elevates expression of the 35-kDa Fra-1 protein in both clones; however, only in the P⁻ clone is the 46-kDa Fra-1 protein also induced (Bernstein et al., 1992). It is possible that AP-1 complexes formed with the 46 kDa fra-1 protein act as inhibitors of the TPA response in these cells. Mechanistically, these differences in AP-1 regulation between P⁻ and P⁺ clones could be related to marked differences in antioxidant defenses between the two cell types (Crawford et al., 1989). P⁻ clones are sensitive to oxidant mediated cell killing and have lower levels of superoxide dismutase and catalase than do P⁺ clones (Crawford et al., 1989).

AP-1 and Endogenous Signaling Molecules

The interaction of AP-1 with the muscle-specific transcription factors MyoD and myogenin is an example of a direct effect of AP-1 on the differentiation program of a specific cell type. MyoD and myogenin are members of a family of helix–loop–helix transcriptional activators that act to induce muscle-specific gene expression. Transcriptional activation of the muscle creatine kinase enhancer by MyoD and myogenin is blocked by induction of endogenous c-Fos and c-Jun, and by cotransfection of *c-fos* and *c-jun* expression plasmids (Li et al., 1992). The blocked cells are unable to respond to inducers of muscle differentiation and continue to proliferate (Su et al., 1991). Mutational analysis suggests that the inhibition of MyoD transactivation occurs through a physical interaction between the Jun leucine zipper domain and the helix–loop–helix region of MyoD (Bengal et al., 1992; Li et al., 1992). The role for this type of interaction in carcinogenesis remains to be determined, and may depend on the existence of transcription factors homologous to MyoD and myogenin for other cell types.

Bone hyperplasia and subsequent osteosarcomas in FBJ-MSV-infected and c-Fos transgenic mice may be consequences of AP-1 interaction with hormone receptors in bone-forming cells. Proliferating cultures of normal osteoblasts express c-Fos and c-Jun and have high levels of AP-1 activity, which is downregulated when differentiation and matrix mineralization is induced by vitamin D (Owen et al., 1990). Cotransfection of *c-fos* and *c-jun* expression vectors into a bone cell line inhibits transcriptional activation of the osteocalcin promoter by vitamin D (Schüle et al., 1990) through binding of AP-1 to a TRE within vitamin D response elements of the osteocalcin promoter (Owen et al., 1990; Schüle et al., 1990). Similar inhibition occurs with the alkaline phosphatase gene, another marker of differentiation induced by vitamin D (Lian et al., 1991). Thus, constitutive overexpression of Fos in virally infected and transgenic osteoblasts could block the normal induction of differentiation by vitamin D and lead to hyperplasia of the bone precursor cells. In this sense vitamin D is a promoter that behaves differentially on affected cells overexpressing Fos. Since 75 to 85% of the transgenic animals that get early bone lesions and tumors are male (Rüther et al., 1989), it is possible that androgens also contribute to tumor formation through a similar interaction with the AP-1 pathway.

The molecular interaction of the AP-1 complex and hormone receptors in the mammary gland provides a basis for understanding hormonal promotion in this tissue. The growth of most experimentally induced and spontaneous rat and mouse mammary carcinomas depends on estrogen and other female sex hormones, such that pregnancy and parturition result in cycles of growth and regression (Russo et al., 1983). Human mammary carcinomas have high AP-1 activity (Biunno et al., 1988). When cultures of normal mammary epithelial cell lines are stimulated to proliferate, AP-1 binding activity is comparable to their oncogene-transformed counterparts (Jehn et al., 1992). Estrogen is required for growth of several human breast carcinoma cell lines, and c-Fos expression is induced after treatment with this hormone (van der Burg et al., 1992; Sutherland et al., 1992), suggesting that proliferation is associated with high levels of AP-1. In contrast, induction of terminal differentiation in mammary epithelial cell lines by glucocorticoid and prolactin results in a decrease in AP-1 activity (Jehn et al., 1992). Transformation of mammary cells with *ras*, *mos*, and *src* oncogenes abrogates the reduction in AP-1 activity in response to lactogenic hormones, and blocks the induction of differentiation-specific proteins. Coexpression of transfected *c-fos* and *c-jun* in normal mammary cells also blocks glucocorticoid mediated-transactivation of differentiation-related gene expression (Jehn et al., 1992). The inhibition of glucocorticoid receptor appears to be cell type and promoter specific (Jonat et al., 1990; Schüle et al., 1990; Yang-Yen et al., 1990; Shemshedini et al., 1991) and can involve protein–protein interaction between the AP-1 complex and the glucocorticoid receptor as well as competition for overlapping binding sites. Thus, while normal mammary epithelial cells differentiate in response to specific hormones, initiated cells with increased levels of AP-1 activity would be blocked in a proliferative state and continue to grow in a hormonal milieu that induces differentiation in normal cells.

Anti-promotion may represent a pharmacological extension of the antagonistic interaction of endogenous hormone receptors with AP-1. A classical model for anti-promotion is the inhibition of TPA-mediated promotion by retinoids in two-stage chemical carcinogenesis of the mouse epidermis. Glucocorticoids and vitamin D analogues are also active inhibitors in this system (Chida et al., 1985; Jonat et al., 1990). In the presence of retinoic acid, transcriptional activation by c-Jun of a collagenase promoter AP-1 CAT construct is blocked by cotransfection with RARα, RARβ, and RARγ (Schüle et al., 1991; Yang-Yen et al., 1991). Repression is mediated by a protein–protein interaction involving the RAR DNA binding domain (Schüle et al., 1991; Yang-Yen et al., 1991). This is specific for the RAR subfamily since the RXR does not repress activation (Schüle et al., 1991). Interestingly, the PML-RARα fusion protein that results from the t(15;17)(q22;q12-21) translocation of acute promeylocytic leukemia acts as a retinoic acid-dependent activator rather than inhibitor of AP-1 activity (Doucas et al., 1993).

MALIGNANT CONVERSION AND PROGRESSION

Malignant conversion and neoplastic progression involve marked changes in the phenotype of the premalignant cells and in the interaction of tumor cells with other cells in the target tissue. As discussed previously, elevated levels of AP-1 may enhance genetic damage, providing a pathway for Fos and Jun to influence later stages of carcinogenesis. Additionally, *in vitro* studies with the *c-fos* promoter have shown that it is downregulated by both the Rb and wt p53 proteins (Robbins et al., 1990; Kley et al., 1992). Thus, loss of these tumor suppressor genes during malignant progression could further amplify genetic change through upregulation of Fos expression, or directly cause a progressive change in the neoplastic phenotype. Evidence that quantitative changes in AP-1 levels are relevant for progression is seen in experimental carcinogenesis of the mouse epidermis. Introduction of *v-fos* into benign skin tumor cells, or coinfection of primary keratinocytes with *v-fos* and *v-ras* retroviruses, results in malignant conversion (Greenhalgh et al., 1988, 1990). The carcinomas produced by introducing *v-ras* and *v-fos* into normal cells express the same tumor markers as chemically induced carcinomas (Greenhalgh et al., 1990; Tennenbaum et al., 1992). Additionally, AP-1 activity as measured by gel shift, transactivation of reporter constructs, and expression of AP-1 responsive metalloprotease genes, was constituitively increased in a radiation induced malignant subclone of the 308 papilloma cell line, while in the parent line AP-1 activity was TPA dependent (Domann et al., 1994). v-*fos* can also cooperate with an E1a/T antigen chimaeric gene and HPV-

18 to neoplastically transform human keratinocytes (Lee et al., 1993; Pei et al., 1993). In contrast to these results, targeted expression of both v-ras^{Ha} and v-fos to the epidermis by a keratin 1 promoter in a double transgenic mouse line, resulted only in benign tumors, although the tumors arose earlier and grew larger than the single transgene parental strains (Greenhalgh et al., 1993b). One interpretation is that additional changes other than elevation of AP-1 activity are required for malignant conversion in the epidermis. However, these results are confounded by the known down-regulation of keratin 1 during premalignant progression of epidermal cells. Overexpression of v-Jun in benign tumor cells does not convert papillomas to carcinomas (Tsang et al., 1992a, 1992b; Glick unpublished). However, similar experiments with mammalian c-Jun should be done before the final interpretation of these results.

The spectrum of genes regulated by the AP-1 complex is particularly relevant to this stage of carcinogenesis. A number of reports have indicated that overexpression of v-Fos or c-Fos in malignant cell lines increases their metastatic potential (Kawano et al., 1987; Nakamatsu et al., 1989). In some cases, the acquisition of metastatic activity correlates to increased cell mobility or invasiveness (Taniguchi et al., 1989). The expression of a number of proteases is regulated by AP-1 activity, including stromelysin, collagenase I, and urokinase (Angel et al., 1987; McDonnell et al., 1990; Nerlov et al., 1991). Furthermore, procathepsin L secretion is enhanced in rat fibroblasts transfected with v-fos (Taniguchi et al., 1990). Together, these studies indicate that AP-1-mediated expression of secreted proteases contributes to later stages of tumor progression. At least one study proposes a mechanism where high expression of c-Fos may reduce metastatic potential. Lewis lung carcinoma cells are induced to express the H2-K MHC class I antigen when c-Fos or v-Fos is overexpressed, thus increasing their immunogenicity and reducing metastasis (Kushtai et al., 1990). This biological change would only be detected in an *in vivo* assay in the proper host but must be considered in any evaluation of Fos and Jun in multistage carcinogenesis.

CONCLUSIONS

Given the central role of the AP-1 complex in regulating gene expression relating to all aspects of cell physiology, it is not surprising that changes in AP-1, through measurements of activity or expression of its constituent proteins Fos and Jun, have been implicated in all steps of the neoplastic process. The immediate cellular repair response to chemical and physical carcinogens results in increased AP-1 activity, but this repair mechanism may, in turn, enhance survival of neoplastic cells, as well as generate further mutations necessary for tumor progression. Accumulating molecular and cell culture data indicate that increased AP-1 activity resulting from activating mutations in dominant oncogenes, and loss of suppressor genes, could produce the altered response to proliferation and differentiation signals that is characteristic of the initiated cell. Elevated AP-1 activity is likely to upregulate genes such as proteases and cytokines, as well as directly block differentiation signals mediated by steroid hormone receptors and other transcription factors. All of these alterations in gene expression contribute to the clonal outgrowth of initiated cells as well as the autonomous and invasive growth properties of cells in the later stages of the neoplastic process.

In spite of the clear *in vitro* relationship of AP-1 and carcinogenesis, only a limited number of human cancers have elevated expression of c-Fos, and information on expression of c-Jun is sparse. *In situ* methods should be pursued to more accurately define the extent of this type of regulation of AP-1 activity in human tumors. However, with the potential synergistic and antagonistic interactions that can occur between members of the Fos and Jun families and between AP-1 and other transcription factors, as well as the posttranslational modifications that can alter AP-1 activity, it is clear the measurements of AP-1 activity in human and experimental animal systems should be the relevant biomarker. Transgenic mice with an AP-1 element linked to either CAT or ß-galactosidase should allow analysis of changes in AP-1 activity at discrete stages of carcinogenesis.

Given the seemingly central role of AP-1 activity in all steps of the neoplastic process it is surprising that in two transgenic mouse models overexpressing oncogenic forms of Fos and Jun in a wide range of tissues, only the bone and thymus were affected without experimental intervention. Although AP-1 levels have not been reported in these animals, the results suggest that there are powerful mechanisms within normal cells for controlling AP-1 activity. While much important biochemical and molecular analysis remains, these mice provide a powerful model to demonstrate the *in vivo* relevance of the influence of AP-1 activity on carcinogenesis *in vitro*. Chemical carcinogenesis studies on these transgenic animals should provide useful information on the role of Fos and Jun in multistage carcinogenesis. Mice with individual members of the Fos and Jun families deleted could be useful for carcinogenesis studies if there is not a high degree of functional overlap between family members. Transgenic animals generated with dominant negative versions of Fos and Jun (Ransone et al., 1990) might provide a better source of "AP-1-deficient" animals. Finally, transgenic animal models will be necessary to demonstrate the fundamental importance for carcinogenesis of the AP-1–steroid hormone receptor interaction *in vivo*.

REFERENCES

Abate, C., Patel, L., Rauscher, F. J., III, and Curran, T., Redox regulation of Fos and Jun DNA-binding activity in vitro. *Science*, 249, 1157, 1990.

Amstad, P., Crawford, D., Muehlematter, D., Zbinden, I., Larsson, R., and Cerutti, P., Oxidants stress induces the proto-oncogenes, c-*fos* and c-*myc* in mouse epidermal cells, *Bull. Cancer*, 77, 501, 1990.

Angel, P., Imagawa, M., Chiu, R., Stein, B., Imbra, R. J., Rahmsdorf, H. J., Jonat, C., Herrlich, P., and Karin, M., Phorbol ester-inducible genes contain a common cis element recognized by a TPA-modulated trans-acting factor, *Cell*, 49, 729, 1987.

Balmain, A. and Pragnell, I. B., Mouse skin carcinomas induced in vivo by chemical carcinogens have a transforming harvey ras oncogene, *Nature*, 303, 72, 1983.

Balmain, A., Ramsden, M., Bowden, G. T., and Smith, J., Activation of the mouse cellular Harvey-ras gene in chemically induced benign skin papillomas, *Nature*, 307, 658, 1984.

Ben-Ari, E. T., Bernstein, L. R., and Colburn, N. H., Differential c-*jun* expression in response to tumor promoters in JB6 cells sensitive or resistant to neoplastic transformation, *Mol. Carcinogenesis*, 5, 62, 1992.

Bengal, E., Ransone, L., Scharfmann, R., Dwarki, V. J., Tapscott, S. J., Weintraub, H., and Verma, I. M., Functional antagonism between c-Jun and MyoD proteins: A direct physical association, *Cell*, 68, 507, 1992.

Bernstein, L. R., Bravo, R., and Colburn, N. H., 12-O-tetradecanoylphorbol-13-acetate-induced levels of AP-1 proteins: A 46-kDa protein immunoprecipitated by anti-fra-1 and induced in promotion-resistant but not promotion-sensitive JB6 cells, *Mol. Carcinogenesis*, 6, 221, 1992.

Binétruy, B., Smeal, T., and Karin, M., Ha-Ras augments c-Jun activity and stimulates phosphorylation of its activation domain, *Nature*, 351, 122, 1991.

Biunno, I., Pozzi, M. R., Pierotti, M. A., Pilotti, S., Cattoretti, G., and Della Porta, G., Structure and expression of oncogenes in surgical specimens of human breast carcinomas, *Br. J. Cancer*, 57, 464, 1988.

Boyle, W. J., Smeal, T., Defize, L. H., Angel, P., Woodgett, J. R., Karin, M., and Hunter, T., Activation of protein kinase C decreases phosphorylation of c-Jun at sites that negatively regulate its DNA-binding activity, *Cell*, 64, 573, 1991.

Caignard, A., Kitagawa, Y., Sato, S., and Nagao, M., Activated K-*ras* in tumorigenic and non-tumorigenic cell variants from a rat colon adenocarcinoma, induced by dimethylhydrazine, *Jpn. J. Cancer Res.*, 79, 244, 1988.

Chida, K., Hashiba, H., Fukushima, M., Suda, T., and Kuroki, T., Inhibition of tumor promotion in mouse skin by 1, 25-dihydroxyvitamin D3, *Cancer Res.*, 45, 5426, 1985.

Chiu, R., Angel, P., and Karin, M., Jun-B differs in its biological properties from, and is a negative regulator of, c-Jun, *Cell*, 59, 979, 1989.

Cohen, D. R. and Curran, T., *fra*-1: A serum-inducible, cellular immediate-early gene that encodes a fos-related antigen, *Mol. Cell. Biol.*, 8, 2063, 1988.

Crawford, D. R., Amstad, P. A., Foo, D. D., and Cerutti, P. A., Constitutive and phorbol-myristate-acetate regulated antioxidant defense of mouse epidermal JB6 cells, *Mol. Carcinogenesis*, 2, 136, 1989.

Cresteil, T. and Jaiswal, A. K., High levels of expression of the NAD(P)H:quinone oxidoreductase (NQO1) gene in tumor cells compared to normal cells of the same origin, *Biochem. Pharmacol.*, 42, 1021, 1991.

Curran, T., Peters, G., Van Beveran, C., Teich, N. M., and Verma, I. M., FBJ murine osteosarcoma virus: Identification and molecular cloning of biologically active proviral DNA, *J. Virol.*, 44, 674, 1982.

Datta, R., Hallahan, D. E., Kharbanda, S. M., Rubin, E., Sherman, M. L., Huberman, E., Weichselbaum, R. R., and Kufe, D. W., Involvement of reactive oxygen intermediates in the induction of c-jun gene transcription by ionizing radiation, *Biochemistry*, 31, 8300, 1992.

De Togni, P., Niman, H., Raymond, V., Sawchenko, P., and Verma, I. M., Detection of *fos* protein during osteogenesis by monoclonal antibodies, *Mol. Cell. Biol.*, 8, 2251, 1988.

Devary, Y., Gottlieb, R. A., Lau, L. F., and Karin, M., Rapid and preferential activation of the c-*jun* gene during the mammalian UV response, *Mol. Cell. Biol.*, 11, 2804, 1991.

Devary, Y., Gottlieb, R. A., Smeal, T., and Karin, M., The mammalian ultraviolet response is triggered by activation of Src tyrosine kinases, *Cell*, 71, 1081, 1992.

Domann, F. E., Levy, J. P., Finch, J. S., and Bowden, G. T., Constituitive AP-1 DNA binding and transactivating ability of malignant but not benign mouse epidermal cells, *Mol. Carcinogenesis*, 9, 61, 1994.

Dong, Z., Birrer, M., Watts, R. G., Matrisian, L. M., and Colburn, N., Blocking of tumor promoter-induced AP-1 activity inhibits induced transformation in jB6 mouse epidermal cells, *Proc. Natl. Acad. Sci. U.S.A.*, 91, 609, 1994.

Dony, C. and Gruss, P., Proto-oncogene c-*fos* expression in growth regions of fetal bone and mesodermal web tissue, *Nature*, 328, 711, 1987.

Dotto, G. P., Gilman, M. Z., Maruyama, M., and Weinberg, R. A., c-*myc* and c-*fos* expression in differentiating mouse primary keratinocytes, *EMBO J.*, 5, 2853, 1986.

Doucas, V., Brockes, J. P., Yaniv, M., de Thé, H., and Dejean, A., The PML-retinoic acid receptor α translocation converts the receptor from an inhibitor to a reinoic acid-dependent activator of transcription factor AP-1, *Proc. Natl. Acad. Sci. U.S.A.*, 90, 9345, 1993.

Farber, E., The sequential analysis of liver cancer induction, *Biochim. Biophys. Acta*, 605, 149, 1980.

Farber, E., Cellular biochemistry of the stepwise development of cancer with chemicals: G. H. A., Clowes memorial lecture, *Cancer Res.*, 44, 5463, 1984a.

Farber, E., Pre-cancerous steps in carcinogenesis their physiological adaptive nature, *Biochim. Biophys. Acta*, 738, 171, 1984b.

Finkel. M. P., Biskis. B. O., and Jinkins. P. B., Virus induction of osteosarcomas in mice. *Science*, 151, 698, 1966.

Fisher. C., Byers. M. R., Iadarola, M. J., and Powers. E. A., Patterns of epithelial expression of Fos protein suggest important role in the transition from viable to cornified cell during keratinization. *Development*, 111, 253, 1991.

Fürstenberger. G., Schurich, B., Kaina. B., Petrusevska. R. T., Fusenig. N. E., and Marks. F., Tumor induction in initiated mouse skin by phorbol esters and methyl methanesulfonate: Correlation between chromosomal damage and conversion ('stage I of tumor promotion') *in vivo. Carcinogenesis*, 10, 749, 1989.

Garcia. M. and Samarut. J., Cooperation of v-*jun* and v-*erb*B oncogenes in embryo fibroblast transformation in vitro and in vivo. *J. Virol.*, 64, 4684, 1990.

Gignac. S. M., Buschie. M., Pettit. G. R., Hoffbrand. A. V., and Drexler, H. G., Expression of proto-oncogene c-jun during differentiation of B-chronic lymphocytic leukemia. *Leukemia*, 4, 441, 1990.

Greenhalgh. D. A. and Yuspa. S. H., Malignant conversion of murine squamous papilloma cell lines by transfection with the fos oncogene. *Mol. Carcinogenesis*, 1, 134, 1988.

Greenhalgh. D. A., Welty. D. J., Player, A., and Yuspa. S. H., Two oncogenes, v-fos and v-ras, cooperate to convert normal keratinocytes to squamous cell carcinomas. *Proc. Natl. Acad. Sci. U.S.A.*, 87, 643, 1990.

Greenhalgh. D. A., Rothnagel. J. A., Wang. X. J., Quintanilla. M. I., Orengo. C. C., Gagne. T. A., Bundman. D. S., Longley. M. A., Fisher. C., and Roop. D. R., Hyperplasia, hyperkeratosis and benign tumor production in transgenic mice by a targeted v-*fos* oncogene suggests a role for *fos* in epidermal differentiation and neoplasia. *Oncogene*, 8, 2145, 1993a.

Greenhalgh. D. A., Quinatanilla. M. I., Orengo. C. C., Barber. J. L., Eckhardt. J. N., Rothnagel. J. A., and Roop. D. R., Cooperation between v-*fos* and v-*ras*Ha induces conversion. *Cancer Res.*, 53, 5071, 1993b.

Grigoriadis. A. E., Schellander. K., Wang. Z.-Q., and Wagner. E. F., Osteoblasts are target cells for transformation in c-*fos* transgenic mice. *J. Cell Biol.*, 122, 685, 1993.

Harris. C. C., Chemical and physical carcinogenesis: Advances and perspectives for the 1990s. *Cancer Res.*, 51, 5023s, 1991.

Hashimoto. Y., Tajima. O., Hashiba. H., Nose. K., and Kuroki. T., Elevated expression of secondary, but not early, responding genes to phorbol ester tumor promoters in papillomas and carcinomas of mouse skin. *Mol. Carcinogenesis*, 3, 302, 1990.

Hass. R., Brach. M., Kharbanda. S., Giese. G., Traub. P., and Kufe. D., Inhibition of phorbol ester-induced monocytic differentiation by dexamethasone is associated with down-regulation of c-fos and c-jun (AP-1). *J. Cell Physiol.*, 149, 125, 1991.

Heintz. N. H., Janssen. Y. M., and Mossman. B. T., Persistent induction of c-*fos* and c-*jun* expression by asbestos. *Proc. Natl. Acad. Sci. U.S.A.*, 90, 3299, 1993.

Herrlich. P. and Ponta. H., 'Nuclear' oncogenes convert extracellular stimuli into changes in the genetic program. *Trends Genet.*, 5, 112, 1989.

Holbrook. N. J. and Fornace. A. J., Jr., Response to adversity: Molecular control of gene activation following genotoxic stress. *New Biol.*, 3, 825, 1991.

Hollander. M. C. and Fornace. A. J., Jr., Induction of *fos* RNA by DNA-damaging agents. *Cancer Res.*, 49, 1687, 1989.

Hsu. S. M., Xie. S. S., el-Okda. M. O., and Hsu. P. L., Correlation of c-fos/c-jun expression with histiocytic differentiation in Hodgkin's Reed-Sternberg cells. Examination in HDLM-1 subclones with spontaneous differentiation. *Am. J. Pathol.*, 140, 155, 1992.

Jehn. B., Costello. E., Marti. A., Keon. N., Deane. R., Li. F., Friis. R. R., Burri. P. H., Martin. F., and Jaggi. R., Overexpression of Mos, Ras, Src, and Fos inhibits mouse mammary epithelial cell differentiation. *Mol. Cell. Biol.*, 12, 3890, 1992.

Johnson, R. S., Spiegelman, B. M., and Papaioannou, V., Pleiotropic effects of a null mutation in the c-*fos* proto-oncogene, *Cell*, 71, 577, 1992.

Jonat, C., Rahmsdorf, H. J., Park, K. K., Cato, A. C. B., Gebel, S., Ponta, H., and Herrlich, P., Antitumor promotion and antiinflammation: Down-modulation of AP-1 (Fos/Jun) activity by glucocorticoid hormone, *Cell*, 62, 1189, 1990.

Kawano, T., Taniguchi, S., Nakamatsu, K., Sadano, H., and Baba, T., Malignant progression of a transformed rat cell line by transfer of the v-fos oncogene, *Biochem. Biophys. Res. Commun.*, 149, 173, 1987.

Kley, N., Chung, R. Y., Fay, S., Loeffler, J. P., and Seizinger, B. R., Repression of the basal c-fos promoter by wild-type p53, *Nucleic Acids Res.*, 20, 4083, 1992.

Klimpfinger, M., Zisser, G., Ruhri, C., Putz, B., Steindorfer, P., and Höfler, H., Expression of c-myc and c-fos mRNA in colorectal carcinoma in man, *Virchows Arch. B Cell Pathol.*, 59, 165, 1990.

Koo, A. S., Chiu, R., Soong, J., deKernion, J. B., and Belldegrun, A., The expression of C-jun and junB mRNA in renal cell cancer and in vitro regulation by transforming growth factor beta 1 and tumor necrosis factor alpha 1, *J. Urol.*, 148, 1314, 1992.

Kushtai, G., Feldman, M., and Eisenbach, L., c-fos transfection of 3LL tumor cells turns on MHC gene expression and consequently reduces their metastatic competence, *Int. J. Cancer*, 45, 1131, 1990.

Lee, M.-S., Yang, J.-H., Salehi, Z., Arnstein, P., Chen, L.-S., Jay, G., and Rhim, J. S., Neoplastic transformation of a human keratinocyte cell line by the v-*fos* oncogene, *Oncogene*, 8, 387, 1993.

Lee, W. M., Lin, C., and Curran, T., Activation of the transforming potential of the human *fos* proto-oncogene requires message stabilization and results in increased amounts of partially modified *fos* protein, *Mol. Cell. Biol.*, 8, 5521, 1988.

Li, Y. and Jaiswal, A. K., Regulation of human NAD(P)H:quinone oxidoreductase gene. Role of AP1 binding site contained within human antioxidant response element, *J. Biol Chem.*, 267, 15097, 1992.

Li, L., Chambard, J. C., Karin, M., and Olson, E. N., Fos and Jun repress transcriptional activation by myogenin and MyoD: The amino terminus of Jun can mediate repression, *Genes Dev.*, 6, 676, 1992.

Li, Y. C., Seyama, T., Godwin, A. K., Winokur, T. S., Lebovitz, R. M., and Lieberman, M. W., MTrasT24, a metallothionein-ras fusion gene, modulates expression in cultured rat liver cells of two genes associated with in vivo liver cancer, *Proc. Natl. Acad. Sci. U.S.A.*, 85, 344, 1988.

Lian, J. B., Stein, G. S., Bortell, R., and Owen, T. A., Phenotype suppression: A postulated molecular mechanism for mediating the relationship of proliferation and differentiation by Fos/Jun interactions at AP-1 sites in steroid responsive promoter elements of tissue-specific genes, *J. Cell. Biochem.*, 45, 9, 1991.

Lin, C. S., Goldthwait, D. A., and Samols, D., Induction of transcription from the long terminal repeat of Moloney murine sarcoma provirus by UV-irradiation, x-irradiation, and phorbol ester, *Proc. Natl. Acad. Sci. U.S.A.*, 87, 36, 1990.

Linardopoulos, S., Malliri, A., Pintzas, A., Vassilaros, S., Tsikkinis, A., and Spandidos, D. A., Elevated expression of AP-1 activity in human breast tumors as compared to normal adjacent tissue, *Anticancer Res.*, 10, 1711, 1990.

Lipkin, M. and Deschner, E., Early proliferative changes in intestinal cells, *Cancer Res.*, 36, 2665, 1976.

Maki, Y., Bos, T. J., Davis, C., Starbuck, M., and Vogt, P. K., Avian sarcoma virus 17 carries the *jun* oncogene, *Proc. Natl. Acad. Sci. U.S.A.*, 84, 2848, 1987.

McDonnell, S. E., Kerr, L. D., and Matrisian, L. M., Epidermal growth factor stimulation of stromelysin mRNA in rat fibroblasts requires induction of proto-oncogenes c-fos and c-jun and activation of protein kinase, *Mol. Cell. Biol.*, 10, 4284, 1990.

Miller, A. D., Curran, T., and Verma, I. M., c-*fos* protein can induce cellular transformation: A novel mechanism of activation of a cellular oncogene, *Cell*, 36, 51, 1984.

Miller, E. C., Some current perspectives on chemical carcinogenesis in humans and experimental animals, *Cancer Res.*, 38, 1479, 1978.

Minna, J. D., Schütte, J., Viallet, J., Thomas, F., Kaye, F. J., Takahashi, T., Nau, M., Whang-Peng, J., Birrer, M., and Gazdar, A. F., Transcription factors and recessive oncogenes in the pathogenesis of human lung cancer, *Int. J. Cancer*, Suppl. 4, 32, 1989.

Müller, R., Curran, T., Müller, D., and Guilbert, L., Induction of c-*fos* during myelomonocytic differentiation and macrophage proliferation, *Nature*, 314, 546, 1985.

Nagai, M. A., Habr-Gama, A., Oshima, C. T., and Brentani, M. M., Association of genetic alterations of c-*myc*, c-*fos*, and c-Ha-*ras* proto-oncogenes in colorectal tumors, *Dis. Colon Rectum*, 35, 444, 1992.

Nakabeppu, Y. and Nathans, D., A naturally occurring truncated form of FosB that inhibits Fos/Jun transcriptional activity, *Cell*, 64, 751, 1991.

Nakagawa, K., Saijo, N., Tsuchida, S., Sakai, M., Tsunokawa, Y., Yokota, J., Muramatsu, M., Sato, K., Terada, M., and Tew, K. D., Glutathione-S-transferase pi as a determinant of drug resistance in transfectant cell lines, *J. Biol. Chem.*, 265, 4296, 1990.

Nakamatsu, K., Taniguchi, S., Kimura, G., and Baba, T., Enhancement of colony forming ability in the lung by transfer of the v-fos oncogene into a ras-transformed rat 3Y1 cell line, *FEBS Lett.*, 257, 422, 1989.

Nerlov, C., Rorth, P., Blasi, F., and Johnsen, M., Essential AP-1 and PEA3 binding elements in the human urokinase enhancer display cell type-specific activity, *Oncogene*, 6, 1583, 1991.

Nishina, H., Sato, H., Suzuki, T., Sato, M., and Iba, H., Isolation and characterization of *fra-2*, an additional member of the *fos* gene family, *Proc. Natl. Acad. Sci. U.S.A.*, 87, 3619, 1990.

Okuda, A., Imagawa, M., Sakai, M., and Muramatsu, M., Functional cooperativity between two TPA responsive elements in undifferentiated F9 embryonic stem cells, *EMBO J.*, 9, 1131, 1990.

Ostrowski, L. E., Finch, J., Krieg, P., Matrisian, L., Patskan, G., O'Connell, J. F., Phillips, J., and Slaga, T. J., Expression pattern of a gene for a secreted metalloproteinase during late stages of tumor progression, *Mol. Carcinogenesis*, 1, 13, 1988.

Owen, T. A., Bortell, R., Yocum, S. A., Smock, S. L., Zhang, M., Abate, C., Shalhoub, V., Aronin, N., Wright, K. L., van Wijnen, A. J., Stein, J. L., Curran, T., Lian, J. B., and Stein, G. S., Coordinate occupancy of AP-1 sites in the vitamin D-responsive and CCAAT box elements by Fos-Jun in the osteocalcin gene: Model for phenotype suppression of transcription, *Proc. Natl. Acad. Sci. U.S.A.*, 87, 9990, 1990.

Papathanasiou, M. A., Kerr, N. C., Robbins, J. H., McBride, O. W., Alamo, I., Jr., Barrett, S. F., Hickson, I. D., and Fornace, A. J., Jr., Induction by ionizing radiation of the *gadd45* gene in cultured human cells: Lack of mediation by protein kinase C, *Mol. Cell. Biol.*, 11, 1009, 1991.

Pei, X. F., Meck, J. M., Greenhalgh, D., and Schlegel, R., Cotransfection of HPV-18 and v-*fos* DNA induces tumorigenicity of primary human keratinocytes, *Virology*, 196, 855, 1993.

Petrusevska, R. T., Fürstenberger, G., Marks, F., and Fusenig, N. E., Cytogenetic effects caused by phorbol ester tumor promoters in primary mouse keratinocyte cultures: Correlation with the convertogenic activity of TPA in multistage skin carcinogenesis, *Carcinogenesis*, 9, 1207, 1988.

Pinto. A., Colletta. G., Del Vecchio, L., Rosati. R., Attadia, V., Cimino. R., and Colombatti, A., c-*fos* oncogene expression in human hematopoietic malignancies is restricted to acute leukemias with monocytic phenotype and to subsets of B cell leukemias. *Blood*, 70, 1450, 1987.

Power, C., Sinha, S., Webber. C., Manson. M. M., and Neal, G. E., Transformation related expression of glutathione-S-transferase P in rat liver cells. *Carcinogenesis*, 8, 797, 1987.

Ransone. L. J., Visvader, J., Wamsley, P., and Verma, I. M., Trans-dominant negative mutants of Fos and Jun. *Proc. Natl. Acad. Sci. U.S.A.*, 87, 3806, 1990.

Riva. P. and Larizza. L., Expression of c-*sis* and c-*fos* genes in human meningiomas and neurinomas. *Int. J. Cancer*, 51, 873, 1992.

Robbins. P. D., Horowitz. J. M., and Mulligan. R. C., Negative regulation of human c-*fos* expression by the retinoblastoma gene product. *Nature*, 346, 668, 1990.

Roop. D. R., Lowy. D. R., Tambourin, P. E., Strickland. J., Harper, J. R., Balaschak. M., Spangler. E. F., and Yuspa. S. H., An activated Harvey ras oncogene produces benign tumours on mouse epidermal tissue. *Nature*, 323, 822, 1986.

Russo. J., Tay, L. K., Ciocca. D. R., and Russo. I. H., Molecular and cellular basis of the mammary gland susceptibility to carcinogenesis, *Environ. Health Perspect.*, 49, 185, 1983.

Rüther, U., Garber. C., Komitowski. D., Müller. R., and Wagner. E. F., Deregulated c-*fos* expression interferes with normal bone development in transgenic mice. *Nature*, 325, 412, 1987.

Rüther. U., Müller. W., Sumida. T., Tokuhisa. T., Rajewsky. K., and Wagner. E. F., c-*fos* expression interferes with thymus development in transgenic mice. *Cell*, 53, 847, 1988.

Rüther, U., Komitowski, D., Schubert. F. R., and Wagner. E. F., c-*fos* expression induces bone tumors in transgenic mice. *Oncogene*, 4, 861, 1989.

Ryder. K., Lanahan. A., Perez-Albuerne. E., and Nathans. D., Jun-D: A third member of the Jun gene family. *Proc. Natl. Acad. Sci. U.S.A.*, 86, 1500, 1989.

Sakai. M., Muramatsu. M., and Nishi, S., Suppression of glutathione transferase P expression by glucocorticoid. *Biochem. Biophys. Res. Commun.*, 187, 976, 1992.

Sandberg. M., Vuorio. T., Hirvonen. H., Alitalo. K., and Vuorio. E., Enhanced expression of TGF-ß and c-*fos* mRNAs in the growth plates of developing human long bones. *Development*, 102, 461, 1988.

Sato. K., Glutathione transferases as markers of preneoplasia and neoplasia, *Adv. Cancer. Res.*, 52, 205, 1989.

Schlager. J. J. and Powis, G., Cytosolic NAD(P)H:(quinone-acceptor) oxidoreductase in human normal and tumor tissue: Effects of cigarette smoking and alcohol. *Int. J. Cancer*, 45, 403, 1990.

Schuh. A. C., Keating. S. J., Monteclaro. F. S., Vogt. P. K., and Breitman. M. L., Obligatory wounding requirement for tumorigenesis in v-*jun* transgenic mice. *Nature*, 346, 756, 1990.

Schuh. A. C., Keating, S. J., Yeung. M. C., and Breitman. M. L., Skeletal muscle arises as a late event during development of wound sarcomas in v-*jun* transgenic mice. *Oncogene*, 7, 667, 1992.

Schüle. R., Umesono. K., Mangelsdorf. D. J., Bolado. J., Pike, J. W., and Evans. R. M., Jun-Fos and receptors for vitamins A and D recognize a common response element in the human osteocalcin gene. *Cell*, 61, 497, 1990.

Schüle. R., Rangarajan. P., Yang. N., Kliewer. S., Ransone. L. J., Bolado. J., Verma. I. M., and Evans. R. M., Retinoic acid is a negative regulator of AP-1-responsive genes. *Proc. Natl. Acad. Sci. U.S.A.*, 88, 6092, 1991.

Schütte, J., Minna, J. D., and Birrer, M. J., Deregulated expression of human c-*jun* transforms primary rat embryo cells in cooperation with an activated c-Ha-ras gene and transforms rat-1a cells as a single gene, *Proc. Natl. Acad. Sci. U.S.A.*, 86, 2257, 1989a.

Schütte, J., Viallet, J., Nau, M., Segal, S., Fedorko, J., and Minna, J. D., *jun*-B inhibits and c-*fos* stimulates the transforming and trans-activating activities of c-*jun*, *Cell*, 59, 987, 1989b.

Shemshedini, L., Knauthe, R., Sassone-Corsi, P., Pornon, A., and Gronemeyer, H., Cell-specific inhibitory and stimulatory effects of Fos and Jun on transcription activation by nuclear receptors, *EMBO J.*, 10, 3839, 1991.

Sherman, M. L., Datta, R., Hallahan, D. E., Weichselbaum, R. R., and Kufe, D. W., Ionizing radiation regulates expression of the c-*jun* protooncogene, *Proc. Natl. Acad. Sci. U.S.A.*, 87, 5663, 1990.

Silverberg, M. and Silverberg, R., Steroid hormones and bone, in *The Biochemistry and Physiology of Bone*, vol. 3, Bourne, G. H., Ed., Academic Press, New York, 1971.

Slamon, D. J., deKernion, J. B., Verma, I. M., and Cline, M. J., Expression of cellular oncogenes in human malignancies, *Science*, 224, 256, 1984.

Smeal, T., Binétruy, B., Mercola, D. A., Birrer, M., and Karin, M., Oncogenic and transcriptional cooperation with Ha-Ras requires phosphorylation of c-Jun on serines 63 and 73, *Nature*, 354, 494, 1991.

Smeal, T., Binétruy, B., Mercola, D., Grover-Bardwick, A., Heidecker, G., Rapp, U. R., and Karin, M., Oncoprotein-mediated signalling cascade stimulates c-Jun activity by phosphorylation of serines 63 and 73, *Mol. Cell. Biol.*, 12, 3507, 1992.

Smeyne, R. J., Vendrell, M., Hayward, M., Baker, S. J., Miao, G. G., Schilling, K., Robertson, L. M., Curran, T., and Morgan, J. T., Continuous c-*fos* expression precedes programmed cell death *in vivo*, *Nature (London)*, 363, 166, 1993.

Stein, B., Rahmsdorf, H. J., Steffen, A., Litfin, M., and Herrlich, P., UV-induced DNA damage is an intermediate step in UV-induced expression of human immunodeficiency virus type 1, collagenase, c-fos, and metallothionein, *Mol. Cell. Biol.*, 9, 5169, 1989.

Stein, B., Angel, P., van Dam, H., Ponta, H., Herrlich, P., van der Eb, A., and Rahmsdorf, H. J., Ultraviolet radiation-induced c-jun gene expression: Two AP-1 like binding sites mediate the response, *Photochem. Photobiol.*, 55, 409, 1992

Stopera, S. A., Davie, J. R., and Bird, R. P., Colonic aberrant crypt foci are associated with increased expression of c-*fos:* The possible role of modified c-*fos* expression in preneoplastic lesions in colon cancer, *Carcinogenesis*, 13, 573, 1992.

Su, H. Y., Bos, T. J., Monteclaro, F. S., and Vogt, P. K., *Jun* inhibits myogenic differentiation, *Oncogene*, 6, 1759, 1991.

Sugio, K., Kurata, S., Sasaki, M., Soejima, J., and Sasazuki, T., Differential expression of c-*myc* gene and c-*fos* gene in premalignant and malignant tissues from patients with familial polyposis coli, *Cancer Res.*, 48, 4855, 1988.

Sutherland, R. L., Lee, C. S., Feldman, R. S., and Musgrove, E. A., Regulation of breast cancer cell cycle progression by growth factors, steroids and steroid antagonists, *J. Steroid Biochem. Mol. Biol.*, 41, 315, 1992.

Taniguchi, S., Tatsuka, M., Nakamatsu, K., Inoue, M., Sadano, H., Okazaki, H., Iwamoto, H., and Baba, T., High invasiveness associated with augmentation of motility in a fos-transferred highly metastatic rat 3Y1 cell line, *Cancer Res.*, 49, 6738, 1989.

Taniguchi, S., Nishimura, Y., Takahashi, T., Baba, T., and Kato, K., Augmented excretion of procathepsin L of a fos-transferred highly metastatic rat cell line, *Biochem. Biophys. Res. Commun.*, 168, 520, 1990.

Tennenbaum, T., Yuspa, S. H., Grover, A., Castronovo, V., Sobel, M. E., Yamada, Y., and De Luca, L. M., Extracellular matrix receptors and mouse skin carcinogenesis: Altered expression linked to appearance of early markers of tumor progression, *Cancer Res.*, 52, 2966, 1992.

Terrier, P., Sheng, Z. M., Schlumberger, M., Tubiana, M., Caillou, B., Travagli, J. P., Fragu, P., Parmentier, C., and Riou, G., Structure and expression of c-*myc* and c-*fos* proto-oncogenes in thyroid carcinomas, *Br. J. Cancer*, 57, 43, 1988.

Toftgard, R., Roop, D. R., and Yuspa, S. H., Protooncogene expression during two-stage carcinogenesis in mouse skin, *Carcinogenesis*, 6, 655, 1985.

Tsang, T. C., Chu, Y. W., Hendrix, M. J. C., and Bowden, G. T., V-Jun oncogene suppresses both the phorbol ester TPA induced tumor cell invasion and stromelysin expression, *Proc. Am. Assoc. Cancer Res.*, 33, (Abstr. 2363), 396, 1992a.

Tsang, T. C., Levy, J., and Bowden, G. T., Viral jun oncogene acts as dominant negative regulator in mouse epidermal tumor cells, *Proc. Am. Assoc. Cancer Res.*, 31, (Abstr. 1915), 323, 1992b.

van den Berg, S., Kaina, B., Rahmsdorf, H. J., Ponta, H., and Herrlich, P., Involvement of *fos* in spontaneous and ultraviolet light-induced genetic changes, *Mol. Carcinogenesis*, 4, 460, 1991.

van den Berg, S., Rahmsdorf, H. J., Herrlich, P., and Kaina, B., Overexpression of c-*fos* increases recombination frequency in human osteosarcoma cells, *Carcinogenesis*, 14, 925, 1993.

van der Burg, B., de Groot, R. P., Isbrucker, L., Kruijer, W., and De Laat, S. W., Direct stimulation by estrogen of growth factor signal transduction pathways in human breast cancer cells, *J. Steroid Biochem. Mol. Biol.*, 43, 111, 1992.

Vogt, P. K. and Bos, T. J., *jun:* Oncogene and transcription factor, *Adv. Cancer Res.*, 55, 1, 1990.

Wakita, K., Ohyanagi, H., Yamamoto, K., Tokuhisa, T., and Saitoh, Y., Overexpression of c-*Ki-ras* and c-*fos* in human pancreatic carcinomas, *Int. J. Pancreatol.*, 11, 43, 1992.

Walker, R. A. and Cowl, J., The expression of c-*fos* protein in human breast, *J. Pathol.*, 163, 323, 1991.

Wang, Z.-Q., Grigoriadis, A. E., Möhle-Steinlein, U., and Wagner, E., A novel target cell for c-*fos*-induced oncogenesis: Development of chondrogenic tumours in embryonic stem cell chimeras, *EMBO J.*, 10, 2437, 1991.

Wang, Z.-Q., Ovitt, C., Grigoriadis, A. E., Möhle-Steinlein, U., Rüther, U., and Wagner, E. F., Bone and haematopoietic defects in mice lacking c-fos, *Nature*, 360, 741, 1992.

Ward, J. M. and Young, D. M., Histogenesis and morphology of periosteal sarcomas induced by FBJ virus in NIH Swiss mice, *Cancer Res.*, 36, 3985, 1976.

Wu, J. X., Carpenter, P. M., Gresens, C., Keh, R., Niman, H., Morris, J., and Mercola, D., The proto-oncogene c-*fos* is over-expressed in the majority of human osteosarcomas, *Oncogene*, 5, 989, 1990.

Xanthoudakis, S. and Curran, T., Identification and characterization of Ref-1, a nuclear protein that facilitates AP-1 DNA-binding activity, *EMBO J.*, 11, 653, 1992.

Xanthoudakis, S., Miao, G., Wang, F., Pan, Y. C., and Curran, T., Redox activation of Fos-Jun DNA binding activity is mediated by a DNA repair enzyme, *EMBO J.*, 11, 3323, 1992.

Yamanishi, D. T., Buckmeier, J. A., and Meyskens, F. L., Jr., Expression of c-jun, jun-B, and c-fos proto-oncogenes in human primary melanocytes and metastatic melanomas, *J. Invest. Dermatol.*, 97, 349, 1991.

Yang-Yen, H. F., Chambard, J. C., Sun, Y. L., Smeal, T., Schmidt, T. J., Drouin, J., and Karin, M., Transcriptional interference between c-Jun and the glucocorticoid receptor: Mutual inhibition of DNA binding due to direct protein-protein interaction, *Cell*, 62, 1205, 1990.

Yang-Yen, H. F., Zhang, X. K., Graupner, G., Tzukerman, M., Sakamoto, B., Karin, M., and Pfahl, M., Antagonism between retinoic acid receptors and AP-1: Implications for tumor promotion and inflammation, *New Biol.*, 3, 1206, 1991.

Yuspa, S. H., Tumor promotion, in *Accomplishments in Cancer Research 1986*, Fortner, J. F. and Rhoads, J. E., Eds., J. B. Lippincott, Philadelphia, PA, p. 169, 1987.

Yuspa, S. H. and Poirier, M. C., Chemical carcinogenesis: From animal models to molecular models in one decade, *Adv. Cancer Res.*, 50, 25, 1988.

Chapter 20

An Introduction to Tumor Progression

J. A. East and I. R. Hart

CONTENTS

TUMOR PROGRESSION

The term *neoplasia* describes "all focal proliferative lesions, benign tumors, primary cancers and metastases, that may affect any given cell system" (Clark, 1991). It has been known for some time that in many tumor systems there is a stepwise process by which the tumor cells become more malignant (Foulds, 1954). This gradual conversion from "bad to worse" occurs because the focal accumulations of cells derived from these sequential changes serve as precursor lesions for the subsequent processes. The periods of selection, which follow each rare event that evokes a cellular alteration, allow the emergence of new focal accumulations of cells and ensure that the whole process may occur over an extended period of time, i.e., months and years rather than days. Termed *tumor progression*, this process generally is responsible for the conversion of cells from normal, through benign, nonmetastatic tumors, to fully metastatic malignancies. While it has been demonstrated for several tumor types (e.g., melanoma, colon, cervix, and stomach), it is by no means a universal phenomenon. Many human tumors do not experience this stepwise increase in malignancy, but seem to be maximally malignant from their inception, e.g., unknown primary tumors (UPTs) and renal, lung, and prostate cancers. These latter tumors have been described as undergoing type 2 progression (Bell et al., 1989), where presentation and diagnosis frequently correlate with the presence of a fully malignant tumor.

Clark, who has spent a lifetime analyzing the histological changes that accompany this process, has suggested that neoplastic lesions can be divided into four classes that describe the pathological characteristics of tumor progression (Clark, 1991). These classes, using the melanocytic and colonic tumor systems as paradigms, are designated as follows. Class IA lesions consist of orderly clonal growth capable of undergoing differentiation and regression. Such tumors are benign, and, in the melanocytic neoplastic system, are the common acquired melanocytic naevi (moles), while in the colonic system they are represented by adenomatous colonic polyps. Class IB lesions exhibit aberrant differentiation accompanied by disorderly growth, and, although regression may occur, most of these lesions are indolent. Examples of these lesions include melanocytic naevi with focal architectural abnormalities, and small persistent tubular adenomas of the colon. The appearance of large atypical cells with prominent and usually hyperchromatic nuclei, which may be indistinguishable from overt cancer cells, signals the progression to Class

0-8493-4573-1/94/$0.00+$.50
© 1994 by CRC Press, Inc.

IC. Regression is still possible and the lesions usually remain indolent, e.g., dysplastic naevi and colonic adenoma with cytological atypia.

Class II comprises intermediate lesions identified by unrestricted temporal growth but which remain confined to the tissue compartment of origin. Class II lesions grow slowly, do not metastasize, but are highly likely to proceed to the next stage of progression. These lesions, by virtue of their appearance, are often called carcinoma *in situ* (e.g., melanoma *in situ*, radial growth phase melanoma, and carcinoma *in situ* within an adenomatous colonic polyp).

The Class III lesions are the primary invasive cancers that display temporally unrestricted growth combined with the ability to grow in more than one tissue compartment; although they may be metastatic, this is not always the case. Examples of this class of tumor are the vertical growth phase melanomas and carcinomas of the colon that have extended into the muscularis propria; the latter event being associated with the ability to metastasize.

The final class of lesion, Class IV, describes metastasis or the growth of malignant cells in the mesenchyme (not the parenchyma) of a secondary site.

MOLECULAR EVENTS DRIVING TUMOR PROGRESSION

It has become clear that each of the stages of tumor progression could arise as a consequence of genetic change, be it activation, mutation, or loss of individual genes (Klein and Klein, 1985). These genes have been categorized into three classes: oncogenes and suppressor genes, which play their major role at the initiation stage, and the modulator genes, which control subsequent biological behavior once transformation has occurred. Cancer remains predominantly a disease of old age, and analysis of age-incidence curves has suggested that six to seven mutation-like changes are involved in the development of carcinomas (Cairns, 1981). Vogelstein and co-workers, in their studies on the colorectal system, have identified a number of molecular events that underlie the initiation and progression of this disease (Fearon and Vogelstein, 1990). Their work has shown that colorectal tumors appear to arise from mutational activation of oncogenes, often accompanied by a mutational inactivation of tumor suppressor genes. Moreover, while the genetic changes frequently occur in a preferred sequence during the course of tumor evolution, this is not obligatory since it is the total accumulation of genetic changes that regulates malignant development, rather than a strict, sequential process (Fearon and Vogelstein, 1990).

Many of the changes in the genome associated with the conversion of benign to fully malignant cells operate at the level of transcriptional regulation, and possible mechanisms have been discussed more fully chapters 8 and 9. However, the assumption that altered gene expression determines the movement of nonmetastatic to metastatic tumors has provided the rationale for differential screening of cDNA libraries in attempts to identify mRNAs varying in abundance between related cell populations (Hart and Easty, 1991). A variety of changes, in extracellular matrix components or expression of metalloproteinases capable of degrading such proteins, has been documented using this approach. Thus, Schalken and his colleagues (1988) found that there was downregulation of fibronectin mRNA in metastasizing, versus nonmetastasizing, prostatic tumors. The upregulation of stromelysin mRNA in squamous cell carcinomas versus noninvasive papillomas (Matrisian et al., 1986) seems to provide an explanation for the increased degradative capacity of the more advanced tumors. However, the use of whole tumor tissue, rather than cell lines, has led to the demonstration that the source of the increased mRNA for metalloproteinases need not be the neoplastic cells themselves but may derive from adjacent, normal stromal cells (Basset et al., 1990). This result suggests that changes in malignant behavior may also be a consequence of changed responses in the stromal cell infiltrate, a frequently ignored component of the cancer lesion. Activating mutations in the dominant oncogenes may function through the transcriptional regulation of the modulator genes to expedite tumor progression, rather than through their better characterized effects on cell growth and transformation. In colorectal cancers, ras gene mutations are frequently observed. Approximately 50% of adenomas greater than 1 cm diam. exhibit such mutated ras genes, whereas small adenomas (<1 cm) have a reduced frequency of ras mutations (<10%) (Vogelstein et al., 1988; Farr et al., 1988). The adenomas with mutated ras genes are more likely to progress to the next stage of the disease than are those without mutated ras genes. Is this a consequence of the selective growth advantage conferred on the mutated ras-containing cells, or could it relate to transcriptional enhancement mediated via the ras pathway? It is interesting to note, in light of the increased expression of protease genes observed in the cDNA library differential screening experiments (Matrisian et al., 1986; Basset et al., 1990), that transfection of recipient cells with activated ras frequently has produced transcriptional enhancement of proteinase expression (Denhardt et al., 1987; Garbisa et al., 1987; Collier et al., 1988).

The most common genetic change detected in human cancer involves the mutation and inactivation of the tumor suppressor activity of the p53 gene (Hollstein et al., 1991). The possible role of p53 in facilitating tumor progression, as distinct from transformation, is less clearcut. Certainly, the p53 protein binds to specific sequences in double-stranded DNA, which, when placed upstream of a minimal promoter and reporter gene, can function as a specific transcription factor (Scharer and Iggo, 1992; Kern et al., 1992). Since this sequence-specific binding capacity is lost by the point-mutated proteins, it is feasible that mutation may result in the downregulation of expression of benign phenotype genes (e.g., the cadherins), and the facilitation of malignant progression. Given that the identity of the genes for which wild-type p53 serves as a natural transcription factor is unknown, such suggestions lie purely within the realm of speculation. It is interesting, however, that an association of enhanced expression of mutated p53 with more advanced tumor progression has been made in some cancers (Van den Berg et al., 1989). Perhaps this association is less a reflection of the transcriptional activity of p53 and more a consequence of its ability to regulate genomic integrity, a function compatible with the suggested role of p53 as "guardian of the genome" (Lane, 1992). The possibility that cells containing inactivated p53 will become genetically less stable and, thus, will accumulate increased rates of chromosomal rearrangements and mutations (Lane, 1992), with increasing selection of malignant clones, provides a molecular basis for the clonal selection theories of tumor evolution (Nowell, 1976; Clark, 1991). With this possible mechanism as a driving force behind the rate of acquisition of genetic aberrations, it seems possible that the often observed amplification of some of the dominant oncogenes in tumors of increased metastatic capacity may be a reflection of enhanced growth rate promoting the emergence of malignant clones. Thus, amplification of c-myc has been reported in many disseminated tumors (Little et al., 1983), and in small-cell lung carcinoma this lesion often is associated with the highly aggressive variant cell type (Nau et al., 1985, 1986). Amplification of N-myc, often detected in neuroblastoma-derived cell lines (Schwab et al., 1984), appears to correlate with the more advanced stage 3 and 4 disease states (Brodeur et al., 1984). Overall survival and relapse time in node-positive patients with breast cancer may correlate with amplification of the HER-2/neu gene (Slamon et al., 1987). These results tend to suggest that progression may correlate with the increased proliferative capacity provided by oncogene activity. However, the *sine qua non* of malignant capacity is the ability of tumors to invade and metastasize; characteristics that may exist independent of growth rates.

ROLE OF CELL ADHESION MOLECULES IN TUMOR METASTASIS

In order for tumors to metastasize they must complete a series of cell–cell and cell–substrate interactions, which suggests that the cell surface molecules regulating these steps are likely to play a vital role in modulating disseminatory behavior (Hart and Saini, 1992). It is not surprising, therefore, that many changes in levels of expression of cellular adhesion molecules have been reported in more advanced cancers when compared to their benign counterparts. Although some changes are reflected in an upregulation of expression level, others constitute a downregulation of expression. For example, in colon carcinoma a frequently observed late change is allelic loss of chromosome 18q (70% of tumors, 50% of late adenomas) (Vogelstein et al., 1988). The candidate tumor suppressor gene located at this locus is designated DCC (deleted in colon cancer) and encodes for a protein with considerable homology to the neural cell adhesion molecule (Fearon et al., 1990). Recent findings have demonstrated that antisense RNA to this gene transforms Rat-1 fibroblasts and modifies cellular adhesion (Narayanan et al., 1992). Similarly, analysis of a limited series of human tumors has shown that expression of the homophilic cell–cell cadherin adhesion molecules is downregulated in more progressed cancers (Schipper et al., 1991). Such findings suggest that the reduced ability of malignant cells to cohere to one another is a consequence of alterations in expression of specific genes and that these alterations lead to an enhanced ability to invade (Fahraeus et al., 1992). While downregulation of cell–cell adhesion may constitute an important component of the initial detachment of cells from the primary tumor, it is conceivable that later steps in the metastatic sequence will benefit from an upregulation of adhesion receptor expression. The heterodimeric integrin receptors are a major family of cell surface proteins involved in determining interactions between cells and the adhesive proteins of the extracellular matrix (Hynes, 1987). While experimental studies have suggested that cellular transformation is associated with the downregulation of integrin expression (Plantefaber and Hynes, 1989), it is apparent that, at least in some tumor types, the progression toward the more aggressive phase of malignancy is correlated with increased integrin expression. Thus, in cutaneous melanomas, transition from radial growth phase to the vertical/metastatic growth phase is correlated with enhanced expression of the $\alpha_v\beta_3$ vitronectin receptor (Albelda et al., 1990). Analysis of the regulation of expression of cell adhesion molecules in evolving and progressing tumors may well shed light on their altered behavioral characteristics.

CONCLUSION

The process of tumor progression was first defined and characterized by Foulds (1954), who suggested that the phenomenon could be defined by the following principles: (1) there is an individual progression of individual tumors; (2) there is an independent progression of individual characteristics; (3) the process is independent of continued growth; (4) the process could be continuous or discontinuous; (5) the process could follow one of several developmental paths; and (6) an endpoint need not be reached within the life span of the host (a tenet which appears to relate more to Fould's analysis of transplantable tumors than to clinical oncology). Responsible for the development and evolution of most common solid cancers, the molecular basis of tumor progression remains little understood. Elucidation of the molecular processes behind the development of malignancy could lead to improved interventional strategies to prevent or reverse the phenomenon.

REFERENCES

Albelda, S. M., Mette, S. A., Elder, D. E., Stewart, R., Damjanovich, L., Herlyn, M., and Buck, C. A., Integrin distribution in malignant melanoma: Association of the β3 subunit with tumor progression, *Cancer Res.*, 50, 6757, 1990.

Basset, P., Bellocq, J. P., Wolf, C., Stoll, I., Hutin, P., Limacher, J. M., Podhajcer, D. L., Chenard, M. P., Rio, M. C., and Chambon, P., A novel metalloproteinase gene specifically expressed in stromal cells of breast carcinomas, *Nature*, 348, 699, 1990.

Bell, C. W., Pathak, S., and Frost, P., Unknown primary tumors: Establishment of cell lines, identification of chromosomal abnormalities, and implications for a second type of tumor progression, *Cancer Res.*, 49, 4311, 1989.

Brodeur, G., Seeger, R. C., Schwab, M., Varmus, H. E., and Bishop, J. M., Amplification of N-*myc* in untreated human neuroblastomas correlates with advanced disease stage, *Science*, 224, 1121, 1984.

Cairns, J., The origin of human cancer, *Nature*, 289, 353, 1981.

Clark, W. H., Human cutaneous malignant melanoma as a model for cancer, *Cancer Metastasis Rev.*, 10, 83, 1991.

Collier, I. E., Wilhelm, S. M., Eisen, A. Z., Marmer, B. L., Grant, G. A., Seltzer, J. L., Kronberger, A., He, C., Bauer, E. A. A., and Goldberg, G. I., H-*ras* oncogene-transformed human bronchial epithelial cells (TBE-1) secrete a single metalloprotease capable of degrading basement membrane collagen, *J. Biol. Chem.*, 263, 6579, 1988.

Denhardt, D. T., Greenberg, A. H., Egan, S. E., Hamilton, R. E., and Wright, J. A., Cysteine proteinase cathepsin L expression correlates closely with the metastatic potential of H-*ras*-transformed murine fibroblasts, *Oncogene*, 2, 55, 1987.

Fahraeus, R., Chen, W., Trivedi, P., Klein, G., and Obrink, B., Decreased expression of e-cadherin and increased invasive capacity in ebv-lmp-transfected human epithelial and murine adenocarcinoma cells, *Int. J. Cancer*, 53, 834, 1992.

Farr, C. J., Marshall, C. J., Easty, D. J., Wright, N. A., Powell, S. C., and Paraskeva, C., A study of *ras* gene mutations in colonic adenomas from familial polyposis coli patients, *Oncogene*, 3, 673, 1988.

Fearon, E. R. and Vogelstein, B., A genetic model for colorectal tumorigenesis, *Cell*, 61, 759, 1990.

Fearon, E. R., Cho, K. R., Nigro, J. M., Kern, S. E., Simons, J. W., Ruppert, J. M., Hamilton, S. R., Preisinger, A. C., Thomas, G., Kinzler, K. W., and Vogelstein, B., Identification of a chromosome 18q gene that is altered in colorectal cancer, *Science*, 247, 49, 1990.

Foulds, L., The experimental study of tumor progression: A review, *Cancer Res.*, 14, 327, 1954.

Garbisa, S., Pozzatti, R., Muschel, R. J., Saffiotti, U., Ballin, M., Goldfarb, R. H., Khoury, G., and Liotta, L. A., Secretion of type IV collagenolytic protease and metastatic phenotype: induction by transfection with c-Ha-*ras* but not c-Ha-*ras* plus Ad2-E1a, *Cancer Res.*, 47, 1523, 1987.

Hart, I. R. and Easty, D., Identification of genes controlling metastatic behaviour, *Br. J. Cancer*, 63, 9, 1991.

Hart, I. R. and Saini, A., Biology of tumor metastasis, *Lancet*, 339, 1453, 1992.

Hollstein, M., Sidransky, D., Vogelstein, B., and Harris, C., p53 mutations in human cancers, *Science*, 253, 49, 1991.

Hynes, R. O., Integrins: A family of cell surface receptors, *Cell*, 48, 549, 1987.

Kern, S. E., Pietenpol, J. A., Thiagalingam, S., Seymour, A., Kinzler, K. W., and Vogelstein, B., Oncogenic forms of p53 inhibit p53-regulated gene expression, *Science*, 256, 827, 1992.

Klein, G. and Klein, E.. Evolution of tumors and the impact of molecular oncology. *Nature*, 315, 190, 1985.

Lane, D. P.. p53, guardian of the genome. *Nature News Views*, 358, 15, 1992.

Little, C. D., Nau, M. M., Carney, D. N., Gazdar, A. F., and Minna, J. D.. Amplification and expression of the c-*myc* oncogene in human lung cancer lines. *Nature*, 306, 194, 1983.

Matrisian, L. M., Bowden, G. T., Krieg, P., Furstenberger, G., Briand, J.-P., Leroy, P., and Breathnach, R.. The mRNA coding for the secreted protease transin is expressed more abundantly in malignant than in benign tumors. *Proc. Natl. Acad. Sci. U.S.A.*, 83, 9413, 1986.

Narayanan, R., Lawlor, K. G., Schaapveld, R. Q. J., Cho, K. R., Vogelstein, B., Tran, P. B.-V., Osborne, M. P., and Telang, N. T.. Antisense RNA to the putative tumor-suppressor gene DCC transforms Rat-1 fibroblasts. *Oncogene*, 7, 553, 1992.

Nau, M. M., Brooks, B. J., Battey, J. F., Sausville, E., Gazdar, A. F., Kirsch, I. R., McBride, O. W., Bertness, V., Hollis, G. F., and Minna, J. D.. L-*myc*, a new *myc*-related gene amplified and expressed in a human small cell lung cancer. *Nature*, 318, 69, 1985.

Nau, M. M., Brooks, B. J., Carney, D. N., Gazdar, A. F., Battey, J. F., Sausville, E. A., and Minna, J. D.. Human small-cell lung cancers show amplification and expression of the N-*myc* gene. *Proc. Natl. Acad. Sci. U.S.A.*, 83, 1092, 1986.

Nowell, P.. The clonal evolution of tumor cell populations. *Science*, 194, 23, 1976.

Plantefaber, L. C. and Hynes, R. O.. Changes in integrin receptors on oncogenically transformed cells. *Cell*, 56, 281, 1989.

Schalken, J. A., Ebeling, S. B., Isaacs, J. T., Treiger, B., Bussemakers, M. J. G., de Jong, M. E. M., and Van de Ven, W. J. M.. Down-modulation of fibronectin messenger RNA in metastasizing rat prostatic cancer cells revealed by differential hybridisation analysis. *Cancer Res.*, 48, 2042, 1988.

Scharer, E. and Iggo, R.. Mammalian p53 can function as a transcription factor in yeast. *Nucleic Acids Res.*, 20, 1539, 1992.

Schipper, J. H., Frixen, U. H., Behrens, J., Unger, A., Jahnke, K., and Birchmeier, W.. E-Cadherin expression in squamous cell carcinomas of head and neck: Inverse correlation with tumor differentiation and lymph node metastasis. *Cancer Res.*, 51, 6328, 1991.

Schwab, M., Ellison, J., Busch, M., Rosenan, W., Varmus, H. E., and Bishop, J. M.. Enhanced expression of the human gene N-*myc* consequent to amplification of DNA may contribute to malignant progression of neuroblastoma. *Proc. Natl. Acad. Sci. U.S.A.*, 81, 4940, 1984.

Slamon, D. J., Clark, G. M., Wong, S. G., Levin, W. J., Ullrich, A., and McGuire, W. L.. Human breast cancer: Correlation of relapse and survival with amplification of the *HER-2/neu* oncogene. *Science*, 235, 177, 1987.

Van den Berg, F. M., Tigges, A. J., Schipper, M. E. I., den Hartog-Jager, F. C. A., Kroes, W. G. M., and Walboomers, J. M. M., Expression of the nuclear oncogene p53 in colon tumors. *J. Pathol.*, 157, 193, 1989.

Vogelstein, B., Fearon, E. R., Hamilton, S. R., Kern, S. E., Preisinger, A. C., Leppert, M., Nakumura, Y., White, R., Smits, A. M. M., and Bos, J. L.. Genetic alterations during colorectal-tumor development. *N. Engl. J. Med.*, 319, 525, 1988.

INDEX

T - #0494 - 101024 - C0 - 222/150/17 - PB - 9781138562165 - Gloss Lamination